IMAGES OF THE STREET

The street has always held a particular fascination. The terrain of social encounters and political protest, sites of domination and resistance, places of pleasure and anxiety, the street is also the focus of many theoretical debates about the city concerning modern and post-modern urbanism.

Images of the Street captures the vitality, excitement and tensions of the street. Drawing on leading historical and contemporary international research in cultural geography, cultural studies, sociology and planning, Nicholas Fyfe focuses contributions into three main sections. *Planning and Design* examines how specific streetscapes are shaped by the interplay between dominant ideas in politics, planning and local economic and political circumstances. *Social Identities and Social Practices* draws on a rich seam of qualitative material, to tease out social differences of peoples' experiences of the street; to examine how social identities are shaped and represented in fiction and film; and to explore the meaning and significance of streets as settings in which social practices are played out. The final section, *Control and Resistance*, focuses on how social life on the street is increasingly regulated, both directly by formal agencies of control, such as the police, and indirectly through architecture and urban design.

Images of the Street subjects the street to sustained critical scrutiny, enriching and extending our understanding of the making and meaning of urban space.

Nicholas R. Fyfe is Senior Lecturer in Geography at the University of Strathclyde.

Contributors: Stuart Aitken, David Atkinson, Jon Bannister, David Bell, Jon Binnie, Tim Cresswell, David Crouch, Gerry Daly, Tim Edensor, Nick Fyfe, Brendan Gleeson, John Gold, Steve Herbert, Peter Jackson, Loretta Lees, Richard Levy, Chris Lukinbeal, Gerry Mooney, Jane Rendell, Gill Valentine

IMAGES OF THE STREET

Planning, identity and control in public space

•

EDITED BY NICHOLAS R. FYFE

London and New York

First published 1998
by Routledge
11 New Fetter Lane, London EC4P 4EE

Simultaneously published in the USA and Canada
by Routledge
29 West 35th Street, New York, NY 10001

© 1998 Nicholas R. Fyfe for the selection and editorial matter; individual chapters, the
contributors

Typeset in Garamond by Florencetype Limited, Stoodleigh, Devon
Printed and bound in Great Britain by T.J. International Ltd, Padstow, Cornwall

British Library Cataloguing in Publication Data
A catalogue record for this book is available from the British Library

Library of Congress Cataloguing in Publication Data
Images of the street : planning, identity, and
control in public space / edited by Nicholas
R. Fyfe
    p.   cm
Papers presented at a conference at the
University of Strathclyde in Glasgow in 1996
1. City planning—Congresses. 2. Streets—
Social aspects—Congresses. I. Fyfe,
Nicholas R.
HT166.I527   1998
307.3   dc21   97–50288

ISBN 0–415–15440–5 (hbk)
ISBN 0–415–15441–3 (pbk)

FOR MY PARENTS

CONTENTS

•

FIGURES

●

TABLES

●

CONTRIBUTORS

•

Stuart C. Aitken is Professor of Geography at San Diego State University.

David Atkinson is Lecturer in Geography at the University of Hull.

Jon Bannister is Lecturer in Social Policy at the University of Glasgow.

David Bell is Senior Lecturer in Cultural Studies at Staffordshire University.

Jon Binnie is Lecturer in Geography at Liverpool John Moores University.

Tim Cresswell is Lecturer in Geography at the University of Wales, Lampeter.

David Crouch is Professor of Cultural Geography at Anglia Polytechnic University.

Gerald Daly is Professor in the Faculty of Environmental Studies at York University, Toronto.

Tim Edensor is Lecturer in Cultural Studies at Staffordshire University.

Nicholas R. Fyfe is Senior Lecturer in Geography at the University of Strathclyde.

Brendan Gleeson is a Research Fellow in the Research School of Social Sciences at the Australian National University.

John R. Gold is Professor of Geography at Oxford Brookes University.

Steve Herbert is Assistant Professor in the Department of Criminal Justice at Indiana University.

Peter Jackson is Professor of Geography at the University of Sheffield.

Loretta Lees is Lecturer in Geography at King's College, University of London.

Richard M. Levy is Associate Professor of Urban Planning at the University of Calgary.

Chris Lukinbeal is a doctoral student in Geography at San Diego State University and the University of California at Santa Barbara.

Gerry Mooney is Senior Lecturer in Sociology and Social Policy at the University of Paisley.

Jane Rendell is Senior Lecturer at Chelsea College of Art and Design.

Gill Valentine is Lecturer in Geography at the University of Sheffield.

PREFACE

•

Soon after coming to live and work in Glasgow I bought a pocket guide to the city's architecture. Its opening sentence surprised and intrigued me: 'Glasgow is more a city of streets than buildings.' Having just arrived from a city of narrow, winding, medieval streets, this observation struck a certain chord as I navigated the long, broad streets of Glasgow's city centre grid. Laid out mainly in the nineteenth century over the undulating topography of a drumlin field, Glasgow's gridded street pattern affords dramatic views over the city and provides long vistas down axial streets which invite exploration. Although evocative of North American city layouts, the building facades fronting the streets also mean the city has a very European feel.

It was in this setting that an international group of geographers, planners and sociologists came together to discuss *Images of the Street* at the Royal Geographical Society–Institute of British Geographers conference at the University of Strathclyde in Glasgow in 1996. The timing of the conference was quite appropriate. The previous year the City Council had published its *City Centre Public Realm Strategy and Guidelines*. This celebrated the city's historic streetscapes and their contribution to the city's unique identity but also outlined an ambitious programme of improving the aesthetic, environmental and social qualities of Glasgow's streets for pedestrians so as to present an image of 'a confident and progressive city'. What gave this *Public Realm Strategy* an added significance was that exactly fifty years earlier in 1945 the City Council had published its *First Planning Report*. This had recommended destroying much of Glasgow's historic street pattern to allow the comprehensive redevelopment of the city centre and the construction of an urban motorway. In contrast to the symbolic importance given to streets in the *Public Realm Strategy*, the *First Planning Report* was preoccupied with their functional role and focused on ways of excluding pedestrians from streets so they did not impede 'fast moving mechanically propelled vehicles'. Although sections of the motorway were constructed, most of the city's streets survived. Nevertheless, the contrasts between these two images of Glasgow's streets, one informed by postmodern sensitivities, the other displaying the bravado of modernist planning, provided an interesting setting for the conference and the discussion of images of the street.

I would like to thank all the contributors to the session at the RGS–IBG conference and the Urban Geography Study Group of the RGS–IBG who sponsored it. Tristan Palmer, then at Routledge, responded positively and very enthusiastically to the suggestion of putting a book together based on this session and his guidance helped enormously in developing a fairly modest book proposal into something more ambitious. I would also like to thank Sarah Lloyd, Sarah Carty and Katharine Day at Routledge for their advice and support. My greatest debt is to the contributors to the book who have met deadlines, dealt constructively with my comments, and waited patiently for me to put the book together.

Finally, I would like to thank Mark Boyle, Neil McInroy and Robert Rogerson who have at various times discussed and guided me around the streets of Glasgow; and Gillian, Alexander and Christopher for insisting that forest tracks and mountain paths are every bit as interesting as city streets.

For permission to use copyright material, Gerry Mooney would like to thank the Ministry of Defence for Figure 3.4; John Gold, the Royal Institute of British Architects for Figure 4.1; and Jane Rendell, the Guildhall Library for Figures 6.1, 6.2 and 6.3.

1

INTRODUCTION

READING THE STREET

Nicholas R. Fyfe

•

Think of a city and what comes to mind? Its streets.

Jacobs, 1961: 39

Streets, as Jane Jacobs reminds us, have always held a particular fascination for those interested in the city. Streets are the terrain of social encounters and political protest, sites of domination and resistance, places of pleasure and anxiety. Located at the intersection of several academic disciplines, the street is also the focus of many theoretical debates about the city concerning modern and, more recently, postmodern urbanism. For modernists the street is a space 'from which to get from A to B, rather than a place to live in', displacing the street 'from lifeworld to system', (Lash and Friedmann, 1992: 10); for postmodernists, the street is a place designed to foster and complement new urban lifestyles, reclaiming the street from system to lifeworld. Exploring these and many other readings of the street, this volume subjects the street to sustained critical scrutiny. An international, cross-disciplinary set of essays, it explores how streets as specific, local landscapes manifest broader social and cultural processes, establishing the strategic importance of the street to wider theoretical questions about the interplay between society and space. What, for example, does the design of streets reveal about dominant ideas in politics and planning? How are social identities and social practices shaped by people's experiences of the street? Does increasing social control signal the end of the street as a 'public' space? These are some of the key issues addressed by contributions to this volume. The street which 'has occupied a cherished place in the lexicon of urbanism' (Keith, 1995: 297) is, of course, no stranger to such scrutiny. Nevertheless, much of our current understanding of the meaning and significance of the street appears dominated by a small number of studies of very particular streets.

URBANISM AND THE STREET: TAKING THE 'GRAND TOUR'

Look through two volumes reprinting what are considered to be some of the most significant contributions to understanding urbanism this century (Kasnitz, 1995; Le Gates and Stout, 1996), and you will find that each offers a very similar 'tour' of city streets. In the company of Walter Benjamin, Le Corbusier, Jane Jacobs and Mike Davis, the reader is taken down broad boulevards and high speed expressways, through communities where residents participate in daily 'street ballets' and on to 'mean streets' where an underclass fight for survival. Although the work of these urban commentators clearly represents only a fraction of what has been written about the street, it is to these classics that many other influential accounts of the city and its streets, as well as many of the contributors to this volume, so often refer (see, for example, Berman, 1983; Sennett, 1990; Sudjic, 1992; Young, 1990). It will therefore be useful to begin by taking the 'grand tour' of city streets before looking in more detail at the individual chapters.

The tour begins in mid- to late nineteenth century Paris with Walter Benjamin's observations on that self-styled 'artist in demolition', Baron Haussmann and his 'constructive destruction' of the city to make way for the straight, wide boulevards of his new urban circulatory system (Benjamin, 1995). The surgical overtones of the phrase 'circulatory system' are not unimportant. As Ellin (1997a) notes, Haussmann viewed the city as a sick organism with his task that of the surgeon cutting out infected areas and opening up clogged arteries. According to Benjamin, however, these surgical metaphors should not obscure 'The true purpose of Haussmann's work', namely to 'secure the city against civil war' (Benjamin, 1995: 54). While it is certainly true that the breadth of streets made the erection of barricades difficult and their straightness provided infantry with a long line of fire, Benjamin's singular reading of Haussmann's boulevards misses other significant implications of their construction. The boulevards had an important economic function, helping to quicken the pace of commerce; socially, large numbers of the working-classes were employed in their construction while the routes of the boulevards caused some dispersion and displace-ment of working class communities; and, symbolically, they provided an unequivocal demonstration of the power of the state to shape the urban landscape in the interests of the bourgeoisie (see Ellin, 1997a: 18–19).

Although the physical legacy of Haussmann's work was enormously important, so too was its clear articulation of that modernist understanding of the street, that 'Streets had been for walking to work or shops and for socialising. Now they were primarily for movement' (Ellin, 1997a: 13). This was to be strongly endorsed by the architect and planner Le Corbusier and it is the streets of his planned 'Contempo-rary City' of 1922 that provide the next stop on this 'grand tour'. To be built on the Right Bank of the Seine, this was to be a city of high towers, open spaces and new kinds of streets. According to Le Corbusier, 'The corridor street "should be toler-ated no longer" because it is full of noise and dust, deprived of light and so 'poisons the houses that border it' (Le Corbusier, 1996: 371). Although this use of a medical metaphor harks back to Haussmann, it is the city as machine which provides the central metaphor of Le Corbusier's urban vision. The corridor street must be replaced by a new type of street which will be 'a machine for traffic' (quoted in Berman, 1983: 167) used exclusively by fast-moving mechanical vehicles, and free from pedestrians and building fronts. Although this would mean the abolition of the street and with it the crowd and many other activities, for Le Corbusier it was a price worth paying. Capturing the modernist spirit, he declared, 'A city made for speed is made for success' (Le Corbusier, 1996: 375). Unfortunately for Le Corbusier, his Contemporary City proposal won him few planning commissions and it was not until the construction of Brasilia in 1960 which drew strongly on his ideas that the implications of a city without streets became apparent. In place of the street, Brasilia substitutes high-speed avenues and residential cul-de-sacs, a configuration which doesn't simply erase a particular type of space (the street) but also undermines particular forms of social and political life. As Holston's (1989) fascinating study reveals, Brasilia is a city without 'street corner societies' where people might gossip informally and exchange information because there are no street corners and people therefore rely more on domestic and private spaces for social interation. And Brasilia is a city without crowds because by abolishing the street the planners effectively destroyed those public spaces where people might meet to express and debate their political beliefs and through which the public sphere of civic life is both represented and constituted (Holston, 1989: 103). If, as many have claimed, 'revolutions entail a taking to the streets' (Mitchell, 1995: 124; but see also Berman, 1983, 1986), Le Corbusier's ideas represent a neat counter-revolutionary strategy.

Next stop on the tour is 1960s New York and the streets of Greenwich Village from where Jane Jacobs (1961; see also Jacobs, 1995, 1996) made her vehement attack on the Corbusian tradition of expressways and tower blocks. Le Corbusier had visited New York some thirty years earlier, delighting in the simplicity with which it was possible to navigate the city because of the regular street grid: 'the streets are at right angles to each other and the mind is liberated' (Le Corbusier, 1995: 100). However, he went on to observe 'an urban no man's land made up of miserable low buildings in poor streets of dirty red brick' and it was precisely from such streets that 'the great refutation of his model of urbanism would be launched' (Kasnitz, 1995: 93). Jacobs describes in vivid detail the rhythms of daily life on Hudson Street in Greenwich Village, arguing that streets play a central role in establishing urban communal life and, in particular, in promoting safety. To achieve this, however, it is essential for the street to be 'multifunctional', not the exclusive domain of traffic, and for there to be 'eyes on the street' belonging to local inhabitants and traders who are able to provide neighbourhood surveillance of activities taking place on the street.

While the contrast with Le Corbusier's vision of the street could hardly be greater, the almost pastoral image of self-regulating street life that Jacobs conjures up (Berman, 1983: 324) also stands in stark contrast to the final destination on this 'grand tour', the streets of contemporary Los Angeles. Displaying 'the gritty street-wise pluck of the truck driver-*flâneur*' (Soja, 1997: 27), Mike Davis guides us round the 'Mean Streets' of LA, pointing out the 'bumproof' benches, sprinkler systems and regular police patrols as evidence of the city's 'relentless struggle to make the streets as unliveable as possible for the homeless and the poor' (Davis, 1995: 362; see also Davis, 1996). Although Davis's account has been criticised by those who feel his 'overheated rhetorical excesses often seem to overwhelm rational discourse' (Legates and Stout, 1996: 158), his disturbing images of 'the inhumanity of Downtown streets' (ibid.: 365; Soja calls them 'sadistic street environments', 1997: 27), do highlight two important and related themes of postmodern urbanism. First, these images underline the way in which 'form follows fear' in the contemporary urban environment (see Ellin, 1996, 1997b); secondly, they point to an increasing erosion of democratic public space (see Sorkin, 1992; Christopherson, 1994).

The importance of this tour of city streets in terms of providing wider insights into urban society should not be underestimated. These studies can be used individually and collectively to illustrate how streets are sites and signs of discipline and disorder, symptoms and symbols of modern and postmodern urbanism. Further, these studies show how streets can be viewed as both 'representations of space', the discursively constructed spaces of planners and architects, and 'spaces of representation', the spaces of everyday life of 'inhabitants' and 'users' (Lefebvre, 1991). Nevertheless, this 'grand tour' clearly has many limitations. Most obviously it is tied to an extremely narrow range of historical, geographical and cultural settings and therefore inevitably fails to engage with the heterogeneity of streets located in different times and spaces. More significantly, this 'grand tour' relies on limited methodological positions. At one extreme is Le Corbusier who sees the street (to borrow Lefebvre's phrase), 'from on high and from afar' (quoted in Gregory, 1994: 404); at the other, the accounts of both Jacobs and Davis involve the 'epistemological privileging of the experience of the *flâneur*', the street-wandering free agent of everyday life' (Soja, 1997: 21). These empirical and methodological limitations, in turn, inevitably circumscribe the theoretical contribution of the 'grand tour'. While the descriptions and analysis of the different streets on the tour can, as suggested above, be used in broader theoretical debates about society and space, individually these accounts are only weakly informed by specific theoretical ideas. It is against this background that the essays in this volume attempt to enrich our understanding of the street.

AN INTRODUCTION TO THE ESSAYS

The contributions are grouped around three broad themes: 'Planning and Design', 'Social Identities and Social Practices', and 'Control and Resistance'. The first section, 'Planning and Design', establishes the importance of seeing streets as environments constructed by knowledgeable agents situated within particular social, political and economic settings. Streetscapes are very much 'a synthesis of charisma and context, a text which may be read to reveal the force of dominant ideas and prevailing practices as well as the idiosyncrasies of a particular author' (Ley and Duncan, 1993: 329; see also Appleyard, 1981; Çelik, Favro and Ingersoll, 1994, and Moudon, 1991). Comprising four essays organised in chronological order, the first by David Atkinson examines the restructuring of Rome under the Fascist regime of Benito Mussolini, focusing on the creation of streets which would both express the ideological agenda and stage the rituals and performances of fascism. Of these streets, the *Via del Mare* (the Road to the Sea) begun in 1926 was one of the most ambitious projects and Atkinson's study provides intriguing insights into the making of totalitarian urban space. Standing in stark contrast to this broad, monumental boulevard constructed in central Rome are the narrow, winding streets of Pollok, a municipal suburb to the south of Glasgow begun in the 1930s. Built to take those displaced by redevelopment in Glasgow's inner city, the design of Pollok, as Gerry Mooney's chapter reveals, drew inspiration from the representations of space produced by the Garden City movement and this is partly expressed in the network of broad, tree-lined streets and narrow, curving roads laid out in sympathy with the local, undulating topography. One of the central themes of Mooney's chapter, however, is that streetscapes rarely reflect a straightforward application of some visionary model (in Pollok the pressures to house more and more people lead to the construction of four- and five-storey tenements on streets originally designed for two-storey cottages) and this theme is reworked in John Gold's chapter in the context of modernist plans to replace the traditional street. The impulse for many of these plans came from an acute sense that the street had become a 'battleground' between competing and conflicting uses: motor vehicles and pedestrians, local and through traffic, commercial and private activities. Although Le Corbusier had vigorously attacked the waste and inefficiency of the *rue corridor*, Gold shows that there remained an important gap between the visions of Le Corbusier (and other modernists) and blueprints for street planning. 'The boulevard might be dead', Gold observes, 'but the urban expressway was yet to arrive'. By considering the work of the British Modern Movement in the 1940s, Gold illustrates how this gap between vision and practice provided scope for considerable experimentation with multi-level, functionally defined circulation systems. Finally, in this first section, Richard Levy provides a glimpse of how current Computer Aided Design (CAD) technology is being used to transform the planning and design of streets. Using animations and virtual reality it is now possible to simulate the pedestrian experience of a planned streetscape, allowing people to see its impact, quite literally, on their view of the environment. Impressive though this technology is, arguably its greatest potential impact is in democratising the design process by allowing anyone with access to a television or computer screen a simulated experience of plans proposed for their community.

In Part II, the focus shifts from the making of streetscapes to explore the meaning and significance of the street in relation to social identities and social practices. By focusing on these themes, the essays in this section contribute to wider debates which question the view that as public spaces streets are universally accessible to a civic public, and provide evidence of how streets can be an active medium through which social identities are created and contested (see Ruddick, 1996: 133–5). Jane Rendell's vivid account of the male rambler on the streets of early nineteenth-century London offers, at one level, an intriguing series of observations on the social, cultural and economic geography of the city, as the rambler takes to the street and guides the reader between sites of leisure and pleasure (theatres, opera houses and parks), consumption and exchange (the main shopping streets, private arcades and bazaars). But these urban explorations have a wider theoretical relevance. The movement of the male rambler through the streets reveals much about the gendering of urban space and his mobility suggests that the relationships between gender and space are more complex than established ideas concerning the 'separate spheres' of the male public realm and the female private realm. In contrast to the mobility of the rambler is the more restricted movement of those with disabilities examined by Brendan Gleeson. Piecing together a fragmentary historical record, Gleeson reveals the strategic importance of the street in the lives of disabled people in colonial Melbourne. Their inability to meet the mobility requirements of industrial capitalism, with its separation of work and home, combined with the desire of those in authority to confine the disabled to the workhouse, asylum or jail, meant that the very presence of disabled people on the streets, as beggars or street-traders, represented a minor victory for those struggling for some sense of inclusion in an exclusionary society. The disabled also feature in the following chapter as one group among the diverse array of people who make up the homeless living on the streets of contemporary Britain and North America. Linking together some of the reasons for homelessness, Gerald Daly unravels a complex chain of events and decisions in which the personal becomes enmeshed in the political and the economic, the local in the global. Daly shows how, once on the street, the homeless enter into the shared experience of a harsh and brutal urban environment in which their movements around the city are mapped out by the locations of hostels, missions, drop-in centres and soup kitchens.

Running throughout Daly's chapter is a strong sense of the physicality of life on the street, a theme developed by David Bell and Jon Binnie in a rather different context in their account of the 'erotics of the street'. Informed by the writings of de Certeau (1984), they argue that walking, looking and being looked at on the street are fundamental aspects of the formation of 'queer consciousness'. By juxtaposing two fictional accounts of urban queer sexualities, they show how the streets can provide sites for very different experiences. In Andrew Holleran's *Dance from the Dance*, the streets are to be cruised in search of love whereas in Stewart Home's *No Pity* the streets are settings of violence and acts of revolutionary agitation. In making sense of these different experiences of the street, it's not simply their different geographical settings that matter (Holleran's novel is set in Manhattan, Home's in London) but also their temporal context, for Holleran's romantic vision predates AIDS while Home's 'nihilistic brutalism' is set firmly within the time of AIDS. The interrelationships between masculinities and the street are also examined by Aitken and Lukinbeal's analysis of the streetscapes found in the films of Terry Gilliam. In film, Aitken and Lukinbeal argue, urban streets often act as 'visual signifiers' of the loss of innocence and the alienation of city life, with the 'mean streets' of the city commonly juxtaposed with the escapism associated with life on the 'open road'. In Gilliam's movies, however, where the streetscapes range from fascist monumentalism to medieval citadels, characters are frequently trapped by and absorbed into the street with little prospect of such escape to the 'open road'. Aitken and Lukinbeal use scenes from Gilliam's films to investigate how street myths, masculinities and representations of hysteria contrive what they call a *mise-en-scène*, a 'continuous space . . . a positioning and positional movement' for multiple male masculinities.

In the next chapter, the focus broadens as David Crouch explores the significance of the street as 'an everyday site of geographical knowledge and leisure practice'. From walking to a football match to going to play in a park or tend an allotment, he shows how people make sense of their lives by way of the street, revealing the rituals and relationships, practices and representations, which are played out routinely on the street. In the following chapter, Peter Jackson is also concerned with the everyday experience of the street but the focus is narrower as he explores the theme of 'domesticating space' in the context of two shopping malls in north London. In contrast to the perceived incivility of city streets, the privately owned and managed but publicly accessible spaces of the mall are environments which allow for the regulation of difference and the promotion of 'the virtues of familiarity'. Such environments don't, however, generate a singular experience of 'consumer citizenship'. Rather, differences of ethnicity, class, age, and gender all mediate and complicate the meanings people attach to the mall. Further, Jackson's chapter alerts us to the dangers of domesticating the street. Purifying and privatising spaces to enhance the consumption experience for some comes at a price of social exclusion and a sense of increasing inequality for others.

The consumption experience is given a more literal twist in Gill Valentine's chapter which examines the interrelationships between food and the street. Historically, she argues, social expectations about eating have determined that the street should be viewed as a cultured space where people exercise self-restraint in the face of natural urges to eat; a public space where intimate, 'private' bodily matters such as eating are not on display; and an ordered space, where the mess of consumption is kept out of sight. Attitudes to eating in the street are changing, however, and Valentine's research indicates many of the past taboos about eating on the street are being broken down. Changing work practices, the growth in canteen style fast foods in schools and on the high street, and the anonymity of contemporary urban life have all contributed to developing more informal social codes of eating. 'Grazing' on city streets thus now appears to be a more accepted form of behaviour than in the past and in turn, Valentine suggests, the street is becoming a more informal and more democratic space, less regulated by codes of 'civility', 'privacy' and 'order'. Of course, taboos about eating on the street are culturally as well as historically specific and the final chapter in this section, Tim Edensor's 'The Culture of the Indian Street', rightly requires us to acknowledge the cultural specificity of the regulated qualities of Western street life. He explores many of the themes already encountered in the book – social practices, movement, regulation, sensual experience – but in the context of an idealised Indian street, destabilising many taken-for-granted notions about the ordering of street practices in the West. This, too, has a wider theoretical relevance for, as Edensor shows, it unmasks the ethnocentricity of Baudelaire's *flâneur*, Foucault's heterotopia and de Certeau's walks in the city.

In Part III, 'Control and Resistance', the four essays pick up on themes of regulation, ordering and the surveillance of street life that have surfaced at various points earlier in the book and address them in more detail. In the frontline of maintaining order on the street are the police, and Steve Herbert's rich ethnographic study of patrol officers in Los Angeles provides an interesting elaboration on Rubenstein's observation on the relationship between the street and police identity: 'For the patrolman [sic] the street is everything; if he loses that, he has surrendered his reason for being what he is' (Rubenstein, 1973: 166). Police efforts to claim sovereignty over the street are always subject to contestation and Herbert, by exploring an infamous 'anti-police' location', shows how attempts to maintain territorial control here are driven by a sense of moralistic fervour and a love of the physicality of the chase as the police try to insert state authority into everyday life on the street. This kind of policing, however, represents only one way of maintaining order on the street. Accounts charting the rise of the 'fortress city' (see, for example, Davis, 1990; Christopherson, 1994) illustrate that issues of security and control are now prominent themes in contemporary planning and urban design as 'form follows fear' in the postmodern city. As Loretta Lees observes in her chapter, there is an increasing use of war rhetoric by many urban analysts as they describe the 'embattled' public spaces of city streets and warn of the 'end of public space'. Although Lees' study of two gentrification projects in Vancouver does lend some support to this view it is, she argues, only part of the story. Alongside attempts at enhanced control in urban design are also concerns with promoting the diversity and vitality of the street, revealing a more complex and ambivalent relationship between gentrification and urban life than notions of the privatised streets of the fortress city allow. Nevertheless, as Nick Fyfe and Jon Bannister argue in their study of closed circuit television (CCTV) surveillance on the streets of Glasgow, the use of public space CCTV (funded partly by private capital) is implicated in a subtle privatisation of public space as commercial imperatives increasingly define what is 'acceptable' behaviour on the street. This reflects broader processes linked to the economic and political restructuring of public space as attempts are made to create a 'downtown as mall'. As Jackson's earlier chapter indicated, however, there is a high price to be paid for such attempts at 'domesticating the street'. Drawing on the work of Sennett, Fyfe and Bannister suggest that purifying the street of disorder and difference may actually deprive people of the ability to handle conflict, leading to problems of violent over-reaction to any social disorder when it does arise. The final essay by Tim Cresswell connects up this theme of disorder on the street with that of resistance by examining the 'subversive scrawls' that appear at night on advertisements, signs and buildings in city streets. Focusing on 'billboard banditry' and the work of the Polish-Canadian artist Krzysztof Wodiczko, Cresswell argues that the intrusion of 'illegal' and unsanctioned texts and images onto the street is important politically because it creates new meanings and messages in public spaces which can destabilize the meanings and messages of officially sanctioned forms of discourse. More generally, however, these 'subversive scrawls' capture a creative tension, one that manifests itself in different forms throughout this book, between the street defined from 'above' as a space of order and discipline and the street as experienced from 'below' as a space of conflict and contestation.

No one volume on streets could hope to capture 'all their hectoring danger, their swirling confusion, and their muddled vitality' (Boddy, 1992, p.153) and this collection makes no claims to offering a comprehensive guide to 'images of the street'. Nevertheless, the essays, individually and collectively, clearly advance our understanding of the street beyond the limited empirical, methodological and theoretical horizons of the 'grand tour' described at the beginning of this introduction. This is reflected in sheer diversity of streets explored by the contributors, the range of methodological strategies and source materials used, and the connections made with different theoretical debates. Capturing the vitality, excitement and tensions of the street, this volume should enrich and extend our understanding of the making and meaning of a key urban space.

ACKNOWLEDGEMENTS

I would like to thank Mark Boyle and Donald McNeill for their helpful comments on this introduction.

REFERENCES

Appleyard, D. (1981) *Livable Streets*, Berkeley, University of California Press.

Benjamin, W. (1995) 'Paris: Capital of the Nineteenth Century', in P. Kasnitz (ed.) *Metropolis: Centre and Symbol of Our Times*, London: Macmillan, pp. 46–57.

Berman, M. (1983) *All That Is Solid Melts Into Air – The Experience of Modernity*, London: Verso.

Berman, M. (1986) 'Taking it to the Streets: Conflict and Community in Public Space', *Dissent*, 333(4): 476–85.

Boddy, T. (1992) 'Underground and Overhead: Building the Analogous City', in M. Sorkin (ed.) *Variations on a Theme Park: The New American City and the End of Public Space*, New York: Hill and Wang, pp. 123–54.

Çelik, Z., Favro, D. and Ingersoll, R. (1994) *Streets: Critical Perspectives on Public Space*, Berkeley: University of California Press.

Christopherson, S. (1994) 'The Fortress City: Privatized Space, Consumer Citizenship', in A. Amin (ed.) *Post-Fordism: A Reader*, Oxford: Blackwell, pp. 409–27.

Davis, M. (1990) *City of Quartz: Excavating the Future of Los Angeles*, London: Verso.

Davis, M. (1995) 'Fortress Los Angeles: The Militarization of Urban Space' in P. Kasnitz (ed.) *Metropolis: Centre and Symbol of Our Times*, London: Macmillan, pp. 355–68.

Davis, M. (1996) 'Fortress L.A.' in R. T. LeGates and F. Stout (eds) *The City Reader*, London: Routledge, pp. 159–63.

De Certeau, M. (1984) *The Practice of Everyday Life*, Berkeley, University of California Press.

Ellin, N. (1996) *Postmodern Urbanism*, Oxford: Blackwell.

Ellin, N. (1997a) 'Shelter from the Storm or Form Follows Fear and Vice Versa', in N. Ellin (ed.) *Architecture of Fear*, Princeton: Princeton Architectural Press, pp. 13–46.

Ellin, N., (ed.) (1997b) *Architecture of Fear*, Princeton: Princeton Architectural Press.

Gregory, D. (1994) *Geographical Imaginations*, Oxford: Basil Blackwell.

Holston, J. (1989) *The Modernist City: An Anthropological Critique of Brasilia*, Chicago: University of Chicago Press.

Jacobs, J. (1961) *The Life and Death of Great American Cities: The Failure of Town Planning*, Harmondsworth: Penguin Books.

Jacobs, J. (1995) 'The Uses of Sidewalks' in P. Kasnitz (ed.) *Metropolis: Centre and Symbol of Our Times*, London: Macmillan, pp. 111–29.

Jacobs, J. (1996) 'The Uses of Sidewalks: Safety' in R.T. LeGates and F. Stout (eds) *The City Reader*, London: Routledge, pp. 104–8.

Kasnitz, P., (ed.) (1995) *Metropolis: Centre and Symbol of Our Times*, London: Macmillan.

Keith, M. (1995) 'Shouts of the Street: Identity and the Spaces of Authenticity', *Social Identities*, 1(2), 297–315.

Lash, S. and Friedmann, J. (1992) 'Introduction: Subjectivity and Modernity's Other', in S. Lash and J. Friedmann (eds) *Modernity and Identity*, Oxford: Blackwell, pp. 1–30.

Le Corbusier (1995) 'New York Is Not a Completed City' in P. Kasnitz (ed.) *Metropolis: Centre and Symbol of Our Times*, London: Macmillan, pp. 98–110.

Le Corbusier (1996) 'A Contemporary City', in R. T. LeGates and F. Stout (eds) *The City Reader*, London: Routledge, pp. 368–75.

Lefebvre, H. (1991) *The Production of Space*, Oxford: Blackwell.

LeGates, R. T. and Stout, F. (eds) (1996) *The City Reader*, London: Routledge.

Ley, D. and Duncan, J. (1993) 'Epilogue', in J. Duncan and D. Ley (eds) *Place/Culture/Representation*, London: Routledge, pp. 329–34.

Mitchell, D. (1995) 'The End of Public Space? People's Park, Definitions of the Public, and Democracy', *Annals of the Association of American Geographers*, 85(1): 108–33.

Moudon, A. V. (ed.) (1991) *Public Streets for Public Use*, New York: Columbia University Press.

Rubenstein, J. (1973) *City Police*, New York: Ballantine.

Ruddick, S. (1996) 'Constructing Difference in Public Spaces: Race, Class and Gender as Interlocking Systems', *Urban Geography*, 17(2): 132–51.

Sennett, R. (1990) *The Conscience of the Eye: The Design and Social Life of Cities*, London: Faber & Faber.

Sorkin, M. (ed.) (1992) *Variations on a Theme Park: The New American City and the End of Public Space*, New York: Hill and Wang.

Soja, E. (1997) 'Six Discourses on the Postmetropolis' in S. Westwood and J. Williams (eds) *Imagining Cities: Scripts, Signs, Memory*, London: Routledge, pp. 19–30.

Sudjic, D. (1992) *The 100 Mile City*, London: Flamingo.

Vidler, A. (1978) 'The Scenes of the Street: Transformations in Ideal and Reality, 1750–1871' in S. Anderson (ed.) *On Streets*, Massachusetts: MIT Press, pp. 29–111.

Young, I. (1990) *Justice and the Politics of Difference*, Princeton: Princeton University Press.

Part I

PLANNING AND DESIGN

2

TOTALITARIANISM AND THE STREET IN FASCIST ROME

David Atkinson

•

INTRODUCTION

Between 1922 and 1943, under the Fascist regime of Benito Mussolini, the city of Rome was transformed. Its population grew from around 660,000 when Mussolini seized power to a figure of 1,500,000 when he finally left Rome (Agnew, 1995; Fried, 1973). To accommodate this growth, the city's perimeter witnessed the relentless construction of new residential districts. The centre of the city was subjected to radical surgery as the regime re-planned and rebuilt the Eternal City in a sustained attempt to articulate its political, cultural and social agenda through the form and use of the cityscape. Indeed, so politicised was this programme of urban interventions and so crucial were public spaces to the regime that Diane Ghirardo argues:

> In fascist Italy, much of the battle for hearts and minds of Italians took place in the public arena, in the streets and squares of the peninsula's cities. . . . Propaganda campaigns carried out in newspapers, books, conferences and parliamentary speeches, and even radio broadcasts, paled in comparison with fascist activities in the streets. Mass civic events became a fascist trope, a means of forging a new, post-democratic collectivity and of inscribing the public character of the new political formation into the urban realm.
>
> (Ghirardo, 1996: 347)

Many commentators have discussed the merits of individual examples of Fascist architecture, whilst an increasing number have expanded the scale of analysis to identify the importance the regime accorded to planning and urban landscapes (Atkinson and Cosgrove, 1998; Cederna, 1979; Ernesti, 1988; Fuller, 1996; Ghirardo, 1990; Kostof, 1973). However, despite their frequent references to the Fascist street and public space, seldom have these studies concentrated in a sustained fashion upon the role of the street under totalitarianism (although see Ghirardo, 1988, 1990). By contrast, we know that Italian cities nurture long-standing traditions whereby the piazza and the street function as a locus of political, civic and social life (Calabi, 1993; Cosgrove 1982, 1984; Muir, 1981; Muir and Weissman, 1989), and Luigi Barzini (1964) and Mario Isnenghi (1994) both affirm that streets and piazzas have remained a central site of Italian life through into the twentieth century. Given that other writers have mentioned the ways in which the Classical and Renaissance notions of ceremonial entrances to the city were reworked in the Fascist period (Atkinson, 1997, Fogu, 1996; Ghirardo, 1996), it is no surprise to find that the street became a focus of the Fascist regime's attention.

This essay considers some of the ways in which the long-standing importance of streets and public spaces in Italy was renegotiated and reworked by the regime as part of its gradual appropriation of the public sphere. From the early days of arbitrary violence in the city's streets and piazzas, through the efforts to police and exclude certain individuals, via the demolition and re-planning of Roman streets and thoroughfares, to the formal ceremonies, parades and processions which were characteristic of the regime, I will argue that a focus upon the street lends us a fruitful insight into the making of totalitarian urban space. I do this by exploring the general processes through which the streets were transformed and – eventually – controlled by the regime towards its totalitarian ends.

EARLY FASCISM AND PUBLIC SPACES

In its earliest form, Italian Fascism was not so much a recognisable political party as a violent and essentially *public* expression of political disaffection at the Government's failure to reward wartime sacrifices with the promised employment, land-reform, extension of the franchise, and economic well-being. It reacted against the old, tired Liberal governments which seemed unable to cope with the instabilities and uncertainties of the post-war world. In contrast, the *Fascisti* imagined themselves capable of building a revived, modern Italy for the new era. Yet, rather than debate political ideas and ideologies in closed rooms or hold orthodox public demonstrations, Benito Mussolini and the motley collection of demobilised soldiers and alienated young men who constituted the *Squadristi* (the fighting squads) of early Fascism, roamed their local city, town or district visiting punitive and random intimidation upon their political opponents and demonstrating a ready penchant for brutal public violence. As Mario Isnenghi (1994) comments, the *Fascisti* understood very clearly the political and symbolic significance of such public confrontations. One Tuscan *Squadrista* wrote in his diary:

> [I]n the heart of those of us who take to the streets, is the conviction that to conquer the old world of the fogeys and the new world of the Asiatic hordes, we must occupy with both the spirit and the body the squares and the streets of the city and hold on to them with fortitude.
>
> (Isnenghi, 1994: 287)

And this sense of mastering public spaces for the redemption of the nation was later echoed by Mussolini himself, who wrote in his autobiography:

> I could not forget that I was also Chief of that Party that for three years had fought in the squares and streets of Italy – not only to gain power, but above all for the supreme task and the supreme necessity of infusing into the nation a new national spirit.
>
> (Mussolini, 1936: 195)

However, the main victims of their killing, beatings and arson attacks – the perceived threat to the redemption of Italy – were usually the left-wing working-class organisations which themselves had mobilised in the hope of fermenting revolution amidst the chaotic aftermath of the post-war period (Lyttelton, 1973). Particularly in the North and centre of the peninsula, the years 1921–2 were marked by violent confrontations between the left and the Fascists, with the latter usually victorious thanks to the tacit support of the police and the land-owning and industrial elites who feared international Bolshevism still more than the domestic phenomenon which was Fascism. By the Autumn of 1922, the left was largely defeated and, despite its lawlessness, Fascism was established as a significant political force, although one which could still have been defeated easily by the Army (Lyttelton, 1973). However, when Mussolini threatened to 'March on Rome' with his militia and seize control to 'save' Italy from anarchy, in an attempt to absorb Fascism into the Italian political system and to 'normalise' its energies, the government acceded to a role for the Fascists and King Vittorio Emmanuele III invited Mussolini to form the next government.

In keeping with their emphasis upon public shows of strength, the future dictator insisted that his self-styled and Black-shirted 'Legionnaires' fulfil their threat to 'March on Rome' and by so doing to demonstrate the transfer of power in a very public fashion. The city of Rome enjoyed a significant mythical importance for the infant Fascist movement and the notion of *romanità* (The Roman Spirit) was a central element of their conservative-nationalism (Visser, 1992). Thus, when the Fascist columns marched into the ancient city through its historical gateways to congregate in the centre, the symbolism of this performance was not lost on their opponents. In order to avoid the violent scuffles which attended the Blackshirt advance, Mussolini himself arrived by train and hurried to meet the king. However, these formalities concluded, Mussolini headed immediately to *Piazza Venezia* where he climbed the steps of the National Monument and, before the assembled Blackshirts, made his first speech as Prime minister (Munro, 1933). In the earliest acts of Fascism, therefore, a direct and immediate concern with the public urban spaces of the city and the need to appropriate these areas for Fascism in a very public and visible manner had already been revealed.

This concern to appropriate Rome's public spaces – including the streets where everyday Roman life continued – was to become a significant element of Fascist government. Just as two years previously the Fascist *squadristi* had fought to 'tame the red piazzas' (Isnenghi, 1994: 302), such violence was frequently translated directly onto the streets of the capital. In his 1926 book entitled *Through Fascism to World*

Power Ion Munro, a British journalist based in Italy, described a typical Roman street-scene in the early days of Mussolini's premiership:

> [y]ou would as likely as not meet a bunch of Fascists – with clubs in their hands that would make Irishmen's shillelaghs look like toothpicks, and more pistols in their belts than in a wild-west film – bearing down on some news-paper kiosk. There you would see them seizing and making a street bonfire of some Opposition publication – the unfortunate vendor wisely keeping his thoughts to himself in a sidestreet. Or you would see a mob with banners pour-ing by, singing *Giovanezza* [a Fascist marching song] and knocking the hats off the heads of those who were not bareheaded before the blackshirt emblems. Then you might see tens of thousands of people pouring into *Piazza Colonna* for a mere glimpse of Mussolini to give him an ovation. And then again you might see, later, the same *squadristi* that you had seen surging like lion-tamers among the citizens, kneeling immobile and in dedicatory homage before the tomb of the unknown warrior and the *Ara Patria* [the National Monument].
>
> (Munro, 1933: 209–10)

Munro's narrative tallies with other accounts of the period when the regime had still to establish itself unequivocally and opposition parties posed a threat, albeit a dimin-ishing one, to Mussolini's government. The streets remained a political battleground as the Fascists continued to provoke brutal public violence – even leading to the blatant intimidation of voters outside polling booths in the April 1924 elections. Giacomo Matteotti, a Socialist deputy in the Italian parliament and the most vocal and unrelenting critic of Fascist rule, attacked the government's election victory as worthless because of this unabashed intimidation. In return, on 10th June, 1924, Matteotti was kidnapped, beaten and killed by a group of Fascist thugs on his way to Parliament. Mussolini's evident complicity in the crime was condemned by a series of significant figures in Italian society and Rome's streets and piazzas were filled by demonstrations against this blatant political assassination.

Many historians identify the Matteotti crisis as a crucial watershed in the Fascist period; had a powerful constituency such as the Monarchy, the Church, the Army or Parliament moved against Mussolini effectively at this point, it is argued, Fascism would probably not have survived. However, the King, the Pope and the generals failed to act against the Fascists, whilst the strategy of the opposition deputies who withdrew from parliament in protest at the political assassination, proved to be an ineffectual democratic response to a new genre of undemocratic politics. Mussolini refused to resign and managed to retain his office. Indeed, whilst the political manoeu-vring continued behind closed doors, Mussolini continued to use the streets to remind ordinary Romans who might protest against him of the violent capacity of Fascism:

> I ordered the Florentine legions of the [Fascist] Militia to parade in the streets of the capital. The armed Militia with its war-songs is an element of great persuasion. It is an argument.
>
> (Mussolini, 1936: 210)

Even at the height of the Matteotti crisis, then, Mussolini maintained a Fascist presence in the streets. Yet at this stage, perhaps not even his most ardent supporters imagined that his influence over these spaces would last for the next eighteen years. They reckoned without the development of the Fascist dictatorship. Towards the end of 1925, under increasing pressure from extremists in his own movement to push the Fascist 'revolution' forward apace, Mussolini took the crucial decision to step beyond the democratic frameworks he had inherited as premier. On 3 January 1925, he addressed parliament and took full responsibility for all the actions of the Fascist squads over the past three years – abandoning any pretence towards democracy. It was effectively a *coup d'état* and in the days which followed, Mussolini and his *Gerarchia* (hierarchy) began systematically to dismantle the democratic Italian state and to establish a dictatorship and the world's first self-consciously labelled 'Totalitarian' state.

TOTALITARIAN FASCISM AND THE STREETS

Mussolini first coined the term 'Totalitarianism' in the aftermath of his 1925 declaration of dictatorship, during a speech in which he declared that all of Italy was to be 'Fascistized' (Morgan, 1995). In retrospect it has become clear that the 'Fascistisation' of Italy never attained the degree of social, cultural and political control which the regime hoped for and which was briefly threatened in 1925 (Aquarone, 1965; De Grand, 1991). For as Forgacs (1986) points out, this 'flawed' totalitarianism was compromised by the relative independence enjoyed by the industrial and financial elites and national institutions such as the monarchy, the Army, and, most significantly, the Catholic Church. The state failed to intervene systematically in the realms of culture and education until the mid-1930s and likewise, the judiciary and the police retained much of their autonomy. In addition, the project was compromised by the uneven impact of the regime's use of communications technologies, the differing socio-economic contexts upon which Fascism attempted to graft its totalitarianism and the backwardness of various regions, especially the south (Ponziani, 1995; Steinberg, 1986). In relative terms, one commentator recently conceded that: 'As major twentieth century dictatorships go, the Mussolini regime was neither especially sanguinary nor particularly repressive' (Payne, 1995: 122).

However, having said all of this, it is clear that the inter-war period witnessed the widespread *attempts* by the regime to crush all political opposition and to control the mass cultures and *everyday* lives of ordinary Italians in its attempt to forge a new, 'fascistised' Italy (Gentile, 1995). The impacts and results of this programme may seldom have been as effective as the Fascists intended, but they nevertheless impacted significantly upon Italian cultural and social life. They certainly transformed democratic politics. In early 1925, elected local government was abolished and the fledgling regime foisted its own mayors upon towns and cities. Opposition political parties were banned in 1926 and the chamber of deputies was abolished two years later (Payne, 1995). Censorship was introduced and unsympathetic newspapers were closed. Laws against secret societies were enacted with the Mafia and Freemasonry as the intended targets.

In the workplace, the union-breaking days of early Fascism were reprised by the legal suppression of labour organisations and unions which were replaced by 'Fascist Corporate Councils' (Abse, 1996). Beyond the workplace, leisure activities and numerous aspects of social and cultural life were increasingly the focus of the regime's strident attempts to order society. In 1929 accommodation was made with the church which had been at loggerheads with the secular state since the incorporation of the papal city into Italy in 1870. And although disputes between church and state continued, especially over the 'moral' training of young minds, the schism between the spiritual realm of Catholicism and the civil domain of the state was not the problem it had been for previous Italian governments (Pollard, 1985). The regime made increasing demands on the remaining leisure time of Italians. A range of neo-militaristic youth organisations and the *Opera Nazionale Dopolavoro* (OND) (National Afterwork Organisation) were designed to occupy Italians with a huge range of 'suitable' activities during the evenings and the specially instituted *Sabato fascista* (Saturday afternoon holidays) (De Grazia, 1981). The development of radio, cinema, exhibitions and mass theatre also attempted to draw the masses into the cultures of Fascism (Brunetta, 1975; Hay, 1987; Schnapp, 1996; Stone, 1993; Monteleone, 1976). Likewise, the exploitation of communications and media technologies was a key feature of the regime's totalitarianism (Thompson, 1991).

Finally, in domestic spaces, legislation reinforced the existing patriarchal cultures of Italy and insisted upon conservative maternal and domestic roles for Italian women (De Grazia, 1992). Despite tokenistic organisations to occupy middle-class women and the spare time of domestic servants and rural housewives (Willson, 1996), Fascist women were primarily supposed to provide a home for the Fascist male and to nurture their brood of young Fascists to fight in future imperial campaigns (Caldwell, 1991). She would contribute to the nation through childbirth, and 'devote herself to maximising her output of children in order to qualify as one of the truly heroic figure of the Fascist age – the "prolific mother"'(Willson, 1996: 80). She was certainly not supposed to enjoy the freedom of the streets (Saraceno, 1991). By contrast, the public spaces already marked by the public violence of Fascism became the domain of the alternative gender permitted by the regime – the idealised Fascist male. This heroic, strong, disciplined and virile individual became one of the central symbolic figures of the state and one who was intended to forge Italy's new place in the world (Spackman, 1990). He occupied the political spaces denied to women and was central to Fascist cultures, as one Fascist admitted:

> Fascist culture must be life and the expression of life; it must create a species
> of man, a new man, a whole man, one who is the same man in his family,
> in society, and in the state.
>
> (Gentile, 1996: 97)

And although commentators are agreed that the regime failed to realise the creation of the '*l'uomo nuovo fascista*' (the new Fascist man) (Corner, 1983; De Grazia, 1981), the streets of Italy were increasingly reworked as the spaces where these Fascist males performed their ascribed roles. As we have seen, public spaces were scarred by political violence during the emergence of Fascism. However flawed Fascist totalitarianism was in practice, the extent to which the pursuit of this idealised, male, Fascist citizen impacted upon the streets of Rome is worth further consideration.

Making Totalitarian streets – order and surveillance

The public violence of *squadrismo* was finally curtailed by Mussolini in late 1925 after a particularly lethal fight in Florence left eight Fascist opponents dead (Payne, 1995). By this stage though, the regime was increasingly ensconced in the structures of the state, and whilst no longer the site of frequent political brutality, the street became no less masculine and oppressive as the preserve of 'The New Fascist Man'. Fascism sought to exercise its authority and to discipline and order Italian life in such public spaces. With opposition parties, unions, and newspapers defeated, the regime began to focus on individuals. New legislation permitted a series of 'subversive' figures to be removed from the public realm. The Public Security decrees of November 1926 allowed local authorities to arrest and punish individuals 'singled out by public rumour as being dangerous to the national order of the state' (Morgan, 1995: 83). Other such 'delinquents' were also targeted, as were the 'wrong' kinds of women for the new Italy. In 1923, prostitutes had been subjected to a programme of regulation which sought to control sexualities which lay beyond the familial units advocated by Fascism. Three years later, under the nascent dictatorship, the new Public Security laws legislated for the eviction of *all* prostitutes from the streets (De Grazia, 1992). So although violent beatings were less frequent, the moral geographies of Fascism were increasingly marked in the public thoroughfares of Fascist society. Even at the most mundane levels, jaywalking was outlawed and the police enforced a one-way system upon the narrow pavements of central Rome – much to the admiration of *National Geographic*'s roving correspondent (Roberts, 1936).

Finally, despite the efforts of the OND and the gradual encroachment of the regime into leisure time, street-cafés and bars remained the hub of Italian social and cultural life. Consequently, they too became sites to be appropriated by the regime. Most of the remaining working-class bars were closed down by the police in 1927 alongside the suppression of other potential opposition meeting places (De Grazia, 1992). At the same time, the Fascist presence became gradually more marked in the more up-market cafes which were untouched by the police crackdown. Their most prestigious, streetside tables became the places which Fascist officialdom occupied to 'show their movement's capacity to occupy all the spaces originally associated with the construction of bourgeoise sociability . . . [and] to display their prowess as men' (De Grazia, 1992: 202). The public spectacle foreshadowed by the March on Rome was an increasingly important aspect of life under Fascism. And Fascist ambitions to enforce its moral geographies and clear the streets of 'unwanted' elements did not end with prostitutes and delinquents. The pathologies of the city were rooted much more deeply than this and resulted in the redesign and reconstruction of entire quarters of Rome.

Planning Fascist Streets – pathologies of the city

In May 1927 *Il Popolo d'Italia* (a leading Milanese newspaper) complained that cities were the 'breeding grounds of all the moral – but not only moral – infections of the people' (Horn, 1994: 104). These sentiments struck a common chord. Italian social scientists of the 1920s frequently interpreted cities as unhygienic harbours of disease and degeneracy but also, when beset by overcrowding and unemployment, as potential sources of political resistance or even hotbeds of bolshevism. Furthermore, Mussolini had declared cities to be a *demographic* problem. Despite the pro-natalism of the regime, the birth rate failed to match Fascist expectations and increasing urbanisation and especially rural-urban migration were blamed (Ipsen, 1996; Treves, 1976):

> [N]ationalists saw the city as threatening the virility and [virtue] of the predominantly rural Italian population by reducing its overall fertility . . . doctors and social scientists identified the city as working to subvert or invert the 'natural' gender distinctions and hierarchies that prevailed in the countryside.
>
> (Horn, 1994: 99)

As David Horn (1994) suggests, the regime responded with a dual strategy. 'Negative Urbanism' prompted laws designed to curb rural–urban migration and to restrict Italians to their 'natural' rural environments where their fertility was assured. The only city exempted from the anti-urban rhetoric of the regime was Rome, the capital, where Mussolini aimed for a population of 20 million by the late twentieth century (Kostof, 1973). Yet Rome was not exempt from 'Positive Urbanism' – a response to the infertility, diseases and potential political resistance of the Romans. Using a host of bodily metaphors, the planners and sociologists who constituted the new discourse of urbanism talked of treating the city and 'correcting' its maladies; theirs' were 'medico-social therapies targeting urban spaces, the home, and the reproductive body' (Horn, 1994: 96). Even to the extent of *sventramento* (disembowelling, or 'gutting' the city) (Cederna, 1979) they wanted to excise the 'dangerous' elements from the metropolis to leave a 'healthy', ordered, disciplined and fertile Fascist city.

The heart of Rome was subjected to perhaps the most spectacular and remark-able clearances of the period. Swathes of the centre were demolished and replaced by broad, modern, monumental avenues which became the sites for Fascist parades and spectacles, as well as the Roman traffic (Kostof, 1973). In tandem with the violent and legislative appropriation of Rome's public spaces, therefore, the regime set about physically remaking parts of central Rome to create streets which would both express the ideological agenda of Fascism *and* stage the rituals and performances of the regime, whilst guaranteeing a more ordered, fertile population. The first major intervention in the city follows as a brief example.

Work began on the *Via del Mare* – the Road to the Sea – in 1926 (Insolera, 1971; Muñoz, 1932). Departing from *Piazza Venezia*, the kernel of the city's road network and sweeping along the north and western flanks of the Capitoline hill, beyond the suburbs the broad, modern avenue eventually became Italy's first *autostrada*

FIGURE 2.1 A Roman courtyard (and some of its residents) of the kind Fascism cleared to make way for the *Via del Mare*. Source: Kostof (1994: 9)

connecting the capital with the Tyhrennian coast some twenty miles distant (Atkinson and Cosgrove, 1998; Vandone, 1929). Much was made of this connection to the coast and Romans were urged to escape the city at the weekend to breath the clean, restorative sea-airs. But in the centre of the city, the creation of a suitably healthy, Fascist environment was also addressed (Ciacci, 1941). The Roman terminus of the *Via del Mare* had previously been one of the most run-down parts of the city. Figure 2.1 shows a block which was part of one of the districts cleared by the Fascists in the 1930s; the regime associated the poorest, working-class Romans who resided there, and the dark maze of narrow streets and small piazzas, with overcrowding and unhygienic conditions (Mulé, 1932). Such streets were clearly at odds with the regime's insistence on the healthy, fertile city:

> According to Fascist theory, straight and wide avenues were indispensable. You could not reconcile tortuous, narrow streets with active traffic and sufficient light and air. . . . The street responded to a principle of public morality of which the state was the interpreter.
>
> (Kostof, 1973: 18)

FIGURE 2.2 The completed *Via del Mare*, looking towards the Theatre of Marcellus. Source: Conte (1934: 640)

Consequently, the regime's agenda, the rhetoric of slum clearance and the need to plan a modern city ensured that the piazzas and streets of the district were razed to the ground to make way for the broad, airy *Road to the Sea*. Figure 2.2, for example, views the avenue from its city-centre terminus (marked by a Roman-style milepost), after the slum clearance. The road can be seen approaching the ancient Theatre of Marcellus, which had been restored especially. The route then swings to the left through a series of newly restored ancient monuments which were isolated and exhibited amidst the kind of manicured parkland which can be seen to the left of the image, at the foot of the Capitoline Hill (Mulé, 1932; Muñoz, 1932). The map reproduced in Figure 2.3 clearly indicates the breadth of the avenue after it departs from the piazza in front of the National Monument (at the top left of the image), through the newly demolished slum districts (the dense street-patterns of which can also be identified) and past the Classical Roman Theatre. A still more spectacular avenue, the *Via dell'Impero* (The Avenue of Empire) was carved between the National Monument and the Colosseum in 1932. This can be seen at the top of Figure 2.3.

The working classes of the inner city were affected in various ways. Some found work on the projects, as the alleviation of unemployment and high-profile public-

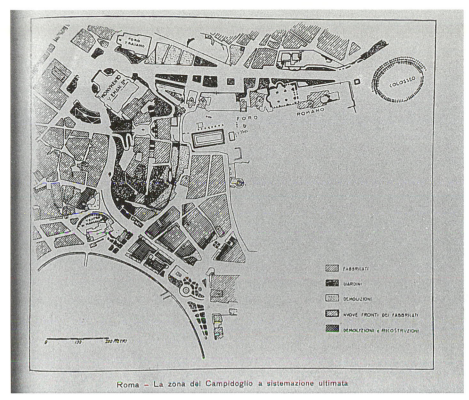

Roma – La zona del Campidoglio a sistemazione ultimata

FIGURE 2.3 Map of the *Via del Mare* and the *Via dell'Impero* cutting through the urban fabric of central Rome. Source: Albertini (1934: 511)

works programmes proved valuable public relations material for the regime. However, the rapid demolition of their residential quarter impacted upon all of them. Most of those displaced were relocated to the eastern outskirts of the city, to the *Borgate* – literally the 'scraps' of the city – which were hastily built settlements at first often lacking basic amenities and infrastructure (Agnew, 1995; Clementi and Perego, 1983; Fried, 1973). Once removed to these supposedly 'rural' *Borgate*, the Romans would be more susceptible to the control of the regime; they would become more fertile and disciplined citizens away from the environment in which they had been vulnerable to disease, vice or Bolshevism. Some residents resisted their evictions briefly (Ridley, 1986), but this was the last opposition the regime would encounter from this district – having cleared it of the 'unhealthy', 'dangerous' classes through the destruction of their urban environment.

The historic monuments which flanked the *Via del Mare* and had been isolated and cleared of the accretions of centuries were central to the regime's cult of *romanità*. This 'symbolic archaeology' (Gentile, 1990) was intended to suggest that the glories and the pre-eminence of the ancient Roman empire would be revived under Fascism (Visser, 1992). Sizeable quarters of the city were re-planned and designed to foster this ideology. However, of more material significance to the Romans forced to move some ten miles to the outskirts of Rome, was the obliteration of their homes, the dissolution of their communities and their difficulties retaining or finding work. In creating these model totalitarian streets, the regime destroyed districts which had developed over the preceding centuries and which hosted long-established communities. The requirements of 'Positive Urbanism', the pursuit of *romanità*, and the drive towards the modern city had all intersected to spell the end of many Roman streets. In their place stood enormous, triumphal avenues, intended to articulate the *Romanità* of Fascist Italy and to host the public spectacle of the regime.

Celebrating Fascist streets – performance and spectacle

By the mid-1930s, on the rare occasions when protesters did take to the streets, they would be met by force and the newspapers (with editorial policies prescribed by the regime) would publicise the latent ability of the regime to quash opposition with violence. *The Daily Telegraph*'s Rome correspondent recorded one such incident:

> The publicity given in the Italian press to a small riot which took place in the province of Aquila . . . when one man was killed and several wounded by the police, was meant as a timely warning that the law against public meetings will be rigorously enforced.
>
> (Salvemini, 1936: 400)

In the face of this appropriation of the streets, the tradition of workers' *sovversivismo* (subversion) retreated to the interior spaces of factories and homes rather than risk the streets (Abse, 1996; Neri Serneri, 1995; Passerini, 1991). And given the lack of any opposition on the streets, the regime celebrated its own triumphs with an unrelenting programme of parades, ceremonies and public spectacles in both the older thoroughfares and newer, monumental avenues. Writing in 1928, the exiled former Prime-minister Francesco Nitti complained that:

> Fascism has not bent its efforts to obtaining results, but to producing [public spectacles]. All over the country there are celebrations, parades, and processions of Blackshirts.
>
> (Nitti, 1928: 420)

Yet these spectacles were more inclusive than Nitti claimed. It is true that the Blackshirted Militia and the military and Fascist heroes of the regime – such as its aviators or successful armed forces – regularly paraded through the capital (Segrè, 1987). However, when huge public-works programmes such as the *Via del Mare* were opened – inevitably by Mussolini and usually on the anniversary of the March on Rome, Fascism's first success on Rome's streets – the parade would often be formed of a selected group from within Fascist society, such as war veterans, the youth movements, or athletes.

Ordinary Italians would not merely watch these spectacles, but would be included in the pageantry and allowed to parade through the great stage set which was the rebuilt Fascist Rome. Participation was extended far and wide. For example, the

1932 *Mostra della Rivoluzione Fascista* was a hugely successful exhibition which celebrated the tenth anniversary of the regime and the modern political, social, cultural and economic system that was Fascism (Andreotti, 1992; Schnapp, 1992; Stone, 1993). Encouraged by the hyperbolic propaganda of the regime and specially discounted train-fares, Italians travelled from all over the peninsula to attend the exhibition. But as they exited the central station and filed down the *Via Nazionale* towards the National Exhibition hall, they would encounter two sights. First, the enormous, temporary facade of red, black and silver steel featuring 25-metre-high *Lictor Fasces* (the symbol of the regime, on the right of Figure 2.4) which masked the front of the nineteenth-century *beaux arts* exhibition hall. Second, positioned upon the broad steps of the building, they would come across a guard of honour, selected from various volunteer groups from all over Italy who vied for the privilege of standing before the ultra-modernist facade. As Emilio Gentile records:

> In the two years it stayed open, there was a ritual changing of the guard several times daily. This was executed not just by the militia but also by family members of the fallen, by mutilees, fighting forces, the *Sansepolcristi* [those present at the first ever Fascist meeting], and organisations of manufacturers and professionals. As [a contemporary writer] pointed out, this demonstrated, 'through a profoundly symbolic act, the closest possible spiritual bond between the citizens and Fascism, the citizen and his government, between each Italian and the consciousness of his welfare and that of the Fatherland'.
>
> (Gentile, 1996: 119)

From children to University students, from teachers to doctors, from dock workers to electricians, at different times almost 11,500 people 'guarded' the building. And this was in addition to the huge processions which included the hall on their parade routes (see Figures 2.4 and 2.5). As Schnapp observes: 'the exhibition immediately became a site for the elaboration of rituals' (Schnapp, 1992: 17).

By the time the *Mostra* closed in October 1934 (its run was extended three times) attendance figures totalled almost 4 million (Schnapp, 1992; Stone, 1993). To provide a suitable setting for this phenomenon, the huge, modernistic facade of the Exhibition of the Fascist Revolution and the constant ceremonials of Fascism had transformed the *Via Nazionale* from a street redolent of the nineteenth-century Liberal government under which it had been constructed, to a site of Fascist pilgrimage. On 15 April 1934, as Figure 2.5 shows, three regimented lines of road sweepers pedalled their official tricycles through Rome to the *Mostra*, demonstrating, in the eyes of one contemporary Italian, 'the almost military character of [their organisation and] the admirable discipline and zeal of its personnel' (Vandone, 1935: 72). They also demonstrated the extent to which the streets of Rome had become a stage set for the performance of Fascist rituals and the performances of idealised, Fascist figures.

My interest revolves around the inclusion of the masses in such rituals and the role of the street as the site where all this occurred. Emilio Gentile (1996) claims that Italian Fascism can be conceptualised as a modern political religion, a national

FIGURE 2.4 A parade along the *Via Nazionale*, past the exhibition of the Fascist Revolution, 12 September 1933. Source: Gentile (1996: between 52–53)

faith replete with myths, rituals and monuments. Other writers confirm that the observance by the masses of: 'the symbols, myths and rituals [of the regime] was meant to signify the populace's acknowledgement of their fascist identity' (Falasca-Zamponi, 1992: 83). Yet, all of this ritual had to take place somewhere – and it seems to me that the streets of Rome played a key role here. Whether temporarily appropriated like the *Via Nazionale,* or permanently transformed, like the *Via del Mare*, it seems that the long-standing traditions of Italian civic, social and cultural life being enacted in the public sphere was renegotiated by Fascism to such an extent that the streets of Italy became some of the key sites wherein the regime articulated its authority, control and – through inclusive spectacles – sought the consensus of the Italian population.

FIGURE 2.5 *Cantonieri* ('road inspectors') parading through Rome on an official visit to the 1932 Exhibition of the Fascist Revolution. Source: Vandone (1935: 71)

CONCLUSION

In early March 1943, factory workers in Turin went on strike, thus initiating the first significant acts of mass public opposition to the Fascist regime. Remarkable because they were unprompted by allied actions but were apparently a spontaneous expression of mass dissent, the strikes were of enormous significance in announcing the downfall of the regime (Abse, 1996; Mason, 1995). However, the strike leaders decided that they would confine their resistance to industrial action *within* the walls of the Turin factories. To demonstrate on the street or in the piazzas of the city was deemed too dangerous – 'pure suicide' in the words of one of the strike's leaders (Mason, 1995: 293). In turn, the main concern of the local police chief was to confine the strikers to the private spaces of the factories, and above all, to prevent any *public* demonstrations of dissent. He concentrated 'on maintaining public order and preventing the strikers from getting out onto the streets' (Mason, 1995: 291). In Turin, at least, it is clear that by the final months of the regime, the street and other public spaces were recognised by the local authorities and their opponents alike as the incontrovertible sphere of a dictatorship which 'treated rule over the piazza as a chief symbol of its public power' (De Grazia, 1992: 203).

In Fascist Rome, like Turin, the totalitarian appropriation of the streets was regarded as a central element of Fascist social control and the regime's attempts to enforce the 'Fascistisation' of Italy. As I have mentioned, the Italian strain of total-itarianism failed to meet the standards it set for itself and, in many respects, it was crucially 'flawed'. Nevertheless, despite these shortcomings, the regime endured for over twenty years and elements of its intended repression and social control did impact significantly on the streets. For the citizens of Rome, particularly those beaten by *Squadristi*, those expelled because they transgressed Fascist moral codes, or those evicted from their homes due to the perceived infertility of the city and the need for monumental avenues, the Fascist concern with Roman thoroughfares must have seemed coherent and effective enough. Moreover, Thompson (1991) argues that, although seldom applied as efficiently as planned, the techniques of totalitarian control first developed by Fascism proved to be both a model and a yardstick for later authoritarian and totalitarian states. Given all this, it seems that any recon-ceptualisation of the street would be well advised to take into account the roles of such public spaces in twentieth-century totalitarian cultures.

NOTE

Unless credited to others, the translations are my own. Unfortunately, the mistakes are all mine as well.

REFERENCES

Abse, T. (1996) 'Italian Workers and Italian Fascism', in R. Bessel (ed.) *Fascist Italy and Nazi Germany. Comparisons and Contrasts*, Cambridge: Cambridge University Press, pp. 40–60.

Acquarone, A. (1965) *L'organizzazione dello stato totalitario*, Turin: Einaudi.

Agnew, J. A. (1995) *Rome*, Chichester: Wiley.

Albertini, C. (1934) 'Urbanismo e viabilità', *Le Strade*, 16(9): 504–25.

Andreotti, L. (1992) 'The Aesthetics of War: The Exhibition of the Fascist Revolution', *Journal of Architectural Education*, 45(2): 76–86.

Atkinson, D. (1997) *Hitler's Grand Tour: the Triumphal Entrance to Fascist Rome*, Royal Holloway Department of Geography Working Papers, Imperial Cities Project paper no. 8.

Atkinson, D. and Cosgrove, D. (1998) 'Urban Rhetoric and Embodied Identities: City, Nation and Empire at the Vittorio-Emmanuele II Monument in Rome 1870–1945', *Annals of the Association of American Geographers*, 88(1): 28–49.

Barzini, L. (1964) *The Italians*, London: Penguin.

Brunetta, G. P. (1975) *Cinema Italiano tra le due guerra*, Milan: Mursia.

Calabi, D. (1993) *Il Mercato e la città: piazze, strade, architetture d'Europa in età moderna*, Venice: Marsilio.

Caldwell, L. (1991) '*Madri d'Italia*: Film and Fascist Concern with Motherhood', in Z. G. Baran-ski and S. W. Vinall (eds) *Woman and Italy. Essays on Gender, Culture and History*, London: Macmillan, pp. 95–116.

Cederna, A. (1979) *Mussolini urbanistica. Lo sventramento di Roma negli anni del consenso*, Bari: Laterza.

Ciacci, F. (1941) 'L'urbanistica fascista in Roma mussoliniana ed Imperiale', *Atti del 5 Congresso di Studi Romani*, vol. IV, pp. 82–90.

Clementi, A and Perego, F. (eds) (1983) *La metropoli 'spontanea'. Il caso di Roma, 1925–1981,* Bari: Dedalo.

Conte, U. (1934) 'Le nuove strade di Roma dal 1922 al 1934', *Le Strade,* 16(11): 635–44.

Corner, P. (1983) 'Consensus and Consumption: Fascism and Nazism Compared', *The Italianist* 3: 72–8.

Cosgrove, D. (1982) 'The Myth and the Stones of Venice: the Historical Geography of a Symbolic Landscape', *Journal of Historical Geography,* 8(2): 145–69.

Cosgrove, D. (1984) *Social Formation and Symbolic Landscape,* London: Croome Helm.

De Grand, A. (1991) 'Cracks in the Facade: The Failure of Fascist Totalitarianism in Italy, 1935–9', *European History Quarterly,* 21: 515–35.

De Grazia, V. (1981) *The Culture of Consent. Mass Organization of Leisure in Fascist Italy,* Cambridge: Cambridge University Press.

De Grazia, V. (1992) *How Fascism Ruled Women, Italy 1922–1945,* Berkeley: University of California Press.

Ernesti, G. (ed) (1988) *La formazione dell'Utopia: Architetti urbanisti nell'Italia fascista,* Rome: Edizioni del Lavoro.

Falasca-Zamponi, S. (1992) 'The Aesthetics of Politics: Symbol, Power and Narrative in Mussolini's Fascist Italy', *Theory, Culture and Society,* 9: 75–91.

Fogu, C. (1996) 'Fascism and *Historic* Representation: The 1932 Garibaldian Celebrations', *Journal of Contemporary History,* 31: 317–45.

Forgacs, D. (1986) 'Introduction: Why Rethink Italian Fascism?', in D. Forgacs (ed.) *Rethinking Italian Fascism,* London: Lawrence and Wishart, pp. 1–10.

Fried, R. C. (1973) *Planning the Eternal City: Roman Politics and Planning Since World War II,* New Haven: Yale University Press.

Fuller, M. (1996) 'Wherever You Go, There You Are: Fascist Plans for the Colonial City of Addis Ababa and the Colonizing Suburb of EUR '42', *Journal of Contemporary History,* 31: 397–418.

Gentile, E. (1990) 'Fascism as Political Religion', *Journal of Contemporary History,* 25: 229–51.

Gentile, E. (1995) *La via italiana al totalitarismo. Il partito e lo Stato nel regime fascista,* Rome: La Nuova Italia Scientifica.

Gentile, E. (1996) *The Sacrilization of Politics in Fascist Italy,* trans. Keith Botsford, Cambridge, Mass.: Harvard University Press.

Ghirardo, D. Y. (1988) 'Architecture and Theatre: The street in Fascist Italy', in S. Foster (ed.) *'Event' Arts and Art Events,* Ann Arbor: UMI Research Press, pp. 231–52.

Ghirardo, D. Y. (1990) 'City and Theatre: The Rhetoric of Fascist Architecture', *Stanford Italian Review* 8, 1–2: 165–93.

Ghirardo, D. Y. (1996) '*Città Fascista*: Surveillance and Spectacle', *Journal of Contemporary History,* 31: 347–72.

Hay, J. (1987) *Popular Film Culture in Fascist Italy, The Passing of the Rex,* Bloomington: Indiana University Press.

Horn, D. G. (1994) *Social Bodies. Science, Reproduction and Italian Modernity,* Princeton: Princeton University Press.

Insolera, I. (1971) *Roma Moderna. Un secolo di storia urbanistica,* Turin: Einaudi.

Ipsen, C. (1996) *Dictating Demography. The Problem of Population in Fascist Italy,* Cambridge: Cambridge University Press.

Isnenghi, M. (1994) *L'Italia in Piazza. I luoghi della vita pubblica dal 1848 ai giorni nostri,* Milan: Mondadori.

Kostof, S. (1973) *The Third Rome: 1870–1950, Traffic and Glory,* Berkeley: University Art Museum.

Kostof, S. (1994) 'His Majesty the Pick. The Aesthetics of Demolition', in Z. Çelik, D. Favro and R. Ingersoll (eds) *Streets: Critical Perspectives on Public Space,* Berkeley: University of California Press, pp. 9–22.

Lyttelton, A. (1973) *The Seizure of Power. Fascism in Italy, 1919–1929,* London: Weidenfeld & Nicholson.

Mason, T. (1995) 'The Turin Strikes of March 1943', in J. Caplan (ed.) *Nazism, Fascism and the Working Class. Essays by Tim Mason,* Cambridge: Cambridge University Press, pp. 274–94.

Monteleone, F. (1976) *La radio italiana nel periodo fascista,* Venice: Marsilio.

Morgan, P. (1995) *Italian Fascism, 1919–1945,* London: Macmillan.

Muir, E. (1981) *Civic Ritual in Renaissance Venice,* Princetown: Princetown University Press.

Muir, E. and Weissman, R. F. E. (1989) 'Social and Symbolic Places in Renaissance Venice and Florence', in J. A. Agnew and J. S. Duncan (eds) *The Power of Place: Bringing Together Geographical and Sociological Imaginations,* Cambridge: Cambridge University Press, pp. 81–103.

Mulé, F. P. (1932) 'La via dell'Impero e la via del mare', *Capitoleum,* 8: 521–36.

Muñoz, A. (1932) *Via dei monti e via del mare,* Rome: Governatorato di Roma.

Munro, I. S. (1933) *Through Fascism to World Power: A History of the Revolution in Italy,* London: Maclehose.

Mussolini, B. (1936) *My Autobiography,* trans. R. Washburn Child, London: Paternoster.

Neri Serneri, S. (1995) *Classe, partito, nazione: alle origini della democrazia italiana, 1919–1948,* Manduria: Laicaita Editore.

Nitti, F. (1928) 'Probabilties of War in Europe', *Atlantic Monthly,* 142 (September 1928): 414–22.

Passerini, L. (1991) *Mussolini immaginario,* Bari: Laterza.

Payne. S. G. (1995) *A History of Fascism 1914–1945,* London: UCL Press.

Pollard, J. F. (1985) *The Vatican and Italian Fascism, 1929–1932: A Study in Conflict,* Cambridge: Cambridge University Press.

Ponziani, L. (1995) *Il fascismo dei prefetti. Administrazione e politica nell'Italia meridionale, 1922–26,* Catanzaro: Donzelli.

Roberts, K. (1936) 'Sojourning in the Italy of Today', *National Geographic Magazine,* September 1936: 355–96.

Ridley, R. (1986) 'Augusti Manes Volitant per Auras: The Archaeology of Rome under the Fascists', *Xenia,* 11: 19–46.

Salvemini, G. (1936) *Under the Axe of Fascism,* London: Victor Gollancz.

Saraceno, C. (1991) 'Redefining Maternity and Paternity: Gender, Pronatalism and Social Policies in Fascist Italy', in G. Bock and P. Thane (eds) *Maternity and Gender Policies, Women and the Rise of European Welfare States, 1880s–1950s,* London: Routledge, pp. 196–212.

Schnapp, J. T. (1992) 'Epic Demonstrations: Fascist Modernity and the 1932 Exhibition of the Fascist Revolution', in R. J. Golsan (ed.) *Fascism, Aesthetics, and Culture,* Hanover: University Press of New England, pp. 1–37.

Schnapp, J. T. (1996) *Staging Fascism,* 18BL *and the Theatre of Masses for Masses,* Stanford: Stanford University Press.

Segrè, C. G. (1987) *Italo Balbo: A Fascist Life,* Berkeley: University of California Press.

Spackman, B. (1990) 'The Fascist Rhetoric of Virility', *Stanford Italian Review* 8(1–2): 81–101.

Steinberg, J. (1986) 'Fascism in the Italian South: The Case of Calabria', in D. Forgacs (ed.) *Rethinking Italian Fascism,* London: Lawrence and Wishart, pp. 83–109.

Stone, M. (1993) 'Staging Fascism: The Exhibition of the Fascist Revolution', *Journal of Contemporary History,* 28: 215–43.

Thompson, D. (1991) *State control in Fascist Italy. Culture and Conformity, 1925–1943,* Manchester: Manchester University Press.

Treves, A. (1976) *Le migrazione interne nell'Italia fascista,* Turin: Einaudi.

Vandone, I. (1929) 'La 'strada automobilistica' Roma-Ostia', *Le Strade,* 11(3): 58–64.

Vandone, I. (1935) 'L'opere dell'Azienda autonoma statale della strada al 30 Giugno 1934 – X11' [editorial], *Le Strade,* 17(2): 66–72.

Visser, R. (1992) 'Fascist Doctrine and the Cult of the Romanità', *Journal of Contemporary History,* 27, pp. 5–22.

Willson, P. (1996) 'Woman in Fascist Italy', in R. Bessel (ed.) *Fascist Italy and Nazi Germany: Comparisons and Contrasts,* Cambridge: Cambridge University Press, pp. 78–93.

3

CHANGING PLACES

PERSPECTIVES ON THE DEVELOPMENT OF A MUNICIPAL SUBURBAN STREETSCAPE

Gerry Mooney

•

INTRODUCTION

The concern of this chapter is to explore the ways in which a particular vision and image of 'the street' was utilised in the planning and development of a large municipal housing estate on Clydeside during the 1930s and 1940s. It is argued here that the street was to be an integral part of the new 'orderly' city which was being created out of the 'disorder' of inner-city slumland housing and chaotic industrial development which characterised much of urban Britain at the time. The image and role of the street in large post-War housing estates owed much to a diversity of influences from the garden city movement and town planners through to advocates of 'neighbourhood planning units'. We will consider some of these influences during the course of this chapter.

The particular housing estate (or '*scheme*' in the Scottish parlance) taken as a case study here is Pollok in the south-west of Glasgow (see Figure 3.1). Lying approximately 6–7 miles from Glasgow city centre, Pollok is usually termed a '*peripheral*' or '*outer*' estate. At it's most basic such a description implies a specifically spatial location on the edge of a large city or built-up urban area. But increasingly these estates are peripheral in a social sense as reflected in high levels of poverty, deprivation and unemployment.

Although Pollok provides us with a rich history through which we can begin to consider the street in such locales, this chapter is not a history of Pollok. Further, much of what is written about Pollok can apply to other large council (or 'social') housing estates of the period and thus the issues and ideas raised in this chapter have a much wider resonance beyond Pollok.

Peripheral estates played a significant role in the reorganisation of the urban landscape of Britain in the inter-war and post-war period. They were, along with the New Towns, the primary means by which the populations decanted as a result

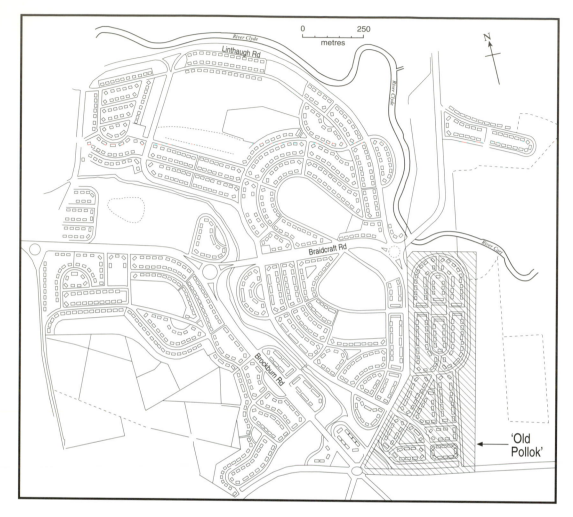

FIGURE 3.1 Lay-out plan for scheme at Pollok, 1935. Source: Corporation of Glasgow Housing Department

of inner-city slum clearance and '*redevelopment*' were to be rehoused. Some of the largest estates housed up to 50,000 people, with a housing stock of around 10,000 dwellings. It has been estimated that around 2 million people live in Britain's 'outer' estates (CES, 1984: 1) and while many estates have experienced '*selective*' demolition and density reductions in the 1980s and 1990s, and consequent population decline, they continue to house a sizeable proportion of Britain's urban population. Yet it can be argued that apart from the interest shown in these estates by sociologists and social anthropologists in the 1950s and 1960s, during a time when 'community studies' were to the fore, these estates have rarely been the central focus of much academic, media and indeed state interest.

One important point to note is that all too often the label 'outer' or 'peripheral' estate is applied with little regard to the specific characteristics of one estate, or to its own particular history and identity or external representation. This is not to deny the existence of similarities and common features but to acknowledge the importance of locating each estate within a particular historical, socio-economic and spatial context. Further, *within* outer estates there is also diversity and heterogeneity, though this is frequently neglected by those who seek to label or stereotype these places as uniform, homogenous deserts of decay and deprivation. It is in relation to this that the early planning and development of Pollok provides a useful and distinctive case study.

Pollok's distinctiveness, at least in relation to Glasgow's other large peripheral housing estates at Castlemilk, Drumchapel and Easterhouse, relates to the two 'periods' in which it was built. Unlike the majority of Glasgow's and Britain's outer estates, work on Pollok began in the mid-1930s and was not completed until the late 1950s or 1960s, the completion date depending on how the boundaries of the estate itself (and neighbouring districts) are drawn. Pre-war Pollok, or 'old' Pollok (Figure 3.1), is distinctive from post-war Pollok in terms of house styles, density of housing, the age profile and background of its population and, importantly for the purposes of this chapter, the image of the street was also different, at least the image of the street used in the urban 'vision' which Pollok was held to represent. In order to understand this we have to consider the plurality of influences which were at work on housing estate planners in the mid-1930s.

PLANNING POLLOK: THE INTER-WAR YEARS

Pollok was to be the success story, *the* model of Glasgow's housing achievements. It was planned to be the largest estate built to date, its size a testament to Glasgow's enormous housing problems in the inter-war period. But in the mid-1930s, with almost a decade and a half's experience of housing estate development behind them, Scotland's local councils, together with the Department of Health for Scotland, the Department responsible at the time for municipal housing, were increasingly concerned about the poor quality of some of the new housing schemes already built. Fears were raised that some of these were rapidly becoming slums themselves. In the main this was attributed to the failure to provide community and recreational facilities and to the creation of 'one-class dormitory estates'. To quote one Department of Health Report at the time:

> We must aim at something more than the provision merely of adequate and healthy internal living accommodation.
>
> (Department of Health for Scotland, 1935: 5)

Among the most critical commentators were organisations such as the Garden cities association. They argued that many of the new estates lacked a 'sense of community' and that the absence of community amenities, a widespread problem at the time, made it more difficult for people, more used to the hustle and bustle of inner-city street and tenement living, to adapt to this new 'suburban way of life'. Mumford summarises many of these sentiments well:

> In the mass movement into suburban areas a new kind of community was
> produced which caricatured both the historic city and the archetypal
> suburban refuge: a multitude of uniform, unidentifiable houses lined up
> inflexibly, at uniform roads, in treeless communal waste, inhabited by people
> of the same class, the same income, the same age-group. . . . Thus the ulti-
> mate effect of the suburban escape in our time, is, ironically, a low-grade
> uniform environment from which escape is impossible.
>
> (Mumford, 1966: 286)

Mumford's comments echoed widespread feelings among social scientists and the planning and housing professions in the inter-War period that municipal estates had failed
in their 'social purpose'. The planners of Pollok then were well aware of the general debate
taking place during this period about housing estate development, and in drawing up
their plans for Pollok some of the concerns raised above were addressed in their proposals.

The choice of the Pollok area is also important. Prior to housing development in
the 1930s, it was mainly an agricultural area (in the County of Renfrewshire) with isolated farms and villages with small-scale mining activities. The area became part of
Glasgow in 1926 when the city boundaries were expanded in anticipation of large-
scale housing development. The land to be used for housing was sold to Glasgow
Corporation in 1934 by the Stirling Maxwell Family. From a prominent landowning
family John Stirling Maxwell felt able to contribute to the debate about town planning.
He was a fierce critic of existing Scottish Housing Schemes and as a keen supporter of
the Garden City movement, he was determined to ensure that the new Pollok housing
estate would be a vast improvement on existing housing estates. In order to achieve
this goal he employed the services of Patrick Abercrombie, then Professor of Civic
Planning at Liverpool University, to assist in the preparation of plans for Pollok.

Among the main recommendations made to Glasgow Corporation by Stirling
Maxwell and Abercrombie was that a mixed variety of cottage-style houses be built;
that housing densities should be very low, indeed well below those employed in garden
suburbs and cities thus far; that houses should be separated by walled-off gardens;
and that the streets should be a mix of wide boulevards and smaller service roads.
The boulevards would also act as 'arterial' roadways dissecting the estate at several
points as it connected Pollok with other part of Glasgow and neighbouring Renfrewshire. Where possible the entire development was to blend in with the many natural amenities which surrounded the area. For example, many of the streets would
follow the contours of the hills and streams in the area and the main roads would also
be tree-lined. Thus the image being created from the start was one very much in
keeping with the ideas of the garden suburb movement. What could be more attractive than images of tree-lined boulevards with neat cottage houses, interspersed at
appropriate points with community retail and recreational facilities? Thus the image
and layout of Pollok, with a network of curving roads laid out in sympathy with the
local topography, stood in stark contrast to the strict grid of central Glasgow with
its long drab streets. The dissemination of such images was aided by the choice of
street names within the scheme. These sought to capture Pollok's rural past and were

derived from the area's many topographical features. The Levern Water, for example, gave rise to Levernside Road and Avenue, the Brock Burn to Brockburn Road and Terrace. Some of the street names were the names of old farms which had been removed for housing, for instance Damshot Road, Calfhill Road and Linthaugh Road.

Some of this rural imagery was used by Glasgow Corporation in the 1940s to present a positive picture of suburban life to those on the lengthy (and getting longer by the day) waiting lists for a new house. Their long wait would be worth it. Under-pinning much of this imagery was an assumption that the better the environment, the better the character of the person who dwelled therein. But this would only happen if 'balanced communities' were created with all 'needs' catered for:

> The large housing estates which ring our big cities cannot be regarded merely as an aggregate of houses built for the sole purpose of providing shelter for certain classes of the population. The spiritual, social and recreational needs of that population must also be met.
>
> (CHAC, 1938: para. 64)

The layout of roads in new schemes typified by Pollok was a key element in this wider social engineering. The roads had a functional as well as aesthetic importance: it is clear that the desire for wide boulevards, together with 'service' roads, would ensure that an 'orderly' suburban landscape would be created. Further, this would also allow for the fragmentation of the estate into smaller component parts, encouraging the emergence of a 'sense of community'. The streets also provided necessary links with the rest of the city in order to serve the occupational needs of the citizens of Pollok. During the mid-1930s the absence of any detailed policy at this time on the relationship between housing development and industrial location was to have continuing implications for Pollok and similar estates. Certainly one of the assumptions employed during the first phase of Pollok's construction was that employment would be found 'elsewhere' in the city, particularly along the Clyde, in the northern parts of the city and in the East End. After all, the argument ran, Pollok was about better housing and a better 'quality of life'. It was not about providing jobs for people who already had employment, this itself telling us much about the 'type' of resident deemed 'appropriate' for the new scheme. The absence then of any employment opportunities – or indeed any economic base as such – within the estate was to become a characteristic of many of Britain's outer estates.

Outer estates then were locales of 'social reproduction', not production, under-pinned by a prevalent discourse about 'normal' family life. Thus the spatial structure of the city is gendered, reflecting assumptions about the full-time male breadwinner and female homemaker. The label 'dormitory estate' was fast to catch on as a result. Not that this was a problem for some commentators of the time:

> The rank and file of workers tend to lose their initiative through too much sameness of environment day after day. The human being likes movement and variety of scene, and a reasonable amount of daily travelling seems reasonable.
>
> (Maxwell, 1937: 9)

The Glasgow Herald newspaper was most fulsome in its praise for the new scheme:

> The Pollok Housing scheme is, by all the portents, to be a brilliant example of the higher art in housing development that is being demanded. . . . The layout is itself eloquent enough to give a good round general impression of the admirable character of the scheme, it looks in every way a very fine piece of planning.
>
> (Glasgow Herald, 6 June 1937)

At this time Pollok Garden Suburb represented the completion of Glasgow Corporation's plans for municipal housing development in the south-west of the city. It also represented, as we have seen, in the eyes of the planners and policy-makers at least, *the* housing estate in contemporary Glasgow. As befitted such a place the rents were higher than those typical in many 'poorer' quality 'rehousing' estates. In the 1930s this served to ensure a significant degree of selection in the choice of tenants. Pollok was not for 'anyone'.

By the commencement of the Second World War in 1939, 'Old' Pollok (Figure 3.1) had been largely completed, with only the foundations in place for house construction to begin in the remaining area of Pollok. This was suspended when the War began. As Figures 3.2 and 3.3 show, what was built by this time was very much in keeping with the planners' vision of an attractive garden suburb.

As the next section unfolds it will become clear however that much of the philosophy which underpinned suburban council estate planning and development in the 1930s was readily dispensed with in the drive to build, to quote Glasgow Corporation's Direct Labour Newsletter in the 1940s, 'the maximum number of houses in the shortest possible time'.

ALL-CHANGE? POLLOK IN THE POST-WAR ERA

The debate which took place in the 1930s regarding the planning and social role of new housing estates gathered pace in the aftermath of the Second World War. The context, however, was rather different in that Britain's housing problems had multiplied enormously as a result of wartime deprivations. Criticisms about the estates which had been built prior to the war continued unabated, and indeed became more vociferous given the new Labour Government's commitment to a massive programme of state housing construction. Once again the new estates had been classed a failure by many observers, a failure for a large number of reasons, but foremost among these were the absence of community facilities; that they had a one-class population; that new residents felt isolated and lonely in the estates and that no 'community spirit' was evident; and that the streetscapes of these areas were bleak places

FIGURE 3.2 Tree-lined boulevard, Braidcraft Road, 'Old' Pollok

FIGURE 3.3 1930s cottage-style housing, 'Old' Pollok

> where all individuality and homeliness have been lost in the endless rows
> of identical semi-detached houses. The depressing appearance of these estates
> is very largely due to monotony in design and layout, and to the repetition
> of the same architectural unit in dull, straight rows or in severe geomet-
> rical patterns which bear no relation to the underlying landscape features.
>
> (Ministry of Health, 1948)

Such sentiments were clearly echoed in Scotland:

> We have reached the general conclusion that there is a great deal of public
> dissatisfaction with the design and character of many of the houses erected in
> Scotland in the inter-War years. The monotony and drabness of many of these
> schemes and the ease with which 'council houses' are identifiable even by the
> least discriminating are matters of frequent comment. ... Housing sites
> should be selected in such a way as to enable subsequent development to take
> a community form either by itself on 'neighbourhood unit' principles or linked
> with existing developments, depending upon the size of the community.
>
> (Department of Health for Scotland, 1944: paras. 264, 304)

While pre-war Pollok escaped much of these criticisms the development of Pollok
in the post-1945 period was to result in streetscapes far removed from the visions
of the 1930s and was vulnerable to precisely these criticisms. Why was this?

In 1945 overcrowding continued to be a major problem in the city with over
700,000 people crammed into an area of 1800 acres, with population densities
reaching as high as 700 persons per acre. In the same year Glasgow's City Engineer,
Robert Bruce, estimated that Glasgow required to build around 200,000 new homes
as part of a fifty-year building programme (Glasgow Corporation, 1945). It was to
the peripheral schemes at Pollok, where completion of the estate had been delayed
since the War commenced in 1939, and to the new large estates at Castlemilk,
Drumchapel and Easterhouse among others that Bruce looked to provide much of
the housing required. Further expansion of the city boundaries had been ruled out
as politically unacceptable and with the acceptance by the Scottish Office in 1949
of Patrick Abercrombie's Clyde Valley Regional Plan, the establishment of a green
belt which encroached over the existing city boundary meant that there was insuf-
ficient space within Glasgow to build all the houses required. Population overspill
(primarily to New Towns) was proposed and accepted as one of the main methods
by which the overcrowded population of central Glasgow could be rehoused. The
other proposal, put forward by Glasgow Corporation amidst ongoing debate and
argument, was to have a devastating impact on the outer estates.

Bruce argued that the low density cottage-style housing (which characterised
estates such as inter-war Pollok) was something that Glasgow could no longer afford.
If his plans for a massive slum clearance and rehousing programme were to be fulfilled
then new peripheral estates would have to be built to much higher densities than
first envisaged, with three- and four-storey flats taking the place of many of the
cottage-style housing initially chosen. All the major outlying estates saw their target
populations increased dramatically. Pollok's population target increased from 34,000

in 1946 to over 40,000 in 1947 alone (Mooney, 1988: 220). Cottages were replaced either by rows and rows of terraced housing, and in some parts of the estate by large numbers of three- and four-storey blocks of flats. But many of the foundations for new houses and the internal road system/estate layout remained in place from 1939 when building work had been postponed. The Scottish Housing Advisory Committee of the Department of Health for Scotland had advised during and immediately after the War, that construction of long blocks of tenements fronting main roads, should be avoided. Again these recommendations were ignored in post-war Pollok (see Figures 3.4 and 3.5) and housing in Brockburn and Linthaugh Roads, two of the longest and busiest in the scheme, were almost completely tenemental, albeit on one side of the road only. While both roads were wide dual carriageways, the resulting effect was rather depressing.

In several parts of the scheme, where the relatively narrow width of the road had been designed for low-rise cottages, the result of flat building along each side was the creation of canyons and wind tunnels. The traditional practice in Scottish tenemental construction was to provide roads of a width at least one and a quarter the height of the tenement blocks, but this was not the case in much of Pollok.

While post-war Pollok represented a marked departure from the plans of the mid-1930s, it is important to bear in mind that the housing and environment represented a major improvement on that which characterised inner-city districts. Many of the new residents enjoyed living alongside tree-lined roads, and the three- and four-storey flats, many of which were to be demolished or rebuilt in the 1980s and 1990s, also proved popular with those eager to escape overcrowding and damp tenements. The following story about one of the new pioneer citizens of Pollok, was typical of many used by Glasgow Corporation at the time to promote suburban life:

> In solitary state and with all the dignity he could muster, he walked to the rear and climbed aboard, the sole passenger at the stop – a ring of the bell and he was being conveyed to Pollok. Although Angus had lived all his life in Glasgow he had never been to Pollok. He imagined it would be something like a modern Gorbals and he was surprised, as the minutes passed and he drew nearer to his destination, to find that he was in the country. Indeed the whole area was very much like the rolling scenery of Kent, in which he had received his battle training.
>
> (Quoted in Mooney, 1988: 238)

In 1945 Glasgow's Housing Department commissioned two films to persuade Glaswegians that the new housing developments springing up throughout the city would transform their lives. In both these 'Progress Reports' newsreel documentary methods were used with the virtues of the countryside very much to the fore in the commentary:

> Mrs. Stuart and the family have been allocated a house at Pollok. This family have lived all their lives in an area you know well, an area of mean streets and dilapidated tenements, skies always continuously clouded by the city's smoke drifts. But from the gloomy backcourts children are escaping day by

FIGURE 3.4 Aerial view of North and West Pollok, 1950, showing mixture of tenement flats and cottage-style housing. Source: Ministry of Defence, Crown Copyright.

day. When all is packed, away to the vans leaving behind the dreary congested areas of the old part of the city, entering a new world where trees are everywhere.

(Quoted in Mooney, 1988: 239)

However, criticisms of the new estate by its new residents were not slow to emerge. Feelings of social isolation and alienation were to the fore here, encouraged by the absence of adequate shopping and community facilities.

The great challenge to the tenants of many of the new Glasgow housing estates is that of remoteness. The men are aware of remoteness from their work; the women of remoteness from shops . . . and perhaps also remoteness from relatives and friends to whom they could formerly turn in times of trouble . . . according to social workers problems in housing schemes

FIGURE 3.5 Tenement flats in Linthaugh Road, Pollok

conform to a plan. Honeymoon of the tenancy – six months . . . first year
– burden of rent and hire purchase; and in the second or third year tenants
are forced – some the hard way – to learn principles of household budgeting.
(Glasgow Herald, 26 December 1956)

Many schoolchildren were still being bussed back to the city centre on a daily basis
for schooling and little in the way of health and welfare provision was made available within the schemes themselves. The vast majority of Pollok's residents came
from overcrowded districts towards the South of the River Clyde, notably from Govan
and the Gorbals. They missed the congested streets which contained a host of pubs,
cafés and small shops. Some just missed the hustle and bustle of street life, which
was in stark contrast to the miles of new roads in the large estates flanked by little
else than housing. Quiet streets with no community or retail facilities did not allow
for frequent casual meetings. Indeed the best chance of such a meeting was on one
of the irregular overcrowded buses taking people out of the estate either to work or
back to family and friends in older areas of the city. That people's places of paid
work were now far removed from their place of residence was also a cause of much
negative comment, not least the added costs which workers now had to bear just to
get to their work. Inadequate transport provision was a major source of discontent
and would continue to be so throughout the post-war period.

Such peripheral schemes were thus a parody of the traditional tenement life of Glasgow. They consisted of tenements indeed, but they were far removed form the urban context in which that mode of life had developed, and incapable of generating their own community life. In spite of the literature of planning, already vast, and containing so many hard learned lessons, these new units were not only devoid of facilities themselves, but were miles from the traditional centres of Glasgow life.

(Checkland, 1978: 68)

On completion in the 1950s Pollok was far removed from the ideals of 1935. The language and imagery of the garden suburb was abandoned, to be replaced by an emphasis on the purely 'functional' role of the estate, as reflected in its planning and house design. The layout of streets served no other purpose than allowing for the free flow of traffic, with no attempt made to segregate housing and traffic. There was little sense of the street being used in any way to replicate community life in older areas of the city. But it was the lack of imagination for the use of streets and public spaces in the new estates which was criticised in the mid- to late 1960s when the peripheral estates stated to be stereotyped as locales of juvenile delinquency.

In one other important sense a particular vision of the street was used in relation to the large peripheral estates. They were to be connected to the 'rest' of the city through a network of 'arterial', subarterial and local roadways (Glasgow Corporation, 1945). Concentric motorways would speed both traffic and population around the city with the minimum of disruption. However, it would not be until the 1990s (December 6 1996 to be precise) before Pollok would be connected to the national motorway network. The opening of the M77 Ayr Motorway, a new 'strategic gateway' to the city, after protracted local protest, represents in part yet another attempt to connect the estate with the wider city economy.

CONCLUSION: IMAGINING THE STREET IN OUTER ESTATES

From being a key part of the solution to housing and 'urban' problems in post-war Britain, peripheral estates have come to be increasingly designated as a part of the 'problem' of the modern city. While these estates were a crucial element in the 1940s vision of a socially engineered city, many of them have progressively become locales for the poor and socially excluded within society. In this chapter it has been argued that we can detect the emergence of a peripheral estate 'problematic' wherein the idea of a peripheral estate becomes symbolic of backwardness and depression.

From the mid-1960s we can also detect a growing 'concern' with the incidence of crime and 'delinquency' in the estates. Given the low-age profile of the typical outer estate population, it was felt that the absence of community recreational facilities and the lack of employment opportunities would encourage younger people to engage in criminal activity. The street gang, whose forte appeared to be 'hanging around' street 'corners', became a frequently used media favourite in reporting crime in Glasgow's outer estates. The street was no longer a symbol of a tranquil garden suburbia but was increasingly emblematic of social disorder.

In exploring the street in outer estates we do not have the hustle and bustle of city life, nor a sense of a 'dynamic', fast-moving city culture to call upon in our attempt to make sense of 'street life'. In outer estates, as in many New Towns and post-war housing development across the length and breadth of Britain, it was the relative absence of street life, as judged in relation to 'traditional communities' which stands out. In part this reflects the increasing privatisation of working-class life throughout the period. In this sense then the outer estate street played a very different role in the lives of people than did the street in older working-class districts of the city. The latter were sites of work, of leisure, of conflict, of struggle but in the peripheral estate the street connected in a very different way to 'work' and to home.

Finally, in making sense of the street in the planning of large council housing estates we need to be careful to avoid depicting this, and indeed planning in general, as a purely technical matter. The role and vision of a housing estate, an estate such as Pollok, is founded on certain assumptions about family life, the class of people who would live there, and the place the estate would occupy in the spatial ordering of the city. Our understanding of the street then derives not from a depoliticised exploration of city planning, but from an understanding of the role of the locality in which it is located in the wider social and geographical ordering of the city.

REFERENCES

CES (1984) *Outer Estates in Britain: Interim Report*, London, CES Paper No 23.

Checkland, S.G. (1978) *The Upas Tree: Glasgow 1875–1975*, Glasgow, Glasgow University Press.

Central Housing Advisory Committee (CHAC) (1938) *Management of Municipal Housing Estates: First Report*, London, Central Housing Advisory Committee, H.M.S.O.

Department of Health for Scotland (1935) *Report on Working Class Housing on the Continent*, Edinburgh, H.M.S.O.

Department of Health for Scotland (1944) *Planning Our New Homes*, Edinburgh, H.M.S.O.

Glasgow Corporation (1945) *First Planning Report*, Glasgow, Corporation of the City of Glasgow.

Maxwell, J. M. Scott (1937) *The Housing Scheme As A Social Training Centre*, Glasgow, John Wylie & Co.

Ministry of Health (1948) *The Appearance of Housing Estates*, London, H.M.S.O.

Mooney, G. (1988) 'Living On The Periphery', Unpublished PhD Thesis, Department of Sociology, University of Glasgow.

Mumford, L. (1966) *The City in History*, Harmondsworth, Penguin.

4

THE DEATH OF THE BOULEVARD
John R. Gold

•

The guiding spirit of every age is crystallized in its monuments – from the Sainte-Chapelle to the Rue de Rivoli. But this splendid heritage was set in the midst of a thoroughly disorganized town, the monuments surrounded and isolated by a tangle of streets. The herculean efforts of Georges-Eugène Haussmann (1809–91), Préfet de la Seine under Napoleon III, drastically altered this situation. It was his desire to provide a splendid framework for the great tradition preserved in Paris.
(Giedion, {1941} 1954: 646)

A titanic achievement. Hats off! Today Paris lives on Haussmann's work.
(Le Corbusier, {1935} 1967: 209)

INTRODUCTION

The rebuilding of Paris during the Second Empire was an important landmark in urban development. Able to proceed without undue concern for issues like displacement of the poor or destruction of the urban fabric, Haussmann's schemes effectively recreated Paris. They opened up vistas, added new sources of monumentality, provided a new water and sewerage infrastructure, and enhanced the city's stock of open spaces and parks. Above all, they instituted a new street pattern. Paris in 1850 retained its centuries-old accretion of narrow roads and alleyways. By 1870, Haussmann's 'regularisation' of the city created hundreds of kilometres of paved and lit streets, including a spinal network of *grandes boulevards*.[1]

For many, the boulevards represented Haussmann's supreme achievement. Cut through the knotted entrails of the medieval city, they set new standards in breadth and directness. The Avenue de l'Impératrice, for example, was approximately 130 metres

wide, with generous provision of pavement space. This width partly reflected considerations of public order, light and circulation of air, but was predicated more on the
boulevards' status as major traffic arteries. Conceived as part of a network, they routed
traffic through the system and supplied rapid connection between the city centre and
the outskirts. They also contained radical experiments in traffic management, as illustrated by the Boulevard Richard Lenoir. Built over an old canal, it had a central lane
reserved for fast moving traffic and two flanking lanes for slower vehicles (Evenson, 1984;
Sennett, 1990: 151).

The boulevards were copied by other cities that were facing similar problems in
meeting the demands of the industrial age. Lyons, Marseilles, Toulouse, Avignon,
Montpelier, Brussels, Rome, Stockholm, Barcelona, Madrid, Washington and Mexico
City all imported ideas developed in Paris (Pinkney, 1958: 4). Yet when Sigfried
Giedion and Le Corbusier hailed Haussmann's achievements, they had something
quite different in mind. As two of the leaders of architecture's Modern Movement,
they saw the boulevards as a source of inspiration not imitation. The boulevards
symbolised the potential of comprehensive and concerted action to effect change,
reinforcing a metaphor of town planning as painful but necessary surgery designed
to improve the health of the body (Ramazani, 1996: 210). They also showed the
value of functional reorganisation of different traffic types. Adopting a second
metaphor, if the city was a machine, then the boulevards were an important component in ensuring its smooth operation.

By contrast, other dimensions of the boulevards were regarded as retrospective
and likely to reinforce the city's problems. Despite their scale, the boulevards were
seen by modern architects as a larger version of the traditional *rue corridor*, with its
rigid line of buildings and intermingling of traffic and pedestrians. They might defer
the day when problems would arise but they were intended for horse-drawn vehicles.
Even their width might be inadequate for peak hour traffic in the age of the car.
Moreover, they still encountered problems at the many crossing points along their
route; places where accidents occurred and circulation was interrupted.

Seen in this light, it was easy to downplay the qualities that Haussmann's boulevards brought to the quality of Parisian life. Tree-lined and fringed with theatres
and restaurants, they were an essential part of the urban experience, as much the
province of the window-shopper and *flâneur* as the delivery-driver and the messenger.
By concentrating on the rigidities and problems brought by the *rue corridor*, modern
architects lost sight of the unique ambience that could be created by the benign
intermingling of pedestrians and vehicles.

This paper looks at one consequence of this style of thinking, namely, experimentation with multilevel, functionally defined circulation systems to replace the
'battleground' of the traditional major urban street. There are three main sections. The
first shows that the ideas about city streets now popularly, and sometimes incor-
rectly, associated with the Modern Movement – the world of the designated urban
expressway and multilevel interchange – emerged only slowly. Despite ransacking
European and North American precedents, modernists found it difficult to develop
consensus about the replacement for the boulevard. The second section uses some

brief case-studies of projects by the British Modern Movement to reveal the degree of plurality and experimentation in modernist thinking about the major urban street even as late as the 1940s. Nevertheless it shows how a new language of urban form and function, of the type encapsulated in Figure 4.1, was steadily created, envisaging new landscapes in which high density limited-access circulation systems were often juxtaposed with tall buildings. The conclusion summarises the major themes before underlining a more general point, namely, the problems that arise from historical interpretations that link emergent trends in post-war city design directly back to visionary architectural prototypes without sufficient consideration of the specific circumstances of reconstruction.

MODERNISM AND THE STREET

No analysis of the Modern Movement's early thinking about the street can omit the influence of North American thinking and prototypes. Between 1880 and 1930, developments in American industrial production, constructional techniques and personal mobility offered clues for ways of transforming the city. Images of mechanised

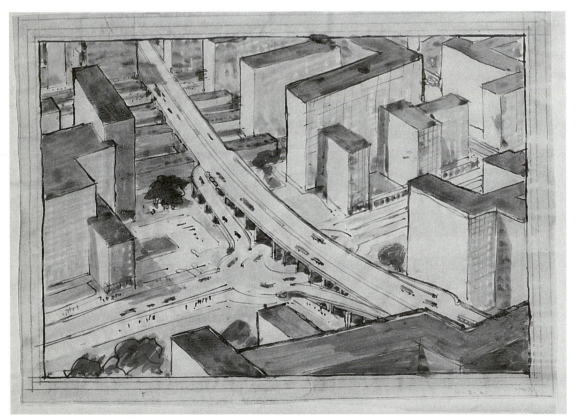

FIGURE 4.1 S. Rowland Pierce, Reconstruction Project, 1942: the abstract language of new urban form. The sketch shows an Inner Circular 'elevated road' of a type proposed by H. Alker Tripp. Source: Drawings Collection, Royal Institute of British Architects

cityscapes and tall buildings were widely depicted in the European press and became taken for granted as a likely future for cities elsewhere. Yet almost as soon as these images appeared, it became clear to observers that this was also the recipe for severe congestion. Sir Raymond Unwin, for instance, calculated that the day population of the Woolworth Building in New York would themselves occupy 1¼ miles of footway 20 feet wide on both sides of the street if walking to work. If just one in ten workers brought their cars to work, then parked cars would occupy the entire width of the 100-foot-wide roadway for the distance of almost one mile (quoted by Tatton Brown and Tatton Brown, 1941: 84).

Although hypothetical and produced for purposes of propaganda, statistics like these highlighted the difficulties faced. The large workforces converging on and leaving the vertical central city at peak hours could bring the city to a standstill. Science-fiction writers and illustrators absorbed this idea into a dystopian view in which the future city could only operate as an oppressive machine, abrogating individual freedoms (see for example Mansfield, 1990). Those more optimistic about the urban future looked for mechanisms to control the untrammelled development of road traffic and alleviate the likely problems.

Not surprisingly, one favoured solution lay in making beneficial use of verticality. Given the lack of room to widen highways and thoroughfares, multilevel circulation systems could accommodate higher density flows. Drawings completed by Harvey Wiley Corbett in 1913, for instance, showed a future New York with pedestrian walkways placed above a triple-decker scheme designed to ease street congestion. This would have vehicles at street-level and subways and goods railways underground. High above this arrangement was a systematic network of overhead bridges linking the skyscrapers (Shanor, 1988: 16–23; Cohen, 1995).

In Europe, limited application of multilevel concepts was seen in a series of planning studies produced for Paris between 1903 and 1909 by a government architect, Eugène Hénard. Besides inventing the roundabout (*carrefour à giration*) to regulate flow at multiple intersections such as the Place de l'Étoile, Hénard suggested a two-level system by which underground concourses would remove the competing presence of pedestrians (Evenson, 1984: 273). Others, however, regarded measures like these as just tinkering with the present dispensation. What was required was a thorough recasting of the relationships between the built forms of the city and its street pattern.

The earliest and most dramatic expression was found in the work of the Italian Futurist movement immediately before the First World War. Antonio Sant'Elia's fragmentary sketches of the New City (*Città Nuova*) from 1914 show a city bisected by giant roadways above which rose tall buildings of reinforced concrete, steel and glass. The conventional street would be replaced by a new multilevel arrangement. The street would 'no longer lie like a doormat at the level of the thresholds, but plunge storeys into the earth, gathering up the traffic of the metropolis connected for necessary transfers to metal cat-walks and high-speed conveyor belts'.[2]

Another radical solution was offered by the *Hochhausstadt* ('Skyscraper City') devised by Ludwig Hilberseimer in 1924. Like other *Neue Sachlichkeit* ('New Objectivity') architects in Weimar Germany, Hilberseimer was interested in the relationship between the city, acting as a molar machine, and its component parts. Devising a plan in which type of dwelling units and amounts of living space were unrelated to class, Hilberseimer's ideal city contained huge slab blocks of flats placed on a grid, with shops and workplaces in five-storey podia beneath them. Pedestrian decks and bridges at sixth-floor level would link the blocks. Through-traffic was handled by the cavernous roadways that separated the blocks, with residual movement by underground railway. The key variable was reduction of movement. Journeys to work or shop simply meant taking the lift. This brought the inherent problem that people would have to move every time they changed jobs, but Hilberseimer did not regard this as a significant drawback. In his view, future urban inhabitants would be happy with a more nomadic existence, satisfied with living in flats equipped as conveniently as hotel rooms and not feeling the need for private furniture (Larsson, 1984: 203; Hays, 1992).

While the sociological and design aspects of such plans brought interest and fierce criticism in equal amounts (see Gold, 1997b), the mounting rhetoric against the street and in favour of multilevel schemes became routine in modernist litera-ture. The most ardent purveyor of that rhetoric was the French-Swiss architect, Le Corbusier. In an article in 1929, he castigated the *rue corridor*:

> It is the street of the pedestrian of a thousand years ago, it is a relic of the centuries: it is a non-functioning, an obsolete organ. The street wears us out. It is altogether disgusting! Why, then, does it still exist?
>
> (quoted in von Moos, 1979: 196)

Four years earlier, Le Corbusier had argued that the street had become: '*a machine for traffic, an apparatus for its circulation*, a new organ, a construction in itself and of the utmost importance, a sort of extended workshop' (Le Corbusier, [1929] 1987: 123). Continuing this theme: 'it is in reality a sort of factory for producing speed traffic. . . . We must create a type of street which shall be as well equipped in its way as a factory' (*ibid.*: 131).

The new street system would have each functionally distinct traffic type occu-pying its own dedicated channel placed at different levels. Heavy traffic would proceed at basement level, lighter at ground level, and fast traffic would flow along great arterial roads placed on immense reinforced concrete viaducts. The viaducts would only have vehicles moving in one direction, with points of access every kilometre or so. The number of existing streets would be diminished by two-thirds due to the new arrangements of housing, leisure facilities and workplaces, with same-level crossing points eliminated wherever possible (*ibid.*: 168–9).

Le Corbusier elaborated a series of models that embraced these ideas, albeit in varying ways. The *Ville Contemporaine* of 1922 contained two major arteries, running north–south and east–west, that formed the axes of the city. The multilevel meeting point at the centre of the city was the hub of not just the street system, but also the surface rail, underground railway and other public transport systems. Later schemes, such as the linear-city inspired *Ville Radieuse* (Le Corbusier,[1935] 1967), eliminated the need for the central crossing. Schematic plans for Rio (1929) and Algiers (1931–4) envisaged viaduct structures with residential accommodation stacked beneath the motorways on their roofs. Innovative though these sketches might have been, the sinuous nature of the viaducts, snaking their way from these ports to their mountainous hinterlands, dramatically revised the insistence on the straight line.

This point illustrates the care necessary when interpreting Le Corbusier on this or any other subject. As the Modern Movement's leading iconographer of the future city, his work is heavily scrutinised for content and nuance, especially since it is central to a Grand Narrative which links the deficiencies of recently designed urban environments back to the *flawed* visions of the pioneers of modern architecture (Gold, 1997b). As such, Le Corbusier's writings and three-dimensional projects offer considerable scope for historical interpretation. His writings were delivered in short staccato bursts with the key expressions already italicised to serve as rallying slogans. Accompanied by powerfully evocative line drawings and photographs, they were always intended for promotional purposes; ideologically charged source material for the international group of architects that adhered to modernism. Yet they were not blueprints. As Ernö Goldfinger noted, 'Le Corbusier did these drawings and you could put them on the wall and admire them. But they did not tell you how to build the houses and streets or where to dig the drains. Not for a real city anyway.'[3]

Confirmation of the gap between broad vision and formulation of coherent principles for street planning in the modern city was shown by the activities of the *Congrès Internationaux d'Architecture Moderne* (CIAM). Founded in 1928 as a forum for the international Modern Movement to meet and discuss matters of shared interest, CIAM devoted its early Congresses to town planning matters. CIAM's early analysis favoured a fourfold classification of the functions of the city – shelter, work, relaxation (leisure) and circulation (transport and communications). Its Fourth Congress (CIAM IV, 1933) took the theme of 'The Functional City' and dealt with the interrelationship between these functions. CIAM had intended to publish the Congress's results, but it took a decade for the two associated volumes (Sert, 1942; Le Corbusier, 1943; see Gold, 1997a) to appear. In the event, neither advanced matters much beyond the level of general principles.

Le Corbusier's sharply polemic *Athens Charter* (1943) drew attention to the new speeds of road vehicles as bringing problems of congestion, safety and hygiene. The solution was to recognise the differences between urban and suburban traffic and classify them according to vehicle speed. On that basis, he argued for major functional transformations that would equip city and region with a road network

> that incorporates modern traffic techniques and is directly proportionate to its purposes and usage. The means of transportation must be differentiated and classified for each of them, and a channel must be provided appropriate to the exact nature of the vehicles employed. Traffic thus regulated becomes a steady function, which puts no constraint on the structure of either habitation or places of work.
>
> (Le Corbusier, [1943] 1973: 98)

Application of multilevel solutions to bring about efficient flow was highlighted by noting: 'Urbanism (town planning) is a three-dimensional, not two-dimensional science.' Adding the element of height would not only solve the problem of modern traffic, it would also assist provision of leisure by allowing use of the open spaces created.

The second volume, José Luis Sert's *Can Our Cities Survive?* (1942), also recognised the interdependence of urban functions. Sert drew attention to the spread of the internal combustion engine as a key element in urban change, drawing people to the heart of cities and later permitting decentralisation towards the outskirts and the open spaces beyond. Street congestion had become severe due to the narrow widths and constrictions of thoroughfares, on-street parking, the frequency of inter-sections, and the fact that railway lines and crossings segmented cities and impeded traffic. Indeed the frequency of obstructions almost neutralised the new possibilities of locomotion (Sert 1942: 171, 174). Modernisation required that each street should be classified according to its function and busy thoroughfares isolated from nearby buildings. Further improvements would come from creation of pedestrian lanes and restriction of on-street parking. Sert recommended that pedestrians would cross busy streets at different levels from vehicular traffic and that traffic was concentrated on to major arteries.

THREE-DIMENSIONAL PLANNING

Taken together, Sert and Le Corbusier succeeded in creating a rhetoric that expressed a growing consensus, even if it was thin fare to show for a ten-year-old project intended to advance the Modern Movement's thinking on town planning matters. Yet there were real problems in moving beyond generalities towards more specific proposals. This point can perhaps be illustrated by considering the work of the British Modern Movement.

At the outset, it is important to stress that their prime area of interest at the urban scale was housing rather than other town planning issues: as Lord Esher remarked, 'it was through housing that we saw cities being rebuilt'.[4] Nevertheless, the image of the congested city thoroughfare was ever-present in the literature of the period as an indictment of urban chaos. Stationary cars and lorries nose-to-tail in the city centre symbolised waste and inefficiency. As the 1930s wore on, various projects were completed that attempted to formulate ideas about what might be done to apply modernist principles to control the land-uses and circulation systems of the city.

The first, the 'Garden City of the Future', was a 1:360 scale model prepared by the partnership of F.R.S. Yorke and Marcel Breuer for the Ideal Home Exhibition in 1936. Built with sponsorship from an interested party, the Cement and Concrete Association, the model measured 2.9 metres by 1.7 metres (Figure 4.2). It showed only a portion of a city's central area rather than the city itself, but this was consid-ered large enough to show the interrelationship of land-uses as well as appropriate transport systems. The model embraced three design principles: geometrically regular layout, strict functional separation of land-uses, and segregation of traffic flows. The choice of housing in 12-storey slabs of flats blended in with the policy on movement within the city. As Yorke and Breuer noted:

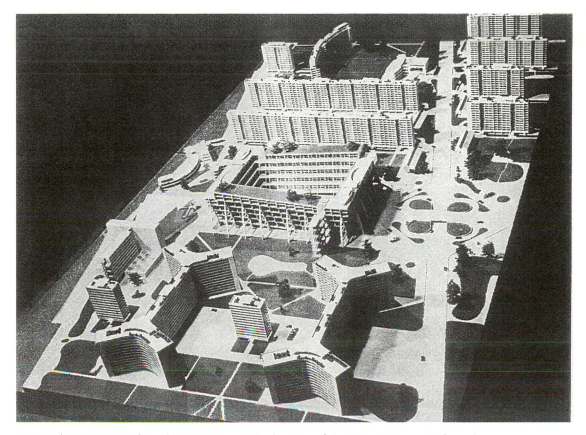

FIGURE 4.2 F. R. S. Yorke and Marcel Breuer, 'Garden City of the Future'. Source: Yorke and Breuer (1937: 182)

> the adoption of higher buildings ensures (that) the distances from point to point are reduced to a minimum. With this type of plan too, the simplification of traffic conditions becomes automatic, there are fewer house doors, fewer streets, and above all, fewer street intersections. . . . The private house in the centre of the city complicates immediately the traffic problem.
>
> (Yorke and Breuer, 1937: 181)

Transport flows themselves would be channelled on to purpose-built, high-capacity routeways of the type exemplified by the axial road that ran across the model. Through-traffic could move unimpeded; traffic serving the megastructural retailing district, the waterfront entertainment district or the office complexes could leave or join the axial road by a modified clover-leaf intersection. Parking spaces or other indications of where vehicles were left when not in use are virtually absent from this conception.

The model effectively portrayed various elements of built form and layout to a public unfamiliar with such ideas. It was intended as a 'demonstration of principle' that showed ideas about form, structure and layout. None the less, despite only representing a part of the city, the model was essentially a pastiche intended to cram as many different features as possible into a small space. There was no underlying prototype of urban form or any explicitly postulated relationship between the form of the street system and user needs.

When the same model was reshown at the Modern Architectural Research (MARS) Group's 'New Architecture' exhibition at the New Burlington Galleries, London (1938), it complemented other exhibits in conveying a loose synthesis of ideas about modern living, design and the city. The MARS Group's torturous efforts to arrive at that synthesis are discussed elsewhere (Gold, 1993), but the exhibition's contents reveal something of the common ground among modern architects at the time. A large introductory screen headed with the words 'It has come to this' contained a photograph of a visually chaotic and highly congested section of Oxford Street (London) superimposed over a view of English parkland, with the following caption:

> The mischief is done. The monstrous town enmeshes our life and wealth. We regret. We condemn. But what can we do? This: we can reconsider the aims of building: establish a new standard of integrity and realism in architecture, so that as we build we recreate.
>
> (MARS Group, 1938: 7)

Compared with the power of this image, related attempts to illustrate any alternative were, at best, sketchy. The catalogue asserted that: 'Transport demands an architecture modelled on a planned flow of traffic, an architecture eliminating stress and confusion at terminal points, clarifying the transition from foot to wheel, from wheel to air or water' (*ibid*: 13). Eight different media were identified and depicted on a revolving wheel construction, with pictures to illustrate the opportunities that each presented for modern architecture. No direct impression was given at this stage as to how broader questions of segregation of flows or system interchange were to be handled, other than reference to the adjacent Town Planning exhibit.

The latter exhibit was the MARS Group's first Plan for London. This was a lightly revised version of a project originally prepared for the fifth Congress of CIAM in 1937 by Aileen and William Tatton Brown in collaboration with the architectural publisher Hubert de Cronin Hastings. The plan primarily comprised ideas for handling urban growth stimulated by projected rises in car ownership. Although these increases had the capacity to destroy existing cities through congestion, they offered prospects of new forms of urban development at the city fringe, creating a new texture through redevelopment in the city itself.

The conceptual basis was a 'Theory of Contacts', which asserted that the city's prime function was to maximise opportunities for human contacts (or transactions), whether intellectual, social or commercial. Efficient communications were essential in achieving that goal and this, in turn, required recognition of the different speeds of traffic. Four categories were proposed: pedestrian movement (below 10 km/hr), local traffic (up to 100 km/hr), rapid highway traffic (more than 100 km/hour), and air travel. The centrepiece of the plan was a notion for creating new developments by locating 'neighbourhood units' along the routes of spinal high-speed road arteries. Housing would be screened from noise and pollution by placing the through-routes at sunken levels, with additional concealing earth-works. There would be multilevel intersections at one-mile intervals, frequent enough to provide access but not so frequent as to impede the flow of through traffic (Gold, 1995).

This style of development was held to offer considerable advantages over conventional development regarding both directness of route and length of roads required. As shown in Figure 4.3, this was shown by contrasting a one-mile square grid cell containing a new neighbourhood unit against the layout for Hampstead Garden Suburb. Although the densities would have been roughly the same (40 per acre), the amount of roads required for the former was calculated at only 6 miles, compared with 19 miles for Hampstead Garden Suburb. Naturally, the choice of the latter as an example was not accidental. Built in 1905–9 and closely associated with Raymond Unwin and the Garden City movement, this unflattering comparison was intended as propaganda against the claims of an alternative urban vision.

When shown in the Exhibition, the Town Planning Exhibit was carefully marked as 'A preliminary survey of London by a *section* (my emphasis) of the MARS Group' – accurately implying that this paper exercise did not enjoy unequivocal support from the organisers. The MARS Group itself went on to produce a second full plan, published in 1942, which was based on reorganisation of London into linear strips of development centred on spinal railways (Korn and Samuely, 1942; see also Gold, 1995). The Tatton Browns, however, regarded the new emphasis on railways rather than roads as a retrograde step and decided instead to develop their ideas about the car and the street further. The end-product was their 'Three Dimensional Planning' scheme (Tatton Brown and Tatton Brown, 1941, 1942).[5]

Appearing during the early years of the war, their project addressed issues of reconstruction, although its essential elements grew out of work completed before 1939. It was partly based on Alker Tripp's (1938) ideas on traffic management and precinct planning, which the Tatton Browns sought to extend by recasting them in three-dimensional form. The authors argued that the by-laws regulating development needed to be fixed in three dimensions:

> It is not enough to control the space between buildings. We must envisage the possibility of controlling the space above and below them in order to create new levels, on which traffic of different kinds can operate. Regulations enforcing a uniform height might be used to create a new ground level for pedestrians; in the same way regulations requiring the construction of basements of standard depth (if necessary sub-basements also) could be used to obtain new parking places at a level when much space is wasted at present.
>
> (Tatton Brown and Tatton Brown, 1941: 84)

A specimen district plan showed 24-storey blocks straddling large highways. The blocks were widely spaced to allow light and the passage of air, with the spaces between interspersed with new parkland and low buildings. Powers to ensure that the lower buildings had flat roofs of similar height meant that they could carry

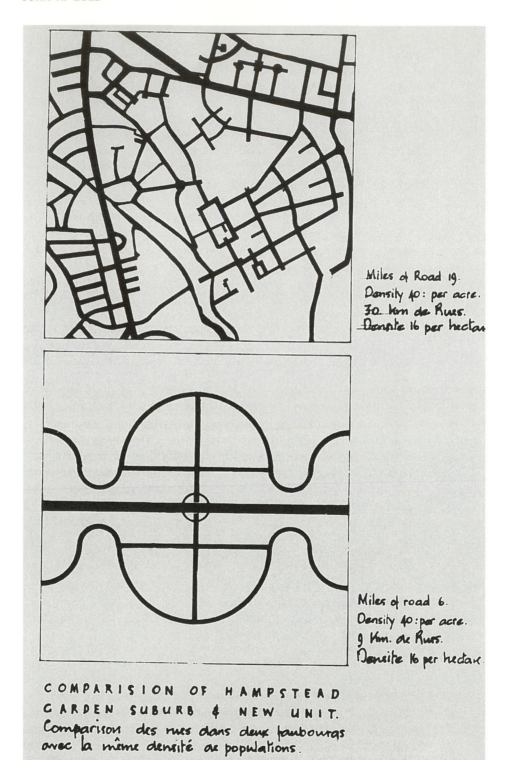

FIGURE 4.3 Comparison of road layouts, Hampstead Garden Suburb and a New Neighbourhood Unit. Source: William Tatton Brown, personal papers

walkways, with pedestrians able to approach buildings from roof level downwards rather than from below. Bus stops were located close into the tall buildings and away from the fast roadways. By placing parking and garaging spaces below ground, the amount of roadway required would be less than in normal plans given the absence of obstructions to traffic flow. Major intersections contained both roundabouts and underpasses, in arrangements envisaged as the urban equivalent of the clover-leaf crossings possible in less restricted spaces.

As a scheme for a new site rather than an existing city – even one that had suffered severe bombing – there were no costings for clearance and renewal. This, like other similar projects before it, was for a hypothetical city. While understandable given that this was a spare-time project that had no major input of research, engagement with the complexities of actual city reconstruction lay in the future. Nor was there yet recognition of the potential problems of pollution and microclimate caused by this intermingling of high-capacity roadways and tall buildings. As William Tatton Brown noted: 'the traffic noise, of course, would have been terrific and it would have had to be air-conditioned . . . we (also) never realised how windy this (roof-deck) would be and how little it is used, even at the Barbican.'[6]

CONCLUSION

Seen with benefit of hindsight, the projects considered in the previous section hint at several broader themes. First, they show the existence of a loose constellation of shared ideas. Certainly many modern architects interested in town planning were excited by the principle of multilevel arrangements for streets and interchanges, in much the same way as, say, they eagerly anticipated the potential of prefabrication for transforming house design. It was an article of faith, rather than a closely researched alternative. Understandably, the analysis was rooted in the times, with much lighter road usage and lower car-ownership. No-one was then questioning the environmental consequences of these ideas, nor considering possible *adverse* consequences of functional separation of traffic systems. For instance, areas of the town would be placed outside the experience of one group or another (see too Mooney, this volume). The multilevel interchange replaced the crossroads or, viewed another way, places where no-one crossed anyone's path replaced places where people met (Augé, 1995). Whatever the postulated functional arguments for that development, the net impact of the change was not wholly advantageous.

Related to this, it needs little emphasis by now to state that any agreement about broad trends did not mask the absence of more detailed ideas that might be readily applied in recreating the form or layout of streets in the future metropolis. The fundamental relationships between streets, buildings and land-use had still to be properly researched and formulated, let alone costed. For all the bold line drawings and confident allusions to science and rationalism, these were essentially spare-time, low-budget projects. Modern architects remained at an early stage in devising convincing principles for reconstructing the major urban thoroughfare. Almost two decades after Le Corbusier had fulminated against the iniquities and illogicalities of the *rue corridor*, there was still little fixed idea about the shape of its successor. The boulevard might be dead but the urban expressway was yet to arrive.

This leads to a final point. It was mentioned earlier that the work of Le Corbusier and other pioneers of modernism are now deeply implicated in a narrative that links emergent trends in post-war city design directly back to visionary architectural prototypes. The findings of this paper show that it is possible to be misled by polemics essentially intended to promote the cause of modernism by journalistic devices. These were the early days of the Modern Movement. Ideas were fluid and ideologies malleable. There were no overarching visions of either the future city or its component elements at this stage. The eventual shape of reconstructed streets, as I have argued elsewhere (Gold, 1997b), stemmed from the specific circumstances of reconstruction and not from blind application of visionary prototypes from a previous era. To make a leap that connects the postwar reshaping of the street pattern of cities to these fragmentary prototypes is to attach too much importance to the beguiling appeal of imagery and ignore the broader context of which that imagery was part.

NOTES

1   While Haussmann's new thoroughfares effectively created the image of what is now normally understood by the word 'boulevard', the word existed well before his time. It was coined in the fourteenth century to apply to defensive works and was first applied to wide routeways in the early nineteenth century (CNRS, 1975: vol. 4, 795–6).
2   Sant'Elia, A. (1914) *Messagio*, quoted in Banham (1960: 129). There is considerable dispute about the true authorship of this document; see de Costa Meyer (1995, 141–68).
3   Interview with Ernö Goldfinger, 16 December 1986. Transcript reference, [T6/5].
4   Interview with Lord Esher, 15 April 1987. Transcript reference, [T4/6].
5   This was published first in *Autocar* (11 July, 18 July and 25 July 1941). As William Tatton Brown noted: 'I was waiting to be called up. I was hard up for money – my practice had folded – and *Autocar* were hard up for material, after all, this was during the war and there were no new motorcars to write about. Thus, they switched on to town planning, as part of a broad concern with streets and street planning.' Interview with William Tatton Brown, 5 December 1986. Transcript reference, [T15/7].
6   Interview with William Tatton Brown, 5 December 1986. Transcript reference, [T15/7].

REFERENCES

Augé, M. (1995) *Non-Places: An Introduction to an Anthropology of Supermodernity*, London: Verso.
Banham, R. (1960) *Theory and Design in the First Machine Age*, London: Architectural Press.
CNRS (Centre National de la Recherche Scientifique) (1975) *Trésor de la langue française*, Paris: Centre National de la Recherche Scientifique.
Cohen, J. L. (1995) *Scenes of the World to Come: European Architecture and the American Challenge, 1893–1960*, Paris: Flammarion.
de Costa Meyer, E. (1995) *The Work of Antonio Sant'Elia: Retreat into the Future*, New Haven, CN: Yale University Press.
Evenson, N. (1984) 'Paris, 1890–1940', in A. Sutcliffe (ed.) *Metropolis, 1890–1940*, London: Mansell, pp. 259–87.
Giedion, S. (1941) *Space, Time and Architecture: The Growth of a New Tradition*, Cambridge, MA: Harvard University Press [3rd Edition (1954) consulted here].

Gold, J. R. (1993) '"Commoditie, Firmenes and Delight": Modernism, the MARS Group's "New Architecture" Exhibition, 1938, and Imagery of the Urban Future', *Planning Perspectives*, 8: 357–76.

Gold, J. R. (1995) 'The MARS Plans for London, 1933–1942: Plurality and Experimentation in the City Plans of the Early British Modern Movement', *Town Planning Review*, 66: 243–67.

Gold, J. R. (1997a) 'Creating the Athens Charter: The Functional City, Town Planning and the Ideology of the Modern Movement', unpublished manuscript.

Gold, J. R. (1997b) *The Experience of Modernism: Modern Architects and the Future City, 1928–53*, London: E. & F. N. Spon.

Hays, K. M. (1992) *Modernism and the Posthumanist Subject: the Architecture of Hannes Mayer and Ludwig Hilberseimer*, Cambridge, MA: MIT Press.

Korn, A. and Samuely, F. J. (1942) 'A Master Plan for London', *Architectural Review*, 91: 143–50.

Larsson, L. O. (1984) 'Metropolitan Architecture', in A. Sutcliffe (ed.) *Metropolis, 1890–1940*, London: Mansell, pp. 191–220.

Le Corbusier (1929) *The City of Tomorrow and its Planning*, London: John Rodker. Originally published (1925) as *Urbanisme*, Paris: Editions Crès. [Version quoted here is the Revised Edition (1987) London: Architectural Press.]

Le Corbusier (1935) *La ville radieuse*, Boulogne-sur-Seine: Vincent, Freal and Cie [translated (1967) as *The Radiant City*, by P. Knight, E. Levieux and D. Coltman, New York: Orion Press].

Le Corbusier (1943) *La charte d'Athènes*, Paris: La Librairie Plon [translated by A. Eardley, 1973, as *The Athens Charter*, New York: Grossman].

Mansfield, H. (1990) *Cosmopolis: Yesterday's Cities of the Future*, New Brunswick, NJ: Centre for Urban Policy Research, Rutgers University.

MARS (Modern Architectural Research) Group (1938) *New Architecture*, London: New Burlington Galleries.

Pierce, S. R. (1942) 'Control: an Exploratory Survey', *Architect and Building News*, 24 (April): 55–7.

Pinkney, D. H. (1958) *Napoleon III and the Rebuilding of Paris*, Princeton, NJ: Princeton University Press.

Ramazani, V. K. (1996) 'Writing in Pain: Baudelaire, Benjamin, Haussmann', *Boundary 2*, 23(2): 199–224.

Sennett, R. (1990) *The Conscience of the Eye*, London: Faber & Faber.

Sert, J. L. (1942) *Can Our Cities Survive? An ABC of Urban Problems, Their Analysis, Their Solutions, Based on the Proposals Formulated by the CIAM (International Congresses for Modern Architecture/Congrès Internationaux d'Architecture Moderne)*, Cambridge, MA: Harvard University Press.

Shanor, R. R. (1988) *The City that Never Was*, Harmondsworth: Penguin.

Tatton Brown, A. and Tatton Brown, W.E. (1941, 1942) 'Three-dimensional Town-planning', parts 1 and 2, *Architectural Review*, 90: 82–8 and 91: 17–21.

Tripp, H. A. (1938) *Road Traffic and its Control*, London: Edward Arnold.

von Moos, S. (1979) *Le Corbusier: Elements of a Synthesis*, Cambridge, MA: MIT Press.

Yorke, F. R. S. and Breuer, M. (1937) 'A Garden City of the Future', in J. L. Martin, B. Nicholson and N. Gabo (eds) *Circle: International Survey of Constructive Art*, London: Faber & Faber, pp. 181–3.

5

THE VISUALISATION OF THE STREET
COMPUTER MODELLING AND URBAN DESIGN
Richard M. Levy

•

THE ARCHITECTURE OF STREETS

Computer modelling and visualisation as powerful tools in understanding the full impact of urban design proposals, are invaluable to planners and members of the community concerned with sensitive design issues. The design of our streets is one such issue. However, streetscape design is not only of contemporary interest. We are fortunate that the concern for the appearance of city streets was recorded by artists and architects of the Renaissance including Perruzzi, Bramante, Michelangelo, and Piranesi (Bacon, 1969).

Although we realise today that design alone cannot correct or compensate for social ills, it is an important tool in shaping our sense of place. In purely commercial terms pleasant surroundings translate into a variety of gains for both the resident and the tourist. Attractiveness can enhance a community's tourism potential, draw shoppers and new businesses, and improve the value of commercial and residential real estate (Brambilla and Longo, 1977; Brower, 1984; Keister, 1990; Krohe, 1992). However, it must be recognized that good urban design goes beyond the facade (Barnett, 1995; Bressi, 1996; Knack, 1994). Good urban design is synonymous with the physical form of a community from the building to the city (Calthorpe, 1993; Lynch, 1971, 1981, 1982).

Cities are ultimately the consequence of design decisions accumulated over decades and centuries. Though some of these decisions were made with a sensitivity to a community's appearance and function, many decisions were at best *ad hoc*. While topography sets the stage, dictates of law and cultural influences work together in creating the form of a city. Unique paths are followed by each community that contribute to a sense of place. Understanding 'place' requires a sensitivity to the geometry of form.

The proportion and scale of city blocks shape the length of our streets. Adding the size and location of buildings to this equation, the three-dimensional form of our cities is defined (Bacon, 1969; Evenson, 1979; Krier, 1984; Lynch, 1982).

The relationship between the horizontal and vertical dimensions of a city is critical to our perception of the urban environment (Calthorpe, 1993; Duany and Plater-Zyberk, 1991; Jacobs, 1985; Richert, 1996). A visit to an older community under redevelopment reveals how urban spaces can be adversely affected by merely changing the location of a building (Figures 5.1 and 5.2). For cities built in the nineteenth century, streetscapes were composed of a continuous line of imposing building facades. The introduction of commercial structures invented for the strip malls of the 1970s destroys this type of street by changing our sense of enclosure. With setbacks needed to create on-site parking, an uneven pattern of car parks and one-storey strip malls is introduced into the block. The result is that views of backyards and service areas are exposed, which were never intended to be visible from the street by the town's original founders.

In order to avoid introducing unintended views into an urbanscape, development proposals must be evaluated at the street level or from the pedestrian view. From this perspective, issues of human scale become important in creating attractive and inviting urban spaces. An appropriate sense of scale, distance, proportion, massing, landscaping and even architectural detailing is interwoven into our culture. Familiarity with our environment is established from childhood with memories of a neighbourhood shaped by its architecture (Campbell, 1978; Norberg-Shulz, 1965). However, the visual rhythms established by building facades will often go unnoticed on a walk down a familiar street, unless we are attentive students of architecture. The style, scale and condition of the buildings, the materials of construction, and the placement of windows and doors are important attributes in defining a street elevation (Jacobs, 1985; Rasmussen, 1964: 22–8, 127–33). The experience of a place is dictated by the design of both streets and buildings. Our experience of the street is shaped by the length of the blocks, the width of the pavement and whether its surface is of brick or concrete. The visibility of an entire facade of a three-story building from either side of a street is conditioned by the width of the pavement and that of the street itself. In addition, the placement, age and type of trees planted along the pavement will dramatically affect our perception of the environment. Trees lining a street, besides adding shade on a summer day, create an architectural space reminiscent of the aisle of a cathedral, where the trunks form the column of a vaulted canopy composed of branches and leaves (Arnold, 1993). The character of the street furniture, including light poles, benches and kiosks, all add to the definition of the street scene (Evenson, 1989; Davies, 1982). The quality of any streetscape experience will be dramatically transformed by changes in atmospheric conditions. The time of year affects the patterns of sun and shade, as well as the quality of light, which all work together to transform the appearance of a street. Understanding the interaction of the environment with design contributes to creating places that are both attractive and functional (Rasmussen, 1964, chap. 4).

FIGURES 5.1 and 5.2 Gaspar Block, Geneva, NY, 1990, 1993. Photographs reveal the impact of new development on a city built in the nineteenth century

URBAN DESIGN: LIFE'S ACTIVITIES

To understand urban space one must go beyond a mere appreciation of architectural form. The uses of buildings found along a block determine the level of pedestrian activity and define the quality of the 'street life'. Urban planners, mainstreet coordinators and retail associations have always known that in areas where lawyers, accountants and real-estate and insurance agents inhabit first-floor shopfronts, few shoppers will be present. The clustering of retail and commercial enterprises will significantly affect the character of the area. A large concentration of bars on a block may encourage night-time traffic but may do little to attract shoppers during the day. Similarly, the location of community buildings such as post offices and government offices will draw steady traffic and encourage the aggregation of commercial uses, but only if opportunities exist for development. Especially in older communities, finding space adequate for today's retailing can be a challenging exercise. Older buildings have interior bearing walls and narrow proportions, which are difficult to adapt to the demands of modern retailers. For these reasons, revitalising a city centre requires a sensitivity to the space requirements of tenants who can potentially benefit from being in close proximity to existing government buildings, hospitals and transportation terminals (Barnett, 1995; Gallagher, 1991).

The density and pattern of uses in a city generate different levels of automobile traffic. In contrast to city engineers who must consider standards to support a specific level of traffic in a street design, urban planners must consider the experience of 'getting there'. City streets provide the links between commercial, residential and industrial areas. Neo-traditional planning has awakened designers to the issue of streetscape design and its impact on the quality of a residential neighbourhood (Appleyard, 1981; Arendt, 1994; Calthorpe, 1993; Duany, 1991, 1992; Fernandez, 1994; Knack, 1993; Mohney and Easterling, 1991). Plans that define place by using a street hierarchy of boulevards, streets, lanes, bike paths and public parks have been designed to address the needs of both the pedestrian and automobile. For example, standards sympathetic to the pedestrian, rather than to the speed of the automobile, can create an environment reminiscent of neighbourhoods found in many of our older communities. In response to concerns for the safety and security of home-owners developers have begun to create communities with narrow, tree-lined streets. Thus, street standards are critical in achieving a hierarchy of scale within a community from block to neighbourhood centre (Appleyard, 1981; Calthrope, 1993; Duany, 1991; Fernandez, 1994; Gibbons and Oberholzer, 1992).

THE GOALS OF COMPUTER VISUALIZATION IN URBAN DESIGN

In urban design, cities and towns must adopt strategies that can translate a community's vision of itself into action. Computer modelling can assist the planner who wants to facilitate goal setting for a community. A master plan must answer such concerns as usable open space for recreation, streets which are appealing to retailers and attractive residential neighbourhoods. A CAD (Computer Aided Design) based approach to urban planning promotes the examination of density, zoning, sun and shade, open space and views as part of the design process (Kwartler, 1989; NEA, 1993; Shirvani, 1981). By actually visualising possible development scenarios, communities have the means to evaluate potential action against vision (Charles and Brown, 1992; Jacobson, 1994; Liggett and Jepson, 1993; Littlehales, 1990; Macleod, 1992; McCullough, 1995). It is important to know if a proposed bylaw, with its specific setbacks and heights for buildings, will place a public park in shade during the winter months, or in the case of waterfront development, if new construction will block existing views along the water's edge. These are some of the visualisation questions that all modelling efforts should be designed to answer.

Devising a strategy for the use of CAD in urban planning must take into account issues of policy and urban design. Clearly, changes in a zoning bylaw can effect the physical form of our streetscapes (Kwarlter, 1989; Shirvani, 1981). Examining the potential physical consequences of a proposed bylaw amendment in a virtual environment may help avoid later surprises. To this, CAD systems need to be integrated into decision making. It is simple enough to use CAD to insure that buildings do not exceed heights and setbacks or to evaluate space allocated for a specific use. However, to be effective, planners as well as all other interested parties need access to these virtual environments. Usually, the concerns of the development officer, who must test proposals for compliance with existing municipal law, will be addressed first. Then developers should have access to these environments for analysing the effectiveness of a proposed design concept. The public needs to have access to understand the potential impact of the proposal on their community (Jacobson, 1994; Mahoney and Phillips, 1994; Novitski, 1994a).

The real potential of CAD is that it can offer a common meeting ground for negotiation among all concerned parties: city officials, property developers and members of the community. Ultimately, CAD and visualisation technology can assist in negotiations by focusing on the look and feel of a project rather than on whether it meets every aspect of a bylaw. By providing an environment for testing proposals, it is possible to develop a less prescriptive approach to architectural guidelines. Concepts can be evaluated by experiencing space as a walk-through, rather than as a single artist's rendering that reveal perhaps the one and only flattering view of a project. With computer environments, it is possible to examine a proposal from all visible locations. With animations and virtual reality (VR) the pedestrian experience can be simulated and reviewed by members of the community early in the development process (Liggett and Jepson, 1993; Mahoney, 1994; Novitski, 1994b).

Rather than hindering the development process, a 3-D computer model available to the public domain can help developers avoid making costly investments in proposals that are inappropriate. Developers also thus have the opportunity to present spaces to potential tenants and owners, which can help in the pre-leasing phase of a project and ultimately avoid costly redesign of unsuitable spaces. Furthermore, a more neutral testing environment can actually facilitate agreement between the developer and the local municipality on mutually beneficial design alternatives. Reducing the time needed to secure a permit would certainly encourage communities and developers to participate in underwriting the creation of a comprehensive model for an entire city (McCullough, 1995; Novitski, 1994a, 1994b).

The use of 3-D computer models can create a virtual environment for communities and cities at all scales, from the design of a single streetscape to the massing of an entire city. Whatever the scenario, the link between the scale of the block and the perception of the streetscape must be addressed in the design of a CAD system. Changes to the street introduced by the construction of a new building or changes to existing zoning bylaws can be viewed in context. The height, setback, materials, articulation of facades, size and placement of doors and windows, color schemes and signage can be studied in a virtual environment, giving the community an opportunity for evaluation and criticism (Alley, 1993; Hall, 1995; Liggett and Jepson, 1993; Novitski, 1994b). At the scale of the street, the choice of street furniture, sidewalk materials and tree plantings can affect the 'sense of place'. Proposals for street improvements including, for example, the selection of appropriate species and sizes of trees for a boulevard can be tested along with different styles of lamp-posts, garbage receptacles and benches. This approach could help avoid hiding a retailer's sign from the street by a newly planted tree, generating support for these modelling efforts by town officials. Changes in surface treatments of pavements and the design of car parks and landscaping can be viewed from the pedestrians' vantage point, giving communities an opportunity to consider the relative merits of various design schemes (Levy, 1993, 1995).

Once a CAD model is created, images can be generated to present concepts at any scale or format including elevations, plans and sections as well as animations and VR environments with little additional effort. Computer visualisation provides the designer with an interactive presentation format that can be appreciated by the community (Jacobson, 1994; Liggett and Jepson, 1993; Mahoney, 1994; Novitski, 1994b; Phair, 1996). As access to the internet become increasingly common, anyone with a television or computer will soon be able to experience proposed plans for their community on their community's website as a VR environment (Novitski, 1994b). This development should not be seen as a threat to the designer's role as a creative force. On the contrary, without local support, there is little possibility for conceptual plans to be implemented.

INTEGRATING THE COMPUTER INTO THE DESIGN PROCESS

Architectural and urban designers have numerous clients and users, and their proposals must address community concerns, government regulations, the need for fiscal responsibility and the developer's concern to make a profit. Furthermore, unlike in engineering design there are no optimal solutions, only acceptable ones (Simon, 1970; Ziesel, 1981: chap. 1). Given this context, the design process cannot be left to technicians and experts alone (Forester, 1989). Crucial decisions must be made with the involvement of political officials and community representatives, and a process that takes advantage of computer visualisation makes this involvement more feasible.

To be successful, the CAD and geographic information systems (GIS) must be carefully integrated into a single system. A firm or government department should begin by examining the goals of the new system. Will it serve the needs of urban design, subdivision approval, zoning compliance, and for development permit review? Then the financial resources available for hardware, software, data collection and training must be assessed. Systems need to be flexible. Data should be in enough depth to permit solving problems both today and in the future (Antenucci, 1991; Arnoff, 1989; Franklin and Neubert, 1993; Huxhold, 1991; Obermeyer and Pinto, 1994; Sebastien and Jackson, 1997). In CAD and in modelling there is always the question of what forms the analysis will take. What types of images and documents will be needed for reports and presentations: images, animation, multimedia, VR and other formats? Setting realistic goals will ensure the design of a system that will give immediate return, thus ensuring its future development as well. Though probably not all levels of management will have active users, there must be an appreciation of the capability and limitations of any system throughout the organisation. Training at all levels of management may ensure that newly introduced systems will be part of an overall strategy.

THE NEED FOR A DATA WAREHOUSE

Key to the use of CAD and other visualisation technology in urban design is the recognition that accurate and comprehensive data is primary to any analysis. Whether the study area is a city, a site or a single building, the data must exist at the appropriate level of detail to support the analysis. Modelling might be concerned with formal design issues or with the engineering design of actual structures and municipal systems. Existing and future proposals can be modelled by computer and visualised but only if the data required is available.

Developments in GIS technology mirror those in CAD. Computer applications now enable analysis of topographic, economic and demographic data. Eventually the data query capabilities of GIS will merge seamlessly with the modelling capability of CAD to create a virtual city. This development will allow the investigator to go beyond the appearance of a model and literally walk through a city, ask 'what-if' questions and view the physical outcomes of these inquiries (Jacobson, 1994; Kunze, 1994; Liggett and Jepson, 1993: McCullough, 1995; Phair, 1995).

To gain full benefit from the versatility of CAD and visualisation, an integrated data management approach must be created. The adoption of a 'data warehouse' would give municipalities the opportunity to share information. Then each new investigation might require some additional data, but it would not require the complete reconstruction of baseline data in order to begin the design exploration process. Such a design process tied to data sharing would be part of an integrated approach to problem solving among different departments. Data might come from a variety of sources: photographic and architectural records from an historical archive, a surveyor's database, engineering CAD files, aerial and satellite images (Antenucci *et al.*, 1991; Arnoff, 1989). Having access to data from a variety of sources would also ensure that potential private sector clients, including banks, retailers, real-estate developers and transportation companies, would be able to use and exchange data with municipal databases. Ultimately, a data warehouse would promote greater involvement by the private sector in community planning.

A properly designed data warehouse will allow users to concentrate on solving problems. All users, including municipalities will have to develop strategies for selecting the most appropriate commercially available software. With no single application satisfying all requirements for urban planning and design, there are two strategies in the selection of software. The first is to buy only applications that are guaranteed to be compatible. For example, some of the most popular CAD packages can be purchased for both drafting and engineering, at the same time serving as an umbrella for other applications. Such a package gives the user a single interface for operating other applications such as for architectural modelling or animation and rendering. If data does not have to be transferred between applications, there are fewer delays in the work flow.

Often, however, this is not possible. Suitable applications might be too expensive. As well, some applications do not work within a single environment. Then the user must develop an integrated process that requires the user to transfers files from one application to another. Sometimes, the lack of suitable applications at reasonable prices makes this approach to systems design necessary (Mayall and Hall, 1994; Schutzberg, 1995).

Establishing a successful flow of information among different departments and organisations requires a professional organisational commitment to the use of CAD and computer visualisation technology in urban planning. This commitment includes recognition of a design process involving the participation of architects, urban designers, landscape architects and municipal engineers, and acknowledgement of each discipline's contribution. Without this, confusion and disjointed solutions are likely.

A CASE STUDY IN COMPUTER VISUALISATION: THE TOWN OF BANFF

The implications of development policy on streetscapes can be better appreciated by reviewing a specific project, such as that for the Town of Banff in Alberta, Canada. This case study demonstrates how a 3-D model was used to explain the visual impact that could result from changes in a zoning bylaw. As in many other communities, plans and elevations for a proposed project in Banff are often submitted as part of a development application to the planning board. The purpose of these documents is to establish that proper setbacks, height restrictions, parking needs and landscaping have been met. However, even with the submission of scale models and artist's renderings, it is often difficult to assess the impact of the proposed development on the adjacent properties and area.

In 1992, changes to Banff's bylaw permitted sites in the commercial district to be built at higher densities, up to three stories high. Over the last few years many properties in the town have been redeveloped, with older structures torn down to make way for buildings that fit this maximum envelope. This has led to a growing concern about the impact of such a maximum build-out on the views of downtown and the visibility of surrounding mountain peaks from Banff's downtown. A 3-D computer model could be used to help visualise the impact of future development on the views from downtown.

Because the architectural character of Banff has a significant impact on tourism, modelling had to address the visual impact that proposed development would have on the town and the mountain views (Mitchell, 1994) (Figure 5.3). A 3-D model had to have an accurate sense of the scale of terrain and the character of the city's architecture. To reveal the impact of new construction on view corridors, for planning review, it was critical that images presented on a computer display did not significantly 'paraphrase' a scene, but rendered the architectural and urban context with sufficient detail that viewers of the models would always have an understanding of their location within the town. Views of mountain peaks needed to be taken into consideration in establishing view corridors. Given that the style of a design can create a positive impression, the model had to be able to present different surface materials, textures, and details of signs, windows, doors, and other elements. Trees,

FIGURE 5.3 Photograph of Banff Avenue, 1994

lamp-posts, traffic lights, utility poles, street furniture, cars and pedestrians were introduced to give the viewer a sense of scale and context not usually seen in simple shaded mass models. Recognising that changes in atmospheric conditions affect an observer's perception, different times of the day and year and variations in weather conditions were also represented in the model.

To test the use of modelling as an aid in decision making, a single corner of a major intersection was considered as the site for the maximum allowable development. This site was selected because buildings on the other three corners already were either at or close to the maximum height. By altering only one of the four corners, the visual impact of a maximum build-out could be evaluated without resorting to images that would be unfamiliar to members of the community (Figures 5.4, 5.5 and 5.6). The 1992 bylaw set a maximum height of 12 metres, or three stories. Within this envelope, the first 8 metres of a building must be along the lot line, with any portion of the building above 8 metres set back a minimum of 2 metres from the property line. Furthermore, no more than 90 per cent of the site can be built upon (Town of Banff, 1992). Using the Town of Banff architectural guidelines and images of buildings constructed since the passage of the guideline, a plausible architectural design was created for the site.

A debate common in modelling focuses on the level of detail desireable in architectural models for planning. One argument is that highly detailed models obscure issues of setbacks and massing, with matters of style and surface materials captivating the viewer's attention. The other argument is that without architectural detailing the translation of representation to reality is more difficult. Without such details as windows, doors, awnings, piers, roof forms and materials, it is difficult for the viewer to judge if the form will be appropriate to the site. The design of the proposed building aimed to create a model that residents would not find jarring in terms of its use of materials and architectural elements, while focusing on the mass of the form. By using architectural elements such as awnings, roof gables, piers, and windows from buildings built under the new bylaw, it was hoped that the proposed design would be fairly typical of what actually might be built on the site.

A multimedia presentation containing both still images and animations was used to show possible future cases to the Banff Town Council, the Planning Department and the General Municipal Planning Committee. The animations simulating a drive through Banff were particularly useful in revealing the impact of new construction on the views of the street and surroundings. Because no single artist's rendering is used to highlight the proposal in such a presentation the impression of the site comes from a more experiential portrayal, such as that of walking or driving through an area. The sense of movement through a site makes it more difficult to focus on single issues, such as an awning or an entranceway. When the animations are made even more realistic by adding moving traffic, viewers are distracted away from the major object of concern. An advantage of this is that the impression of the site becomes more like that confronted in our daily experience as a pedestrian or a motorist. Judgement of the proposed development in such a context may actually be fairer because the observer is not making a decision based upon any single view.

FIGURES 5.4, 5.5, 5.6 Computer model, town of Banff. Images reveal the potential impact of the 1992 bylaw

Instead, a series of connected views, mirroring those shared by a community are considered.

In this case study, the issue of street profile was a specific concern. Rather than viewing the demolition of existing structures with disdain, some members of the community regarded a new building on the site as an improvement, bringing the existing cornice line of buildings to the same height on all four corners (Figures 5.4, 5.5 and 5.6). In an age when strip malls and convenience shops are the dominant idiom in commercial development, Haussmannizing a street may be a welcomed improvement. The opportunity for visitors to walk by shops and restaurants under a protected awning is attractive to those promoting tourism. Streetscapes reminiscent of older downtowns in North America now have a greater appeal than those focused on the automobile.

In the Banff study, the architectural elements presented in the images influenced the audience response. Images of streetscapes with elements from existing buildings may have favourably biased some viewers by giving credence to that which is already familiar. However, others had the opposite reaction to the same images. Visual images can solicit strong emotional responses towards a development proposal. A community that feels that more intensive development is contrary to its long-term goals, may have a negative response to architectural elements found in the newer areas of the

city. In contrast, images of a proposal which present a sense of newness may suggest to others that the concept is progressive and highly desirable.

Translating concepts into images can bring to light potential issues which ultimately make demands on the developer. Details may go unnoticed in plan and elevation, but can take on meaning when viewed in a computer-generated walk-though. Conflicts can arise between a municipality and developer over the details of a proposal. The developer who sees any request for change as an imposition on individual property rights, will probably not be supportive of visualisation technology which present detailed views of a proposal from a variety of vantage points. Municipalities which want to avoid conflict in granting needed approvals may also find the ability to visualise the impact of a new development an obstacle. A greater interest in the visual appearance of a project may stimulate the public's demand for change. Consequently, the power to visualise future development may actually work against the acceptance of new computer technology in urban planning.

CONCLUSION: THE FUTURE

There is little doubt that 3-D models will have an increasing role to play in the design of our city streets. The ability to create virtual environments will be important for promoting public discussion of the elements of an urban composition. A 3-D model can be used to present design proposals that potentially may affect the opinions pedestrians will have of their communities. By turning away from a dependency on drawings in plan and elevations, representations that can never actually be observed, communities will be able to measure the impact of new development in the context of the shared urban experience.

Integrated data bases and 3-D models linking social and economic activity to the development of space will be of growing importance in planning. A 3-D GIS could help the urban planner, engineer and architect answer questions about the appropriate use of space in relationship to a range of community issues, including: land-use, downtown revitalisation, open space, public transportation, economic development and demands for public utilities and community services. In this context individual concerns can be considered relative to the context of the site. Visualising alternative proposals in 3-D computer space will thus support an approach that is less prescriptive and more considerate of the totality of a proposal.

Because 3-D models are stored in virtual space, it is possible to share data among all interested parties. The new generation of Web users, who are becoming more visually oriented, will encourage the growing trend in the use of 3-D models in public participatory urban design. To capitalise on this trend, many of the attitudes of those who managed our communities will have to change. Only if government officials, community representatives, architects and urban planners adopt an approach to design that builds models of proposed development in electronic space from shared community data will such a broad exchange of views be possible.

REFERENCES

Alley, J. (1993) 'Using 3-D CAD Perspective Images as Evidence in Planning Hearings', *Plan Canada*, July: 27–30.

Antenucci, J. (1991) *Geographic Information Systems: A Guide to the Technology*, New York: Van Nostrand Reinhold.

Appleyard, D. (1981) *Livable Streets*, Berkeley, Calif: The University of California Press.

Arendt, R. (1994) 'How to Create a Subdivision with Character', *Planning*, 60, 5: 24–7.

Arnoff, S. (1989) *Geographic Information Systems: A Management Perspective*, Ottawa, Canada: WDL Publications.

Arnold, H. (1993) *Trees in Urban Design*, New York: Van Nostrand Reinhold.

Bacon, E. (1969) *Design of Cities*, New York: Penguin Press.

Barnett, J. (1995) 'Shaping Our Cities: It's Your Call', *Planning*, 61, 12: 10–13.

Brambilla, R. and Longo, G. (1977) *For Pedestrians Only: Planning, Design, and Management of Traffic-Free Zones*, New York: Whitney Libary of Design.

Bressi, T. (1996) 'Reveille For Times Square' *Planning*,62,9:4–8.

Brower, D. J. (1984) *Managing Development in Small Towns*, Chicago, Ill.: Planners Press.

Calthorpe, P. (1993) *The Next American Metropolis: Ecology Community and the American Dream*, New York: Princeton Architectural Press.

Campbell, B. G. (1978) 'Evolution and Information,' in S. Kaplan and R. Kaplan (eds), *Humanscape: Environment for People*, North Scituate, MA: Duxbury Press, pp. 23–9.

Charles, C. and Brown, K. (1992) 'Wow 'Em in 3-D', *Planning*, 58, 8: 14–15.

Davies, S. (1982) *Designing Effective Pedestrian Improvements in Business Districts*, Chicago, Ill.: American Planning Association.

Duany, A. and Plater-Zyberk, E. (1991) *Towns and Town-Making Principles*, New York: Rizzoli International Publications, Inc.

Duany, A. and Plater-Zyberk, E. (1992) 'The Second Coming of the American Small Town', *Plan Canada*, May: 6–13.

Evenson, N. (1979) *Paris: A Century of Change, 1878–1978* New Haven, Conn.: Yale University Press.

Fernandez, J. (1994) 'Boulder Brings Back the Neighborhood Street', *Planning*, 60, 6: 21–6.

Forester, J. (1989) *Planning in the Face of Power,* Berkeley, Univeristy of California Press.

Franklin, J. and Neubert, R. M. (1993) 'CAD – To Automate or Not to Automate?', *Progressive Architecture*, April: 59–61.

Gallagher, J. L. (1991) 'The Bronx Is Up', *Planning*, 57, 9: 14–15.

Gibbons, J. and Oberholzer, B. (1992) *Urban Streetscapes: A Workbook for Designers*, New York: Van Nostrand Reinhold.

Hall, T. (1995) 'Visual Reality', *Planning Week*, 16 March: 16–17.

Huxhold, W. E. (1991) *An Introduction to Urban Geographic Information Systems*, New York: Oxford University Press, 1991.

Jacobs, A. B. (1985) *Looking at Cities*, Cambridge, Mass.: Harvard University Press, 1985.

Jacobson, R. (1994) 'Virtual Worlds Capture Spatial Reality', *GIS World*, 7, 12: 36–9.

Keister, K. (1990) 'Main Street Makes Good' *Historic Preservation*, 42, 5: 44–50, 83.

Knack, R. E. (1993) 'Neotrad Meets the Midwest', *Planning*, 59, 4: 29–31.

Knack, R. E. (1994) 'Charleston at a Crossroads', *Planning*, 60, 9: 21–6.

Krier, L. (1984) 'Urban Components' in Leon Krier, *Houses, Palaces, Cities*, ed. Andreas Papadakis, New York: St Martins Press, pp. 42–9.

Krohe, K. Jr. (1992) 'Is Downtown Worth Saving?', *Planning*, 58, 8: 9–13.

Kunze, E. (1994) 'CAD vs. GIS: Mutually Exclusive or a Continuum of Complementary Technologies?, *GIS World*, 7, 6: 36–8.

Kwartler, M. (1989) 'Legislating Aesthetics: The Role of Zoning in Designing Cities', in C. Haar, and J. S. Kayden, (eds) *Zoning and the American Dream*, Chicago, Ill.: Planners Press, pp. 187–220.

Levy R. M. (1993) 'The Role of the Computer in Urban Design Education', *Plan Canada*, September:19–22.

Levy R. M. (1995) 'Visualization of Urban Alternatives', *Environment and Planning B*, Spring, 22: 343–58.

Liggett, R. and Jepson, W. H. (1993) 'An Integrated Environment for Urban Simulation', *International Conference on Computers in Urban Planning and Urban Management*, 565–83.

Littlehales, C. (1990) 'Revolutionizing the Way Cities are Planned', *IRIS Universe*, 19: 15–18.

Lynch, K. (1971) *Site Planning*, Cambridge, Mass.: The MIT Press.

Lynch, K. (1981) *Good City Form*, Cambridge, Mass.: The MIT Press.

Lynch, K. (1982) *The Image of the City*, Cambridge, Mass.: The MIT Press.

Macleod, Douglas (1992) 'A Collective Vision of Dalhouosie', *The Canadian Architect*, May: 25–9.

Mahoney, D. P. (1994) 'Walking Through Architectural Designs', *Computer Graphics World*, 1994: 17, 6: 22–4.

Mahoney, D. P. and Phillips, D. (1994) Cityscapes, *Computer Graphics World*, April: 36–43.

Mayall, K. and Hall, K. G. (1994) 'Integrate GIS and CAD to Visualize Landscape Change', *GIS World*, 7, 9: 46–9.

McCullough, M. (1995) 'Designing Cities on Disk', *Architecture*, 84, 4: 115–19.

Mitchell, A. (1994) 'Banff's Outlook Not a Pretty Picture', *The Globe and Mail*, 45, 244, 24 December: A1, A7.

Mohney, D. and Easterling, K. (ed.) (1991) *Seaside: Making a Town in America*, New York: Princeton Architectural Press.

National Endowment for the Arts (NEA) (1993) 'Looking at Change before it Occurs', video tape produced by Maguire/Reeder Ltd, Alexandria, Virgina: 17min.

Norberg-Shulz, C. (1965) *Intentions in Architecture*, Cambridge, Mass.: The MIT Press.

Novitski, B. J. (1994a) 'Architect–Client Design Collaboration', *Architecture*, 83, 6: 131–3.

Novitski, B. J. (1994b) 'Virtual Reality for Architects', *Architecture*, 83, 10: 121–5.

Obermeyer, N. J. and Pinto, J. K. (1994) *Managing Geographic Information Systems*, New York, New York: The Guilford Press.

Phair, M. (1996) 'The Next Dimension, Virtual Reality is Transforming CAD models into Moving Worlds', *Engineering New Record*, 236, 24: 24–8.

Rasmussen, S. E. (1964) *Experiencing Architecture*, Cambridge, Mass.: The MIT Press.

Richert, E. D. (1996) 'Made to Measure', *Planning*, 62, 6: 16–18.

Schutzberg, A. (1995) 'Bringing GIS to CAD: A Developer's Challenge', *GIS World*, 8, 5: 48–54.

Sebastien, P. and Jackson J. (1997) 'An Informational System for Government Land Use and Development Reguialation: A Client-Oriented Prototype Using GIS and Hyperlinks', *GIS97, Conference Proceedings*: 604–6.

Shirvani, H. (1981) *Urban Design Review: A Guide for Planners*, Chicago, Ill.: Planners Press, American Planning Association.

Simon, H. A. (1970) *The Science of Design: Creating the Artificial*, Cambridge, Mass.: The MIT Press: chapter 3.

Town of Banff (1992) *By-Laws and Architectural Guidelines*.

Ziesel, J. (1981) *Inquiry by Design*, Monterey, Calif.: Brooks/Cole Publishing Company.

Part II

SOCIAL IDENTITIES AND SOCIAL PRACTICES

6

DISPLAYING SEXUALITY: GENDERED IDENTITIES AND THE EARLY NINETEENTH-CENTURY STREET

Jane Rendell

•

This chapter is a theoretical and historical study of the early nineteenth-century street as a spatial representation of gendered identities. My research draws on approaches adopted in various other disciplines, but makes particular reference to the political and methodological concerns of architectural history and feminism. Despite disciplinary differences, feminist analysis of gender and space has tended to focus on critiquing the paradigm of the separate spheres (the binary which describes space as two mutually exclusive and hierarchically placed categories – the male public realm of the city and the female private realm of the home). Feminists involved in 'deconstructing' this binary have shown its ideological underpinnings in patriarchy and capitalism. This chapter is informed by these strategies, but moves further in suggesting that the gendering of space should be conceptualised through urban movement – through display, consumption and exchange.

The early nineteenth century is an important historical period for contemporary feminists investigating gender and space. This period precedes the patriarchal 'fixing' of the ideology of the public–male–city/private–female–home as the most pervasive spatial and gendered ideology in the mid-nineteenth century. I consider examining the early stages of articulation of an ideology an important way of understanding its workings, but it is not my intention to explain why the separate spheres came into being; rather I am interested in looking at how issues of gender were raised in connection with the public spaces of the city. In order to examine different ways in which streets are gendered in early nineteenth-century London, this chapter looks at rambling – a mode of movement which celebrates the public spaces, streets and the excitement of urban life from a masculine perspective.

THE RAMBLE

The verb 'to ramble' describes the exploration of urban space. Rambling is 'a walk without any definite route', an unrestrained, random and distracted mode of movement. As an activity, rambling is concerned with the physical and conceptual pursuit of pleasure, specifically sexual pleasure – 'to go about in search of sex' (see Oxford English Dictionary; Partridge, 1984). The rambler traverses the city, looking in its open and its interior spaces for adventure and entertainment; in so doing, he creates a kind of conceptual and physical map of what the city is. Rambling rethinks the city as a series of spaces of flows of movement rather than discrete architectural elements.

The places of the ramble are sites of leisure, pleasure, consumption, exchange and display in early nineteenth-century London. These include the emerging spaces of urban consumer and commodity capitalism – theatres, opera houses, pleasure gardens, parks, arcades and bazaars. The perpetual movement of the ramble is represented through the public street, and later through the private street, colonnade and arcade. Urban locations are placed in sequential relation, framing social events, activities and rituals in time and space. Social relations are articulated spatially – through movement and containment, and visually – through relations between viewer and viewed, in new urban designs for promenading. Gender differences are played out through the roles of consumer and commodity in the new spaces of consumption – shopping streets, exchanges, bazaars and arcades. This chapter follows the rambler through London, from east to west, by day and night, focusing on a closer examination of a number of public and private streets in the west, including Bond Street, St James's Street, the Haymarket, Regent Street, the Quadrant, the Burlington Arcade and the Royal Opera Arcade.

THE MALE RAMBLER

Tales of urban rambles were published from the seventeenth century onwards,[1] but the decade following the Napoleonic Wars saw the publication of a large number of best-selling books and prints featuring the rambler, as a fashionable and sporting man about town. These semi-fictional urban narratives tell of various country gentlemen's initiations to the adventures of city life under the guidance of a street-wise urban relative.[2] The rambler represents a new kind of urban identity which emerges in this period – the young, single, heterosexual and upper-class man of leisure, fashion and sport. The male 'type' of the rambler is contradicted by and reinforced through other urban masculinities; specifically, the corinthian, or upper-class sporting gentleman; the bruiser, or working-class boxer; and the dandy, or aspiring man of fashion. The urban rambler articulates his masculinity through dress and language codes, and through various kinds of spatialised social activities.

Rambling is a gendered activity concerned with sexual desire and visual display and consumption in an urban setting. The rambler is a pleasure seeker, his aim is to titillate the uninitiated with glimpses of an unknown world of suspense and pleasure

– heterosexual pleasure. Contemporary magazines concerned with sex and whoring described themselves in terms of rambling,[3] and another closely linked activity, ranging, 'intriguing with a variety of women'. All the women the rambler encounters in the streets and public spaces of the city are described as 'cyprians'. The term cyprian includes all public peripatetic women as prostitutes, from the wealthy courtesan to the poorest streetwalker. Female sexual identity is defined through spatial location and movement, rather than through the exchange of sexual favours for financial benefit.[4] Cyprians are represented as stimuli to the ramble. The rambler's desire for, and pursuit of, these female sexual commodities defines his masculinity as urban and heterosexual.

WEST AND EAST

The ramble represents a culturally diverse journey, structured around social and spatial contrasts, from grand interiors to dark streets, from high fashion to popular culture. The most striking juxtaposition is between London's two class zones of the east and the west. From the seventeenth century onwards, the city and the eastern districts surrounding it were commercial and industrial zones, inhabited by the working class, large numbers of immigrants, most numerously the Irish (George, 1992). The west was populated by members of the aristocracy, nobility and wealthy bourgeois class who moved out of the city westwards to new residential squares, first to Covent Garden and Soho and later to St James and Piccadilly (see Smeeton, 1828). The consequence of these movements created a growing gulf within London between the racially mixed and working-class east and the fashionable and upper-class west. In the city, commerce was the traditional source of respect, but in the western squares, the social standing of wealth gained through trade was downgraded in relation to those with titles and land. The west was represented as an exclusively upper-class place, where, despite their presence as servants, the working class were conspicuously absent.

The urban topography of west and east reinforced the social contrasts. Ownership and government legislation, as well as the status and wealth of the residents, affected the kind of urban spaces produced in both areas. The London jurisdictions of the City of London, the City of Westminster, the counties of Surrey and Middlesex, had different legal and institutional structures and attitudes to building (see George, 1992; Low, 1982; Rude, 1971). During the eighteenth century, the east sprawled haphazardly, its fragmented and chaotic growth outside the boundaries of the city partly caused by patterns of property rights. The only areas of planned development occurred in the west, where building patterns were far more uniform due to the nature of the collaboration between the landowners and speculators. The combination of the 1774 Building Act and the various building guide lines set out by the landlords, produced a structured set of urban spaces with footways, street paving, sewers and houses built to a uniform design (see Olsen, 1964; Rasmussen, 1988; Summerson, 1962). As a result, streets in the west were wider, straighter and more regular with new planted squares. In contrast, the streets of the east were irregular and narrow, with rookeries and slum areas.

The early nineteenth century saw the design of a series of new urban spaces of public entertainment, commercial leisure and consumption in London supported by growing public and private investment – theatres, parks, arcades and streets. These major planning improvements, such as Regent Street, or 'Corinthian Path', and prestigious pieces of architecture, such as Carlton House, were built in the west. Such improvement schemes reinforced the development of an associated sophisticated urban culture – classical architecture, plays, concerts, clubs, assemblies, walks and luxury shopping facilities. The representations of this culture through buildings, artefacts and pictorial representations expressed a self-consciousness about the special qualities of urban life. Although using the language of classicism associated with country houses, the fashionable architecture of urban leisure culture, with its public, commercial and cosmopolitan character, its streets, squares, and planned units, constructed an identity based on the status of city life (Borsay, 1990).

MOBILITY

In search of pleasure, the rambler moved freely between the clubs, opera houses, theatres and arcades of the west of London and the taverns of the east. Ramblers' drinking sprees took them through the leisure spaces of the working-class slums, from Covent Garden's Holy Land to the taverns of St Giles and East Smithfield. The rambler was an initiated member of the modern metropolis, wise to the delights and entertainments, the tricks and frauds of the urban realm.[5] The London he represented was an exciting, but frightening place, with an exotic night life, inhabited by criminals and prostitutes. The rambler warned strangers to the metropolis, typically country relations, of physical and moral corruption by familiarising them with criminal language codes and dangerous places, he achieved social integration through donning disguises and earning the respect owed a 'gentleman'. The rambler's ability to mix with a variety of social classes and experience both the west and east of the city represents an important part of his urban identity – his mobility – the prerogative of the dominant upper/middle-class male.

> This day has been wholly devoted to a ramble about London, to look at curiosities. A friend called on me after breakfast, and proposed an excursion; and we accordingly took our way through St. Giles', that paradise of usquebaugh and 'blue ruin', to which the low Irish resort on coming to London. Such a place of filth, and tipsy jollity, and nocturnal rows, and squalid wretchedness, is no where to be found, except on 'Saffron Hill' in the vicinity of Fleet Ditch, where a large portion of the indigenous poverty of the metropolis is congregated.
>
> (Wheaton, 1830: 119)

As London developed into an important centre of leisure and consumption, the role of streets, as spaces allowing the exchange of people and goods, became increasingly important as zones of trade and commerce, administration and entertainment. Street improvements both increased and facilitated the movement of people on the streets, and also provided a social space for visual display and consumption. The eighteenth century saw a number of improvements in lighting, paving and drainage, all of which made pedestrian walking easier (see Bedarida and Sutcliffe, 1980; Cruickshank and Burton, 1990; Johnson, 1991; Porter, 1994). The Westminster Paving Acts of 1761 provided underground drainage, kerbs, a convex carriageway of granite blocks, and removed rubbish and obstructions, such as old shop signs. But the accessibility of streets varied from west to east, the streets of the larger London estates were maintained according to obligations written into the leases, whereas in parts of the east, where occupation was more transient, improvements were harder to initiate as well as maintain. The London Lighting Act of 1761, for example, started to systematise London street lighting by placing the responsibility for lighting streets and other public areas with municipalities and lighting commissions. Improved street lighting gave people the opportunity of using the streets at night, giving ramblers far more potential for urban exploration (see Adburgham, 1981; Burke, 1940; Cruickshank and Burton, 1990).

> The lamps are well disposed. Not a corner of this prodigious city is unlighted. They have every where a surprizing effect and in the straighter streets, particularly at the west end of the town, where those streets cross each other at right angles, the sight is most beautiful.
>
> (Hutton, 1818: 13–14)

DISPLAY

The social ritual of dressing in the correct urban fashions was an important first step in the process of initiation for new ramblers. This set of activities took place in the chambers of the urban bachelor with a tailor in attendance. Dressing was followed by a parade up and down St James's Street and Pall Mall in order to display body and possessions:

> Such was the costume in which he was destined to show off, and thus equipped, after a few minutes they emerged from the house in Piccadilly on the proposed ramble, and proceeded towards Bond Street.
>
> (Badcock, 1822: 102)

In the early nineteenth century St James's Street and Bond Street were the site of male fashion, the large number of barbers, tailors, and hatters in these streets reflecting the bachelors' intense preoccupation with masculinity and self-presentation. Shops which catered for male custom included gun shops, booksellers, theatre ticket agents, sporting prints exhibitions, hatters, tailors, cravats, hairdressers, perfumers, jewellers and other expensive tradesmen. The purchase of new commodities to enhance the self was an important structuring part of the ramble, the desire, 'to look at everything and buy nothing' (MacDonogh, 1819: 40).

It was very fashionable for upper-class young men to display themselves in the ground floor windows of the clubs on St James's Street. Membership to these clubs was extremely select, and White's created a bow window in 1811 in the centre of the front facade, where men could show off their exclusivity to those in the surrounding rival clubs. Boodle's added a bow window in 1821 and Crockford's one in 1827 (Fulford, 1962: 21). The bow windows were in the ground floor coffee rooms, and provided a place to display dress and leisure time within easy gaze of the street. The bow window was a place for ramblers to watch and be watched (see Figure 6.1).

London streets were congested with all kinds of traffic both vehicular and pedestrian (Corfield, 1990: 147). This created a need for urban individuals to appropriate the street by adopting distinctive styles of behaviour and coded spatial practices:

> Walking the streets has been reduced to a system in London; everyone taking the right hand of another, whereby confusion is avoided. . . . The contrary mode is a sure indication of a person being a stranger, or living at the outskirts of town.
>
> (Badcock, 1828: 47–8)

The street played an integral part in producing a public display of heterosexual, upper-class masculinity. The urban male types of the corinthian, bruiser and dandy all

FIGURE 6.1 *'Crockford's, or, male display in the bow window.'* Source: Robert Cruikshank, 'Exterior of Fishmongers Hall, St James's Street, with a view of a Regular Break down, showing "Portraits of the Master fishmonger and many well known Greeks and Pigeons"', in Bernard Blackmantle, *The English Spy*, London, Sherwood, Jones and Co., 1825, vol. 1, p. 373, Fig. XXIV

adopted specific styles of dressing, talking and walking which centred on developing an urban aesthetic and style. Young men were seen as street nuisances, as a result of the clothes they wore and their body posture, they were either criticised for their feminine dress and described as 'ladies' men, who scent thy mawkish way' (Heath, 1822), or alternatively, for their roughness:

> The tandem men are most dangerous; for they generally drive entirely at random. You meet them with their leader looking wildly about him like a stag which has just brooked cover; the driver is either with his eyes fixed between the horse ears, or perhaps as wild and perturbed as the animal itself . . . perhaps with a mail coach horn, on his left hand; they may both be drunk or in high spirits.
>
> (MacDonogh, 1819: 36)

The rambler is represented not only in motion in exterior spaces; the streets of the east and west of London, but also in internal places; the interiors of theatres, assembly rooms, operas, hotels, clubs or bachelor chambers. The area of St James's, described as a male zone, consisted of male spaces of public activity but also numerous places of male privacy, intimacy and domesticity. There were chambers lodgings, like the Albany on Piccadilly, and hotels, such as Long's and Fenton's, whose residents were single men of the nobility and gentry and those in the military who had recently returned from the Napoleonic Wars. Bond Street, St James's Street, Pall Mall and Piccadilly, and many of the minor streets, were lined with coffee houses and exclusive clubs.

SPECTACLE

In the first decades of the nineteenth century in the fashionable, commercial and residential areas around Piccadilly, a scheme was initiated to promote the area west of Regent Street as an upper-class shopping and entertainment zone, in contrast to the city and the east, but also to compete with Bond Street.[6] The urban designs of architect John Nash, treated urban elements, streets and buildings and the flow of traffic, like landscapes, following the picturesque traditions of English town plan-ning, exemplified in the circuses and crescents of Bath.

> It is, in my opinion, a peculiar beauty of the new streets, that, though, broad, they do not run in straight lines, but make occasional curves which break their uniformity.
>
> (Prince Puckler-Muskau, 1832: 47)

Regent Street combined romantic elements with classical, emulating the public works of continental towns, such as Paris and Versailles (see for example Davis, 1960; Summerson, 1935). But Nash's plans precede Hausmann's plans for Paris, which as well as separating production, commerce, residence, recreation and administration, systematising and controlling circulation, also 'orchestrated sites and opened up vistas that marked out a distinct function of the city as spectacle' (Tagg, 1990).

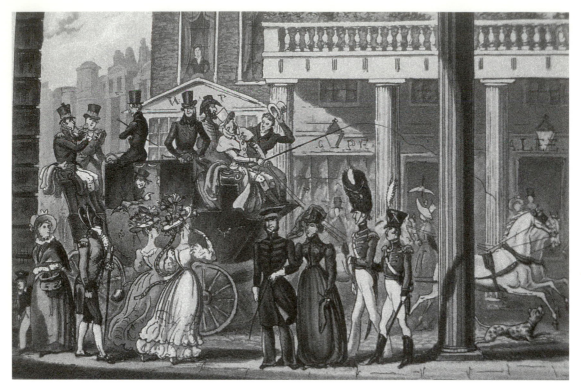

FIGURE 6.2 *'Regent's Street, or, scenes of life and gaiety on "Corinthian Path".'* Source: Robert Cruik-shank, 'The Grand Lounge: Regent Street to wit' in Pierce Egan, *The Finish to the Adventures of Tom, Jerry and Logic in their Pursuits through Life in and out of London*, London, J. S. Virtue and Co., 1828

Nash's plans to improve London and his designs for the Haymarket theatre and the Italian Opera House, including the Royal Opera Arcade, the Quadrant and the colonnade, worked as attractive urban foci. The growth of colonnaded spaces such as these has been connected with the promenade, fashion and display (see Figure 6.2). Regent street was designed with shopping, walking and conversing in mind; it was intended as a scene of life and gaiety. Rambling tales termed Regent Street 'Corinthian Path', indicating its status as the height of fashion (Egan, 1821):

> [T]he Balustrades over the Colonnades will form Balconies to the Lodging-rooms over the Shops, from which the Occupiers of the Lodgings can see and converse with those passing in the carriages underneath, and which will add to the gaiety of the scene.
>
> (quoted in Davis, 1966: 66)

Part of John Nash's urban improvements of 1813 for the Italian Opera House included a single-storey covered colonnade which wrapped around the three sides of the existing theatre, becoming, on the fourth side, the Royal Opera Arcade (see for example Geist, 1983; MacKeith, 1985; Shepperd, 1960). This arcade was the first in England – an enclosed street and continuous with the colonnade, it allowed the opportunity to promenade and view while being protected from the movement of other pedestrians of a different class:

> [I]t was the first time we had met since we were at Eton: he was sauntering an hour away the tedious hour in the Arcade, in search of specific for *ennui*.
>
> (Blackmantle, 1825: 229)

In the early nineteenth century, when Pall Mall and Waterloo Place were taken over by clubs which developed out of the earlier coffee houses, the Royal Opera Arcade operated as a service zone for the men's clubs, with men's drapers and barbers (see Colby, 1964; Geist, 1983). The arcade also functioned as a place around the opera house where people met before and after performances, and gathered during the intervals. After midnight, when the performances were over, the streets around the theatres, especially the Haymarket, were the parade of prostitutes. The Royal Opera Arcade with its spatial connection to the Haymarket, through the theatre lobby or via the external colonnade became the evening promenade for prostitutes looking for clients. As semi-public spaces the arcade and the colonnade offered a concentration of wealthy custom and the possibility of soliciting in shelter on private property without being arrested:

> It came alive in the evening when opera glasses and music scores could be obtained at the shops which were also used as cloak-rooms and for other purposes not so innocent.
>
> (Colby, 1964: 1346)

The occupation of the arcade was represented as gendered; for example, it was noted that female performers were partial 'to sauntering under the facade of the opera house',[7] but the connection of women and colonnaded space was one of sexual deviance:

> We fancy many pretty females traverse the Arcade more from Love than sickness, and too many from necessity? These are different appetites, that crave to be satisfied.[8]

By the 1830s, the spaces of arcades came to be considered as synonymous with prostitution. When the Quadrant Colonnade of Regent Street was found to be a haunt for vice and immorality, prostitution was blamed:

> The Quadrant colonnades were . . . from every point of view except architectural magnificence, a disaster. They were gloomy and dirty; the shops attracted only inferior trades; and, as arcades and colonnades in great cities always do, they became promenades for prostitutes.
>
> (Summerson, 1980: 135)

The architectural form of the arcade or colonnade was itself believed to be the sole determinant in establishing their use as zones of prostitution. The belief was that if the architectural form was altered then prostitution would disappear. However, even after the colonnade was torn down in 1848 there was no noticeable improvement in public morals (Hobhouse, 1975: 72–3).

The arcaded Quadrant Colonnade section of Regent Street was modelled on the Parisian arrangement of mixing shops and flats. The Quadrant provided lodgings for theatrical and operatic people connected to the Italian Opera House, these people were often foreign – French, Italians and Germans. There was also a foreign attendance at the opera (Petty, 1972: 56). This mixture of foreign buildings and people worked as an attraction in terms of the exotic and the different (see for example Hobhouse, 1975; Sala, 1859). However, being foreign at a time when France and England were political rivals was something to be mistrusted. The performance of opera in a foreign tongue, not understood by the audience, stimulated fear (see for example Burney, 1822). The influence of Italian Opera and French ballet was deemed responsible for a lack of morality in the colonnade. Prostitutes were considered to be French, while, conversely, the French women living near the Quadrant were thought to be prostitutes.

VISUALITY

In London's west end, the new spaces of public display were spaces where pleasure and satisfaction were derived from looking; for example, through promenading, shopping and watching performers. The designs of architectural spaces of early nineteenth-century London sought to enhance bodily display. The social importance of the promenade and ramble marked out the importance of the visual in urban experience. A number of cultural commentators have focused on the experience of modernity as dominated by the visual. Georg Simmel identifies the emergence of a new personality due to the fleeting, ephemeral and impersonal nature of urban encounters (Simmel, 1950). Richard Sennet talks of the transformation of public spaces, where social contact was intimate and sensual, into public spaces where the emphasis was on surface appearance, dress, display, fashion and eroticism (Sennett, 1974). Theories of commodity fetishism and reification in social relations developed in Marxist critiques of commodity capitalism are analogous to the increasing emphasis on visual exchange rather than tactile stimulation. Psychoanalytic theories of fetishism, narcissism and voyeurism also begin to link desire and visual relations.

Urban design organises bodies socially and spatially, in terms of positioning, displaying and obscuring. Architecture controls and limits physical movement and sight lines; it can stage and frame those who inhabit its spaces, by creating contrasting scales, screening and lighting (Friedman, 1992). Such devices are culturally determined, they prioritise certain activities and persons, and obscure others according to class, race and gender. Urban space is a medium in which functional visual requirements and imagery are constituted and represented as part of a patriarchal and capitalist ideology. The places of leisure in the nineteenth-century city represent and control the status of men and women as spectators and as objects of sight in public arenas.

A critical aspect of the gendered identity of the rambler is played out through his positionality and visuality. The exchanges of looks and gazes are constructed through gendered relations of equivalence and dominance. In the case of the 'mobile, free, eroticized and avaricious *gaze*' of the Parisian *flâneur* (Pollock, 1988: 79; see also Wilson, 1992; Wolff, 1985), the urban male desires to look at the urban female. The rambler too desires to look, through his social and spatial mobility, the rambler's sexuality gives him access to a dominant mode of vision, a position which the cyprian does not share. Her movement is framed and controlled through patriarchal ideologies concerning women's occupation of the public sphere:

> We have already taken a promiscuous ramble from the West towards the East, and it has afforded some amusement; but our stock is abundant, and many objects of curiosity are still in view.
>
> (Badcock, 1822: 198–9)

The rambler gazes but also requires to be looked at; he demands a visual reciprocity with other men as part of a homosocial system of social exchange. Rambling then is connected with scopophilia or visual pleasure, and also with narcissism. The narcissistic body is an example of a type of mirroring body. It is a body associated with its own surface, on which materials or ornaments can be displayed. In the urban world of consumption, bodies can display 'tastes' which, through relations of imitation, distinction and domination, can mediate gender and class positions. Taste is embodied, it is inscribed onto the body and made apparent in body size, volume, demeanour, sitting, speaking, gesture (Frank, 1991): 'Taste, a class culture turned into nature, that is, *embodied,* helps to shape the class body' (Bourdieu, 1984: 190).

CONSUMPTION

Through maximising the bodily display of taste, streets, particularly the shopping street, are represented as increasingly important spaces of urban life. It is in the street that the rambler displays his leisure time and money, displaying his 'conspicuous consumption' to other men and women. The rambler's activities of consuming, purchasing, spending, enjoying, digesting and displaying took him all over the city. But if men had important roles as consumers, the development of commodity capitalism also encouraged an increasing movement of women outside the family home as workers and consumers (see for example Ryan, 1994; Swanson, 1995; Wilson, 1995).

> In the spring, when all persons of distinction are in Town, the usual morning employment of the ladies is to go a-shopping. . . . This they do without actually wanting to purchase any thing, and they spend their money or not, according to the temptations which are held out to gratify and amuse.
>
> (Espriella, 1807: 121)

The city was the original centre of London's commerce (Adburgham, 1981: 5), but by 1807, although the city was still holding its own as a shopping centre, there were also a number of important shopping streets. Regent Street was the first north–south axis to connect London's three east–west routes, Oxford Street, Piccadilly and the Strand. It was also the first street to be designed as a shopping street as distinct from becoming one. The northern part of the street contained several houses, but lower down towards Piccadilly, the buildings were based on arcades, and a mixture of commerce and residential use (Davis, 1966: 73–4). Within the Quadrant Colonnade, at the meeting point of Regent Street and Oxford Street, shops were to be on the ground floor of the buildings, the shopkeepers and their families on the mezzanine floor, and the floors above to be let as expensive lodgings for visitors or as apartments for bachelors.

> The buildings of this noble street consist chiefly of palace like shops, in whose broad shewy windows are displayed articles of the most splendid description, such as the neighbouring world of wealth and fashion are daily in want of. The upper part of these elegant structures are mostly let as apartments to temporary visitors of the metropolis. . . . It is a continuous style of architecture, with the houses above it, its form is one of the best which could be devised for the purpose; it gives an air of grandeur and space to the streets.
>
> (quoted in Adburgham, 1981: 12)

Controls were exerted over the kind of shops established on the street. There were to be no butchers, greengrocers, domestic trades, hawkers, street vendors or public houses. Cake shops, confectioners and restaurants were allowed since such shops were well decorated, 'fashionable lounging places for the great and titled ones, and the places of assignation for supposed casual encounters – and they were therefore fairly luxurious' (quoted in Adburgham, 1979: 76–7).

ARCADES

The early nineteenth century saw rights of ownership being exerted through government legislation and centralised administration, increasing controls and constraints on public behaviour in the privately owned circulation space of the street. The arcade was a privately owned street of commodity consumption. London arcades were constructed during the early decades of the nineteenth century in the fashionable and wealthy residential areas of the West End around Piccadilly, Bond Street, Oxford Street and Regent Street. The Burlington Arcade was designed by Samuel Ware and built for Sir George Cavendish in 1818 off Piccadilly in London's St James's (see for example, Geist, 1983; Sheppherd, 1960).

But the arcades were intended not only to be spaces of static consumption like the exchanges and bazaars, but also spaces of transition in order to explore new marketing possibilities for the luxury industry. Precedents came from France, along the lines of the Jardins du Palais Royal (1781–6), the Galleries du Bois, Passage Feydeau (1791), and the Passage du Caire (1797–9) (MacKeith, 1985: 16). Ware's early design for the Burlington Arcade was based on the Exeter Change and contained four double rows of shops divided by three open, intervening spaces:

> The Piazza to consist of Two Entrances and Four Double Rows of Shops, separated for selling goods in them, and in the Two Entrances are called 'Stands'; they will be after the principle of those in the Exeter Change.[9]

Arcades provided a new kind of public space, a street environment, which, by being covered over, allowed the opportunity to promenade and view while being protected from the weather. Arcades were the venue for luxury goods, and provided for their wealthy customers a semi-public environment removed from the bodies and movement of pedestrians of a lower class. The Burlington Arcade signified its role as a private street, by a colonnaded screen, which could be closed off. The private ownership of the Burlington Arcade meant that there were regulations concerning spatial behaviour. Lord Cavendish employed members of his ex-military regiment to guard the arcade. Regulations governed the opening and closing times. The arcade was locked at night, and open during the day. There were strict rules governing the kind of movement which could take place in the arcade, these excluded running, pushing a pram, carrying bulky packages or open umbrellas. The rules also controlled the noise level in the arcade; no whistling, singing or playing of musical instruments.[10]

In plans to improve London the arcades worked as an attractive focus or exotic spectacle. The Burlington Arcade (see Figure 6.3) was considered a safe place for women to shop, and built with the intention of providing work for 'industrious females', but in rambling texts it was noted as a place to find 'fresh and fair faced maids' (see for example Geist, 1983).

FIGURE 6.3 *'Burlington arcade, or, an attractive focus or exotic spectacle.'* Source: Thomas H. Shepherd, Wm Tombleson (engraver) 'Burlington Arcade, Piccadilly', London, Jones and Co., 25 October 1828

For many feminists, the position of women in patriarchal society can be compared to that of a commodity. As wives, mothers, virgins and prostitutes, women are the objects of physical and metaphorical exchange among men (Irigaray, 1985). The arcade played an important part in the commerce of prostitution; the windows of the arcade created a fine display of shopgirls and shoppers, the rooms above the shop units were used for part-time prostitution by prostitutes and shopgirls; and as long as the beadles could be bribed, its covered promenade formed a perfect place for prostitutes to pick up rich clients.

CYPRIANS

In streets and arcades, the threat to social order posed by a mixing of classes and genders was cause for middle- and upper-class angst, so too was the worry that female forms of male property – mothers, wives and daughters – would be visually and sexually available to other men. In such public places where the masking of social identity through deliberate disguise or class emulation, lead to fears of working- and middle-class contamination of the public realm; such fears were represented through concerns with female sexuality in these spaces. The body of the urban female was the site of conflicting concerns; those of public patriarchs seeking to control female occupation of the city and of consumer capitalists aiming to extend the roles of women as cheap workers and consumers of household and personal commodities. The representation of women as sexual commodities in the rambling narratives, as cyprians in upper-class venues and prostitutes in working-class taverns, articulated male concerns with female sexuality. The female body represented exchange, and operated as a sign of display and consumption.

The figure of the public woman, synonymous with the prostitute or the cyprian, represents the blurring of public and private boundaries, and the uncontrollable movement of women and female sexuality. The movement of middle-class women outside the control of the private patriarch resulted in the extension of patriarchal control into public spaces of the city, through such codified legislation as the Vagrancy Acts of 1822, which exerted social and moral control over female urban movement. New importance was placed on upper- and middle-class family privacy and the spatial confinement of bourgeois women to the private and controlled sphere of the home, where their moral duty was to take care of the family. The private and good 'woman' provided a means of controlling class deviancy; she was a signifier of social stability, respectability and domesticity for the lower classes to look up to.

All women in the streets of the public realm were the woman-as-sign who represents private intimacies in public. But if all women in the city were represented as prostitutes, it is important to recognise different kinds of prostitute, and to deal with prostitution as a real issue, rather than as a piece of rhetoric, 'debating points about the horror of women's oppression' (Rubin, 1994). There is a difference in wealth between streetwalkers or working-class prostitutes and courtesans or cyprians. There is also a difference in their occupation of space. The wealthy cyprians lived off large incomes in houses rented for them, for this class of prostitute the street was a place in which they chose to display themselves. For streetwalkers, the public street was a place occupied out of economic necessity.

The dialectical relationship of male rambler and female cyprian, precursors to the Parisian *flâneur* and prostitute, represent gendered urban spaces in early nineteenth-century London. Rambling is an activity framed in relation to gender, class, sexuality and nationality. It is also a patriarchal construction representing male control of urban space through mobility. The female cyprian represents women's use and experience of the urban realm, confined both spatially and temporally by the activities of the rambler. Females who strolled through the streets and shopping arcades at certain times were considered to be of loose morals, confused with prostitutes, and represented as cyprians. By considering the interrelation of urban spaces, the ramble suggests a configuration of gender and space that is far more complicated than the separate spheres ideology suggests. Thinking about the dialectical relation between gendered identities and urban space starts to allow a more complex analysis, one which pays greater attention to the fluidity of relations between different representations of gender and urban space by focusing on movement and temporality.

NOTES

1 See for example *A Ramble through London*, 1738; *The Country Spy or a Ramble through London*, 1750; *The Modern Complete London Spy*, 1760; *A Sunday Ramble*, 1774; G. Andrewes, *The Stranger's Guide or the Frauds of London Detected etc.*, London, J. Bailey, 1808; G. Barrington, *Barrington's New London Spy for 1805*, London, Barrington, 1805 and 1809.

2 See for example Badcock, 1822, 1828; Blackmantle, 1825; Egan, 1821; Heath, 1822; Smeeton, 1828.

3 See for example, *The Rambler's Magazine or Annals of Gallantry, Glee, Pleasure and Bon Ton*, 1783; *The Rambler's Magazine or Fashionable Emporium of Polite Literature*, 1822; *The Rambler's Magazine or Fashionable Companion*, London, T. Holt, 1824–5; *The Rambler's Magazine or Annals of Gallantry, Glee, Pleasure and Bon Ton*, London, J. Mitford, 1828.

4 Cyprian: 'belonging to Cyprus, an island in the eastern Mediterranean, famous in ancient times for the worship of Aphrodite or Venus', and as 'licentious, lewd, in the eighteenth and nineteenth centuries applied to prostitutes'. See *Oxford English Dictionary*, CD ROM, 2nd edn.

5 This semi-narrative structure first appears in Ned Ward, *The London Spy*, 1698–1700.

6 See for example, Marshall, 1956: 74–5; Davis, 1960, 1966; Hobhouse, 1975: 72–3; Mansbridge, 1991; Summerson, 1935, 1980.

7 *Memoirs of the Life of Madame Vestris of the Theatre Royal Drury Lane and Covent Garden*, London, Privately Printed, 1830, p. 33.

8 *Memoirs of the Life Public and Private Adventures of Madame Vestris*, London, John Duncombe, 1836, p. 127.

9 Report from S. Ware, 'A Proposal to Build Burlington Arcade', 16 March 1808, London, The Royal Academy of Arts Library, Collection of Lord Christian.

10 Interviews with Beadles, John Simpson, 5 September 1995, Michael Lock, 24 August 1995.

REFERENCES

Adburgham, A. (1979) *Shopping in Style*, London, Thames and Hudson.

Adburgham, A. (1981) *Shops and Shopping 1800–1914*, London, George Allen & Unwin.

Badcock, J. (1822) *Real Life in London*, London, Jones & Co.

Badcock, J. (1828) *A Living Picture of London for 1823*, London, W. Clarke.

Bedarida F. and Sutcliffe, A. 'The Street in the Structure and the Life of the City: Reflections on Nineteenth Century London and Paris', *Journal of Urban History*, 6: 379–96.

Blackmantle, B. (1825) *The English Spy*, London, Sherwood, Jones & Co.

Borsay, P. (1990) 'Introduction', *The Eighteenth Century Town 1688–1820*, Harlow, Longman, pp. 1–38.

Bourdieu, P. (1984) *Distinction: A Social Critique of the Judgement of Taste*, London, Routledge & Kegan Paul.

Burke, T. (1940) *The Streets of London*, London, B. T. Batsford Press.

Burney, F. (1822) *Evelina*, London, Jones & Co.

Colby, R. (1964) 'Shopping off the City Streets', *Country Life*, 136 (19 November): 1346.

Corfield, P. (1990) 'Walking the City Streets, an Eighteenth Century Odyssey', *Historical Perspectives*, 26: 132–74.

Cruikshank D. and Burton, N. (1990) *Life in the Georgian City*, London, Viking.

Cruikshank, R. (1828) *The Grand Lounge: Regent Street to Wit* in P. Egan, *The Finish to the Adventures of Tom, Jerry and Logic in their Pursuits through Life in and out of London*, London, J. S. Virtue & Co.

Davis, T. (1960) *The Architecture of John Nash*, London, Studio.

Davis, T. (1966) *John Nash: The Prince Regent's Architect*, London, Country Life Ltd.

Egan, P. (1821) *Life in London*, London, Sherwood, Neely & Jones.

Espriella, M. A. (1807) *Letters from England*, London, Longman, Hurst, Rees, Ormes.

Frank, A. W. (1991) 'For a Sociology of the Body: An Analytic Overview', *The Body: Social Process and Cultural Theory*, London, Sage Publications, pp. 36–102, p. 63 and p. 67.

Friedman, A. (1992) 'Architecture, Authority and the Gaze: Planning and Representation in the Early Modern Country House', *Assemblage*, 18 (August): 40–61.

Fulford, R. (1962) *Boodle's*, London, Privately Printed for Members of the Club.

Geist, J. F. (1983) *Arcades: the History of a Building Type*, Cambridge, Mass., MIT Press.

George, M. D. (1992) *London Life in the Eighteenth Century*, London, Penguin.

Heath, W. (1822) *Fashion and Folly: The Bucks Pilgrimage*, London, William Sams.

Hobhouse, H. (1975) *A History of Regent Street*, London, Queen Anne Press.

Hutton, W. (1818) *A Journey to London*, London, William Hutton.

Irigaray, L. (1985) *This Sex Which Is Not One*, Ithaca, Cornell University Press.

Low, D. A. (1982) *Thieves Kitchen: The Regency Underworld*, Totowa, New Jersey, Biblio Distribution Centre.

MacDonogh, F. (1819) *The Hermit in London*, London, Henry Colburn.

MacKeith, M. (1985) *Shopping Arcades: a Gazetteer of British Arcades 1817–1939*, London, Mansell.

Mansbridge, M. (1991) *John Nash*, Oxford, Phaidon.

Marshall, D. (1956) *English People in the Eighteenth Century*, London, Longmans.

Olsen, D. (1964) *Town Planning in London: The Eighteenth And Nineteenth Centuries*, New Haven, Yale University Press.

Partridge, E. (1984) *Dictionary of Slang and Unconventional English*, London, Routledge & Kegan Paul.

Petty, F. C. (1972) *Italian Opera in London 1760–1800*, UMI Research Press.

Pollock, G. (1988) *Vision and Difference: Femininity, Feminism and the Histories of Art*, London, Routledge.

Porter, R. (1994) *London*, London, Hamish Hamilton.

Prince Puckler-Muskau, (1832) *Tour in England, Ireland, and France*, London, Effingham Wilson (5 October 1826), vol. 3.

Rasmussen, S. E. (1988) *London: The Unique City*, London, MIT Press.

Richardson, A. E. (1931) *Georgian England*, London, B. T. Batsford.

Rubin, G. (1994) 'Sexual Traffic: Interview with Judith Butler', *Differences: A Journal of Feminist Cultural Studies*, 6, 2/3: 62–99.

Rude, G. (1971) *Hanoverian London 1714–1808*, London, Secker and Warburg.

Ryan, J. (1994) 'Women, Modernity and the City', *Theory, Culture and Society*, 11, 4 (November): 35–64.

Sala, G. (1859) *Twice Around the Clock, or Hours of the Day and Night in London*, London, Houlston & Wright.

Sennett, R. (1974) *The Fall of Public Man*, New York, Vintage Books.

Shepperd, F. H. W. (ed.) (1960) *The Survey of London. The Parish of St. James's Westminster, South of Piccadilly*, London, The Athlone Press, University of London, vol. 29, part 1, pp. 248–9, figs. 38 and 39.

Simmel, G. (1950) 'Metropolis and Mental Life' in Kurt Wolff (ed.) *The Sociology of Georg Simmel*, Glencoe, Ill., Free Press, pp. 409–24.

Smeeton, G. (1828) *Doings in London; or Day and Night Scenes of the Frauds, Frolics, Manners and Depravities of the Metropolis*, London, Southwark, Smeeton.

Summerson, J. (1935) *John Nash: Architect to King George IV*, London, George Allen & Unwin.

Summerson, J. (1962) *Georgian London*, London, Pelican.

Summerson, J. (1980) *The Life and Work of John Nash Architect*, London, George Allen & Unwin.

Swanson, G. (1995) 'Drunk with the Glitter: Consuming Spaces and Sexual Geographies', in Sophie Watson and Katherine Gibson (eds), *Postmodern Cities and Spaces*, Oxford, Blackwell, pp. 80–99.

Tagg, J. (1990) 'The Discontinuous City: Picturing and the Discursive Field', *Strategies*, Los Angeles, no. 3.

Wheaton, N. S. (1830) *A Journal of a Residence During Several Months in London: Including Excursions through Various Parts of England and a Short Tour in France and Scotland in the Years 1823 and 1824*, London.

Wilson, E. (1992) 'The Invisible *Flanêur*', *The New Left Review*, 191: 90–110.

Wilson, E. (1995) 'The Rhetoric of Urban Space', *New Left Review*, 209: 146–60.

Wolff, J. (1985) 'The Invisible *Flâneuse*: Women and the Literature of Modernity', *Theory, Culture and Society*, 2, 3: 36–46.

7

THE SOCIAL SPACE OF DISABILITY IN COLONIAL MELBOURNE

Brendan Gleeson

•

INTRODUCTION

The theme and setting

This chapter explores the social geography of disability in one important industrial capitalist city, colonial Melbourne.[1] Specifically, my aim is to situate the street life of disabled people within this unique historical social space. I shall also broaden the picture's frame by drawing upon some British accounts of everyday street life in the industrial city. The focus is class specific, centring on the urban proletariat (including its most marginal strata), for whom the term 'street life' evokes much of their historical social experience. Within this, I am interested in the quotidian experience of physically disabled proletarians; meaning individuals who lacked a limb, part of a limb, or who had a defective organism or mechanism of the body (see for example Oliver, 1990).

Melbourne is a worthy exemplar of the industrial city (Davison, 1978). The capital of the Colony (now State) of Victoria, Melbourne had an 1891 population of nearly half a million. By the late nineteenth century Melbourne was regarded as one of the premier cities of the British Empire with a rateable value surpassed only by London and Glasgow (Briggs, 1968). By the early 1890s, the city's extensive manufacturing sector employed about thirty per cent of the male labour-force. Most industrial establishments and the proletarian labour-force were located in the inner ring of suburbs circling the Central Business District (CBD) (Lack, 1991). The fragmentary historical records of life in this industrial, proletarian core suggest the presence of a considerable, if marginalised, population of disabled people.

Historiographical considerations

Two specific historiographical problems condition the study of disablement in past Western societies. First, disability is a pervasive social phenomenon which the social sciences have, astonishingly, long ignored – only in recent times has this deficiency begun to be corrected (Abberley, 1987). Of most concern for the present analysis is the lack of historical analyses of disablement. Oliver has remarked that '[o]n the experience of disability, history is largely silent' (1990: xi). There is therefore an epistemological problem – theoretical underdevelopment – that immediately confronts any attempt to *explain* some aspect of the lived experience of disability.

Second, and related to the above, there is a very serious absence of historical records concerning disabled people in previous societies. The reasons for this empirical silence are not clear. It may simply be attributable to the fact that in previous European societies, disability was an unremarkable, not to say pervasive, fact of social life, which did not therefore qualify as an observation variable in the official eye. Thus, there is also an ontological difficulty to be confronted when attempting to *describe* the past lived experience of disabled people. The selection of sources for this analysis reflects this difficulty, being mostly fragments drawn from a few surviving documentary records, various secondary analyses, and my own reasoning based on these materials.

Are these two historiographical problems connected? Perhaps they both reflect an enduring marginalisation and therefore social invisibility of disabled people, which extends to the present day. Several observers (for example Abberley, 1987; Oliver, 1990) have remarked upon the silence of historians on the question of disability, implying that this discipline has helped reproduce the silences of past forms of oppression. Although constrained by limited empirical sources, it is hoped that this analysis will contribute to the broadening of scholarly interest in the historical experience of disabled people.

The chapter is in two parts. The first situates structurally the phenomenon of disability in the industrial city, arguing that the street life of disabled people reflected a broader socio-spatial marginalisation, which they both succumbed to and resisted. The following part sketches the street life of disabled people in the industrial city, focusing on the case of colonial Melbourne.

DISABILITY AND THE INDUSTRIAL CAPITALIST CITY

Whilst the discipline of history has virtually ignored the question of disability, there have been a small number of social scientific analyses that have explored the historical experience of disabled people in capitalist societies (see for example Finkelstein, 1980; Oliver, 1990, 1993; Ryan and Thomas, 1987). My contribution to this emergent literature was an extensive comparative examination of how disabled people lived in feudalism and industrial capitalism (see Gleeson, 1993), centring in the latter case on colonial Melbourne. The following discussion of disability in the industrial city draws upon these more extensive historical analyses of the lived experience of disabled people in nineteenth-century societies.

One disabling feature of the industrial city was the new separation of home and work, a socio-spatial phenomenon which was all but absent in the feudal era. This disjuncture of home and work created a powerfully disabling friction in everyday life for physically impaired people. In addition, industrial workplaces were structured and used in ways which disabled 'uncompetitive' workers, including physically impaired people. The rise of mechanised forms of production introduced productivity standards which assumed a 'normal' (namely, usually male and non-impaired) worker's body and disabled all others. As Ryan and Thomas note, the coming of industrialism meant the end of paid work for many disabled people who had been formerly integrated into cottage-based production:

> The speed of factory work, the enforced discipline, the time-keeping and production norms – all these were a highly unfavourable change from the slower, more self-determined and flexible methods of work into which many handicapped people had been integrated.
>
> (Ryan and Thomas, 1987: 101)

As Marx (1981) pointed out at the time, industrialisation and urbanisation produced an 'incapable' social stratum, a mixed estate that could not sell its labour power at the average rate of productivity, and which was therefore consigned to the usual consequences of labour market exclusion: poverty, ill-health, brevity of life, socio-spatial marginalisation, and, for many, dependence upon the informal sector of the economy. We might include here, as Marx did, widows, the elderly, orphans, the sick, and, interestingly, those individuals he called 'the mutilated and the victims of industry', in a clear reference to disabled people. He referred to this heterogeneous social group as the lumpenproletariat. By the late nineteenth century in most industrialised nations, much of this 'impaired' labour power had been incarcerated in what Foucault (1979: 199) termed 'the space of exclusion', a new institutional system of workhouses, hospitals, asylums, and (later) 'crippleages', operated by an extensive private charitable sector and a host of local and central state bodies.

Marx observed that the lumpenproletariat's only alternative to dependence upon public and private (including family) 'benevolence' was a wretched, insecure form of independence, based 'on kinds of work that can only count as such within a miserable mode of production' (Marx, 1981: 366). Amongst the 'miserable' jobs Marx was referring to were the many street trades – hawking goods and services to passers-by – which made the thoroughfares of the industrial city *sites*, rather than merely *conduits*, of economic production.

For disabled people, their economic devalorisation, or 'incapability' to use Marx's term, took on a particular socio-spatial form. This can be understood dynamically, as a general tendency, *but not law*, for marginalisation to specific realms of the city (and sometimes beyond). The motive forces for this were both centrifugal and centripetal in nature, and were sourced in three key sites in a 'social space of disability' (Figure 7.1).

The paid workplace was the principal centrifugal force of marginalisation: the site where the devalorisation of disabled labour power was actually practised; a node

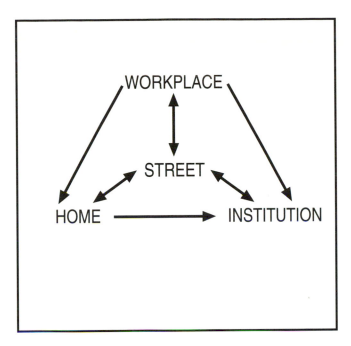

FIGURE 7.1 The social space of disability in the industrial city

of repulsion for disabled people. The economic centrality of that exemplary site of industrialism – the factory – gave this centrifugal force a breadth in the industrial city that cannot be underestimated. The key centripetal site for disabled people was the institution. The increasing ubiquity during the nineteenth century of this deliberate, and morally instructive, caricature of the factory (Foucault, 1979) meant that its tentacles reached to most corners of proletarian social space, drawing in redundant labour power for storage in institutional warehousing.

The third key element of social space for disabled people was the proletarian home. Domestic spaces were certainly important sites for the physical production of disability (though, as Marx and many other Victorian commentators observed, it was the factory which produced physical impairments on an industrial scale). However, the home was an ambiguous site: many households quickly, and without sentimentality, rejected their disabled members, either for the institutions or the streets; at other times, in the context of affective domestic relations, disabled people were able to resist the centrifugal and centripetal currents of industrialism.

For the so-called 'incapable stratum', homeworking – usually in sweated (exploitative) conditions – was one common strategy for transcending the centrifugal tendency of the factory to utter devalorisation. Even when a disabled member was unable to make an economic contribution to the household, affective ties were often stronger than the pull of the institution. Many Victorian working-class families harboured disabled relatives, sometimes in a tug of war with the poorhouse.

This three-way typology of workplace, institution, and home is one way of representing the social space of disability in the industrial city. But what of the street?

How does it fit into this social space? Was it simply a conduit which carried the centripetal and centrifugal currents of social power between these key spatial nodes? Was it an important site for disabled people?

In a concrete sense the street – and not just the slum lane – was certainly important to disabled people. Analysis of the nineteenth-century urban commentaries indicates that disabled people were a common sight on the Victorian city street, particularly in major pedestrian thoroughfares (Gleeson, 1993). I say 'sight' rather than simply 'inhabitant' because disabled people were distinguished from the masses of pedestrians: first, by the social inscriptions of difference arising from their disablement, and second, by the nature of their presence on the streets. In the various tableaux of cities constructed by journalists and literary writers, for example, the disabled beggar or trader is usually an element within the kaleidoscopic backdrop of furious, modern street life (Brown-May, 1995). One 'underworld journalist', Thomas Archer, entered the visceral realms of the slum and the workhouse during the 1860s – many disabled people figure within his rich portrayals of urban squalor (Archer, 1985). However, disabled people were rarely foregrounded in contemporary urban descriptions, such as Archer's; almost never were they given voice.

Disabled people were not 'pedestrians' – their restricted ambulance makes this category unsuitable to describe their participation in street life. Disabled people were often on the street for very immediate economic reasons, engaging either in begging or petty street trading, thus distinguishing them from strolling consumers, people in circulation, idlers or others for whom the street was not the immediate source of their existence. The street was a place of subsistence as much as it was a stage that constantly retold the story of their social difference and exclusion. If disabled people were present on the Victorian city street, as both agents of petty commerce – street traders – and symbols of anti-commerce – beggars – we might say that this indicates both the failure and success of the oppressive structures which bore down on them. How so?

By clinging to society on the streets, some disabled people resisted the 'duty to attend the asylum' (as Foucault would have it), that weighed increasingly heavily upon them as the century progressed. Alternatively, others, those remembered as 'crippled beggars', were indeed a public revelation of the crushed and lowest stratum of industrialism. For these, the street was really a 'non-place', the site of a truly wretched existence which served only as a waiting room before the inevitable moments of institutionalisation or death.

The word 'abjection' describes both the action of casting off, of excluding a person or group, and the experience of being cast down, of degradation. Drawing upon Kristeva's (1982) work, Sibley has developed a geographic notion of abjection as both the 'unattainable desire to expel' those things which threaten the socio-spatial boundaries of normality and 'that list of things and threatening others' (Sibley, 1995: 18). Sibley's notion of abjection illuminates the experience of disabled people in the nineteenth-century capitalist city. Through their 'incapability' disabled people threatened the Victorian social order which was framed by economic class structures. However, both the diffuseness of oppressive power and the determination of many disabled people to resist exclusion, meant that these 'threatening others' could not be totally expelled from the public view – from public streets – and placed within the safe institutional boundaries of the 'space of exclusion'. The presence of disabled people on the Victorian city street was, as Sibley would have it, *a ritual of abjection*, a sort of uneasy (and unstable) truce between the oppressor and the oppressed.

The street then was a place where both abjection was experienced and resistance practised against the forces of abjection. It did not simply conduct the charges of centripetal and centrifugal power that sought variously to expel and attract disabled people. Rather, the street reveals these modalities in tension with the various biographies of disabled people, with their various capacities for resistance and subversion. It was, roughly speaking, a public equivalent of the home, which was also a site of resistance and abjection.

This conceptual prism of social space helps refract some understanding of disablement from the various images of Victorian street life which have been left to us. In general it seems that the street was a place of both abjection and resistance for disabled people, but maybe not much in between.

COLONIAL MELBOURNE

Three views of social space

What about Melbourne specifically? What was the social space of disabled people in this colonial metropolis? More specifically, how did disabled people experience the city's street life? My analysis of the social space of disability in colonial Melbourne (see Gleeson, 1993) showed that disabled people from the city's lower economic strata experienced socio-spatial marginalisation, meaning that many were forced to undertake marginal economic activities, such as street trading, in order to avoid forms of public and private dependency, or worse. The Melbourne study drew upon three exemplary data sources which each shed light on the three distinct dimensions of the social space of disability – workplace, institution and home (see Figure 7.1).

The first window on this social space was the set of factory records left by Guest and Company, a large biscuit and cake manufacturing concern whose principal plant was located in the CBD for most of the Victorian period (Figure 7.2). These records – principally the employment engagement books (1889–91) – reveal both a mechanised labour process that enforced an average productivity standard and the employer's intolerance of 'slow' and impaired workers. There are many recorded dismissals of workers for being 'too slow', 'useless', 'careless' and 'unsteady'. Speed, dexterity and obedience were demanded of the workers. There is one recorded instance in which impairment is cited as a reason for dismissal in the study period. On 4 June 1889, the foreman noted the departure of a 15-year-old boy with the following remark: 'no good, paralysed hand'. It is doubtful that the internal work process at Guest's ever admitted impaired labour-power, at least not for any significant length of period.

There is every reason to believe that these characteristics were general to factories during the industrial era. Both contemporary observers, such as Marx (1981),

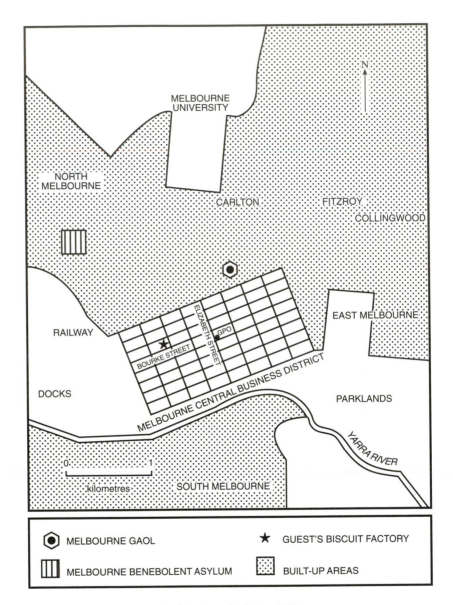

FIGURE 7.2 Melbourne's CBD and adjoining suburbs by 1880

and more recent analysts, such as Doray (1988), Rabinbach (1990) and Ryan and Thomas (1987), have argued that the labour regimes of mechanised production made it nearly impossible for disabled people to work within the factory system. As McCalman (1984: 31) has observed, Melbourne's 'factory system had no place for slow workers'.

Apart from the internal labour rhythms of the factory, there were critical external considerations which prevented most disabled people from joining industrial labour forces. Analysis of the Guest and Company employee data for the early 1890s showed that the average daily return work trip for workers was 6 kilometres. Clearly this sort of mobility expectation was a further exclusionary factor for disabled people that added to the internal centrifugal force of the labour regime. We may assume that disabled people hardly figured amongst the hordes of workers who streamed through inner Melbourne's streets at the beginning and end of each day.

My examination of Melbourne's institutional landscape also confirmed the centripetal pull of the asylum and the poorhouse. By the late century, the city had seventeen major institutions, and a host of other smaller places of indoor charity. A source of great pride for the city was its showpiece poorhouse, the Melbourne Benevolent Asylum (MBA), located just a few kilometres to the North of the CBD (Figure 7.3). Institutions, like the Melbourne Asylum helped support the ideological construction of Victoria as an affluent and civilised fantasyland, where even paupers

FIGURE 7.3 The Melbourne Benevolent Asylum

got to live in palatial homes (see Gleeson, 1995). But the reality was that conditions in the asylums were nothing less than barbarous. Life for those impoverished
disabled people who were 'lucky' enough to gain admittance to the 'benevolent' was
generally wretched and short.

Many thousands of individuals passed through the gates of the MBA after they
opened in 1851: some many times, as though through revolving doors; others only
once, the moment of institutionalisation being their permanent exit from public
space. The Asylum was above all a place of social and physical death. My analysis
of the admissions records for the MBA between 1860 and 1880 revealed that a
substantial number (597) of its inmates during this time were disabled (Table 7.1).

As the city's institutional archipelago grew, so too was the heterogeneity of street
life progressively narrowed, as more subclasses of the lumpenproletariat were confined
within its carceral landscape. The first, and enduring, receptacle for 'street refuse'
was the gaol. My study uncovered evidence to show that the judiciary had a regular
practice of sending street vagrants, many of them described as 'cripples', to prison
as a humane gesture aimed at providing sustenance to the 'physically incapable' (see
also Lynn, 1990).

'Helpless' persons – those amongst the poor with disabilities or psychiatric
illnesses – were frequently imprisoned by judicial authorities for want of institutional

TABLE 7.1 Numbers of physically impaired males and females by stated
impairment type, 1860–80

	No. of males	No. of females	Total persons
Impairment type			
Disabled*	14	6	20
Loss of Limb(s)	11	7	18
Palsy	2	3	5
Hemiplegia	1	2	3
Paraplegia	4	0	4
Paralysis	264	52	316
Part paralysis#	31	6	37
Impairment to:			
Shoulder	1	0	1
Arm(s)	9	4	13
Hand(s)	7	3	10
Spine	20	16	36
Side	26	11	37
Hip(s)	17	9	26
Leg(s)	44	8	52
Knee(s)	6	8	14
Foot/Feet	4	1	5
Total impaired persons	461	136	597

* Includes those inmates described as 'disabled', 'crippled', or 'lame'
# Includes those cases only described as 'partially paralysed'. In the case where paralysis
of a specific part of the body was indicated, the observation has been included in the
relevant impairment category

Source: Melbourne Benevolent Asylum Registers of Applicants and Inmates, 1856–90

alternatives. The usual pathway from the street to prison for the disabled poor was via an arrest for vagrancy. This was not always intended as a punitive measure; police commonly used their powers under the vagrancy statute on compassionate grounds when a needy indigent was brought to their attention. The problem was that, in the face of perennial institutional crowding, magistrates had little option but to commit the vagrant poor to the city's gaols. In 1863, an anonymous correspondent to a major city newspaper (very possibly a Member of Parliament) stated the problem succinctly:

> Victoria might well be proud of her public institutions, considering her youth. . . . However much our Government has done, there are some unfortunate classes unprovided for. The maimed, the diseased, and the unfortunate widows and destitute children are insufficiently cared for. It is scarcely right that an unfortunate cripple should be treated as a vagabond, and sent to prison under the Vagrant Act, merely to provide sustenance. Yet it is the only humane way for the bench at present. It is really too bad that no comprehensive legislative measures have been made to provide for those who are physically incapable of earning their living.[2]

Another, and important, window on the social space of disability were the voluminous case records left by the Melbourne Ladies' Benevolent Society (hereafter, 'the Society'), the principal source of outdoor – that is, 'home delivered' – charity in the city. The Society was made up entirely of 'lady visitors' who attended the homes of the poor to dispense doles and advice on every aspect of home management. Members were drafted from women of the lower or middle strata of the bourgeoisie, usually the wives of doctors, businessmen, and minor clergymen.

From 1855, the Society's field of operation settled upon the CBD and four adjoining suburbs; the whole divided into forty smaller districts, each with its own lady visitor. By the 1890s, the Society's operating area was home to about 150,000 persons. These areas contained extensive slum tracts and a considerable lumpenproletariat of widows, deserted families, the aged, the sick and disabled people.

The Society's lady visitors have left us extremely rich accounts of the domestic and public life of the industrial working class. The records covering 1849–1900 reveal that the Society aided many families with disabled relatives: some 1004 disabled individuals were identified (doubtless an undercount) (Table 7.2). Interestingly, the impairment types closely paralleled those recorded at the Benevolent Asylum; indeed a few individuals figured in both sets of data indicating the centripetal pull between the institution and home. As meagre as it was, outdoor charity such as this helped many families and disabled individuals to resist the centripetal pull of the institution, if only for periods.

Clinging to the home, and sometimes family, none the less often meant exclusion from public space for disabled people. The Society records confirm that many disabled people engaged in sweated homework to sustain themselves: such people were rarely seen in the street. One of many recorded examples reported thus: 'elderly widow, with very bad leg, found her working at shirts, with her leg up on bed'.[3] This picture is confirmed in other surviving fragments which record the visceral world of sweating.

TABLE 7.2 Numbers of physically impaired children and adults by stated impairment type, 1850–1900

	No. of children	No. of adults	Total persons
Impairment type			
Crippling condition	60	107	167
Disabling condition	3	33	36
Loss of limb(s)*	3	76	79
No use of limb(s)*	1	36	37
Weakness of limb(s)*	–	4	4
Deformity	8	5	13
Paralysis	10	179	189
Lameness	8	58	66
Disease of spine, etc.#	31	21	52
Club foot	1	1	2
St Vitus's Dance	1	–	1
Permanently invalided	–	7	7
Long term injury+	–	18	18
Total impaired persons	126	545	671
Possible impairment^	3	330	333
Total impaired and possibly impaired persons	**129**	**875**	**1004**

* Includes persons having lost part of limb(s)
# Includes persons with diseases of leg(s) and hip(s)
+ Includes persons disabled for at least six months by injury
^ Includes persons with 'bad' limb(s) or part thereof

Source: MLBS Minutes, 1850–1900

In his 1891 report to Parliament, for example, the Chief Inspector of Factories relayed the following pathetic case of two outworkers: 'These girls live with parents, and pay them for keep. . . . One is a cripple and is laughed at by factory hands'.[4] The Chief Inspector remarked that the crippled sister had rarely left the home after being humiliated by factory workers.

Alternatively, some disabled people used the home as a base for street trading and begging as supplements to charity and the aggregate household income. The Society smiled on any attempt to please those Victorian gods of Thrift and Independence, and it sometimes helped disabled people to establish themselves in various street trades, as the following case notes reveal.

In 1873, the Society helped a recently disabled man establish himself as a produce hawker: 'respectable couple, husband has had a broken leg, not properly set, and is anxious to get a stand in the market'.[5] Seventeen years later, the practice was still common, as one lady visitor's report on the plight of one woman makes clear: 'husband wooden leg, and wants pounds to start with vegetables . . . four children'.[6] For women in such a position, the procurement of a basket to facilitate the selling of fruit or flowers was sometimes a means to some form of economic independence: '[name] is desirous of obtaining a basket to sell fruit, one arm being disabled preventing her taking a situation'.[7]

Other impaired people attempted the life of a street musician. The Society sometimes helped with the purchase of an instrument in such cases. In 1875, for example,
it assisted a 'widow, with a grandson, who is a cripple, and a musician – is anxious
for aid towards the purchase of a flute'.[8] Some impaired people seem to have survived
by combining street trading and charity with a measure of old fashioned, venial
villainy. In March 1891, a visitor reported assisting a Fitzroy woman whose 'husband
is a cripple, but has a coffee stall. They are dirty thriftless people – brawling and
noisy'.[9] One can almost hear the clicking of tongues which greeted this disturbing
news and it was duly recorded that aid was to be discontinued in this case. But, in
practice, the ladies were rarely as stern as their recorded pronouncements, and the
visitor clearly relented in this case, because two months later she was forced to make
the following distressing report concerning the case of the obstreperous coffee vendors:

> on visiting found everything cleared out of the house. The coffee stall was
> in the yard. Mrs [name] learned from Sergeant of Police that they had
> nothing to pay for the stall, and that a warrant was out for [name]'s
> husband.[10]

I was able to conclude with certainty that thirty-six of the disabled people assisted
by the Society were in some form of paid work (Table 7.3). Many of these trades
were clearly either some form of street hawking or outworking.

The view from the street

Sadly, apart from these fragments, there are few surviving records of what life was
like for Melbourne's disabled street hawkers. However, the commentaries of one
English observer of street life have recorded the voices of other disabled traders, and
these speak powerfully of the abjection which must have overshadowed the lives of
the industrial lumpenproletariat. Henry Mayhew (1968a, 1968b), renowned slum
journalist of mid-century London, was one of the few observers of the Victorian city
to foreground disabled street people, if only for a few instants. Mayhew regarded
disabled people as 'one of the classes *driven* to the streets by utter inability to labour'
(his emphasis, 1968a: 329). His surveys record fragments of these unique street lives
in the form of sketches, snatches of conversation, and his own commentary.

On one of his meanderings through London's slum streets and rookeries, Mayhew
asked a crippled bird seller why he was working. The reply: 'Father didn't know
what better to put me to . . . I liked the birds and do still. I used to think at first
that they was like me, they was prisoners and I was a cripple' (Mayhew, 1968b: 68).
He then tells Mayhew that when his father died he succumbed to the centripetal
pull of the workhouse: 'O, I hated it . . . I'd rather be lamer than I am, and be
oftener called Silly Billy – and that sometimes makes me dreadful wild – than be
in the workhouse' (ibid). His view of the future was bleak: 'I feel that I shall be a
poor starving cripple, till I end, perhaps, in the workhouse' (ibid).

Mayhew next encounters a seller of nutmeg-graters whose plaintive appearance
witnesses powerfully to the daily ritual of abjection which many disabled people

TABLE 7.3 Occupations of working impaired persons*

Occupation	No.
Streethawker	5
Needle woman	4
Carter	2
Musician	2
Organ grinder	2
Writer	2
Coffee stall holder	2
Parasol mender	1
Office worker	1
Washer woman	1
Flower maker	1
Flower seller	1
Boot finisher	1
Rag picker	1
Toymaker	1
Mill worker	1
Presser	1
Messenger	1
Shirtmaker	1
Tinsmith	1
Knitter	1
Newspaper seller	1
Caretaker	1
Match seller	1
Total	36

* 'Street occupations' indicated by bold type
  Possible outworkers indicated by italicised type

Source: MLBS minutes, 1850–1900

were compelled to partake of. The seller wears a prominent sign around his neck declaring 'I WAS BORN A CRIPPLE' in recognition of the obdurate suspicion of middle-class Victorians that all disabled street traders and beggars were really well-disguised, 'healthy' vagrants imposing on the sympathy of gullible passers-by. The nutmeg-grater seller tells Mayhew that his relatives despise him for his disabilities and had abandoned him. Mayhew congratulates him for his initiative in making a 'living', and moves on. Later, Mayhew interviews a one-legged crossing sweeper who has little to say other than, 'A man had better be killed out of the way than be disabled' (1968b: 488).

These scenes witness variously to resistance and abjection, and were doubtless reproduced daily in colonial Melbourne. Many of Melbourne's disabled street traders would never have found their way into the Ladies' Benevolent Society's records, as most were surely either homeless or nomads within the liminal accommodation sector of lodging houses, refuges, and the like.

The CBD was stage to a vibrant bourgeois street life, famous for its shopping arcades, galleries, theatres and palatial hotels, with the city being variously named

THE CRIPPLED STREET BIRD-SELLER.

FIGURE 7.4 The crippled street bird-seller

the 'Paris' or 'Chicago' of the South (Davison, 1978). The city was one of the first
in the Empire to receive electric street lighting, and the nocturnal parades of intermingling bourgeois and proletarian streams were a source of both fascination and
concern for Victorian commentators. The flow of street life would constantly negotiate
the many beggars and street traders who populated the streets of the city and
adjoining areas (Kennedy, 1982).

From the 1860s, a succession of public transport types, such as horse (and later,
cable) trams, evolved to carry the growing throngs; to increase the velocity of street
traffic by separating the strollers from the travellers. These were hardly accessible
forms of transport for many disabled people.

THE STREET-SELLER OF NUTMEG-GRATERS.

FIGURE 7.5 The street seller of nutmeg-graters

By the 1880s, Bourke Street, an important commercial thoroughfare in the city
centre, was the centre of the interstitial street economy, and was daily the host of
hawkers selling everything from fruit to matches. (By night, the offerings extended
to bodies.) These petty merchants competed with a brigade of street musicians and
entertainers for prominent positions (Kennedy, 1982); perhaps points of maximum
friction and/or visibility in the flow of street traffic such as street corners (Brown-
May, 1995). Swain (1985) relates the story of Ada, a partially blind single mother,
who survived in the early 1890s by singing and selling matches on city streets until
finally arrested (and separated from her child). Swain notes that 'Ada was not atypical,
for many similar girls were also physically or mentally handicapped and quite alone
in the city' (1985: 99).

John Freeman, in his *Lights and Shadows of Melbourne Life* (1888), describes women beggars displaying their crippled children in order to elicit sympathy and alms. (Some are even said to have 'borrowed' impaired children for the purpose.) Freeman's (1888) prose portrait of street begging and trading also contains several references to impaired hawkers and musicians. One 1887 account records a crowd gathering on a busy street corner to watch a party of showmen: 'the chief attraction is a so called fortune teller called "Gypsy Eliza" and a deformed man' (cited in Brown-May, 1995: 28). Note the anonymity of the disabled showman.

Many amongst the bourgeoisie were clearly alarmed at this unregulated inter-course of classes and moral types in public streets. The Charity Organisation Society (COS), Melbourne's self-appointed guardian of virtue amongst the poor, was partic-ularly concerned at the moral threat posed by this daily street carnival. The 1890 report of the COS fairly recoils at 'the spectacle of old and young, tainted and untainted, commingling and competing in the streets'. Strange that these ruthless champions of *laissez-faire* should find this quintessentially capitalist assemblage so disturbing. On a more amusing level, Andrew Halliday, Mayhew's collaborator, disguised his Victorian squeamishness with voodoo science:

> Instances are on record of nervous females having being seriously frightened, and even injured, by seeing men without legs or arms crawling at their feet.
>
> (Mayhew, 1968c: 433)

Indeed,

> A case is within my own knowledge, where the sight of a man without arms or legs had such an effect upon a lady in the family way that her child was born in all respects the very counterpart of the object that alarmed her. It had neither legs nor arms.
>
> (ibid.)

In the same set of remarks, Halliday exhorts the police to prevent the attempts of 'some of the more hideous of these beggars to *infest the street*' (emphasis added, ibid.). Halliday's metaphorical suggestion that crippled beggars were vermin which directly threatened public health, points to the extreme sense of abjection that the publicly displayed disabled body could conjure in the minds of the Victorian ruling classes. Halliday, of course, was not alone with his anxieties: purification of the street became an ideal for many amongst the late Victorian bourgeoisie. Anxieties about disease converged with fears of social difference in a new campaign for 'street hygiene', the demand that disgusting and contaminating 'objects' be removed to remote institutional spaces in order that the public's health – moral and physical – might be properly safeguarded.

As the century wore on, Melbourne's police duly responded to these sorts of anxieties and other imperatives by tightening general controls on street trading and vagrancy, thereby heightening the centripetal pull of the institution. As Brown-May notes, 'Street life came to be viewed with suspicion, as deviant and pathological, demanding regulation and control' (1995: 30). Although official intolerance of the disabled street beggar hardened as the century progressed, the reverse was probably true of the disabled street trader. Eventually, officialdom came to believe that orderly street trading was a respectable and humane solution to the problem posed by immobile and 'incapable' labour power. Interestingly, it was the COS, Melbourne's moral guardian of charity, which encouraged the acceptance of this 'humane' view:

> In the fullness of time the COS would convert governments, local councils and the police to its viewpoint that street begging should be banned, street vending licensed, and the 'privilege' of street stalls in some locations 'reserved almost exclusively for those under some physical disability'.
>
> (Kennedy, 1985: 209)

During the 1890s, the COS 'established a crippled person in Melbourne's first newspaper kiosk for the disabled' (Kennedy, 1985: 199). This confinement of disabled street traders in well-concealed kiosks was a prelude to the sheltered workshops that were finally to remove disabled people from the public sphere in the next century.

CONCLUSION

Given the fragments of surviving evidence concerning the lives of disabled people in the past, it is difficult to discern precisely their role in the quotidian spaces of the Victorian city. However, we can, I think, identify certain pervasive currents of power that must have shaped, if not determined, the ability of disabled people to participate in these spaces of everyday life. The evidence that we have suggests that disabled people were often drawn from the mainstream into institutional spaces of marginality, but we can also say that this centripetal pull was often struggled against, sometimes successfully.

Indeed, the very presence of disabled people on the streets of an exclusionary society is testimony to this struggle for personal autonomy and social inclusion. Mayhew's crippled seller of nutmeg-graters evoked the determination of many disabled people to resist the dreaded institution at any cost, declaring that he would 'rather die in the streets than be a [workhouse] pauper' (1968a: 332), and it is indeed probable that he succumbed one day in the place of his trade. But we must not end with a cheer for resistance, as these shadows and fragments of autonomy should not be confused with liberation and social inclusion. The street for many disabled people was a place of both struggle and abjection, and we cannot, from this distance, remember these painful biographies without sadness.

NOTES

1 The white invasion of what is now known as the State of Victoria began in a sustained fashion in 1835. The six and a half decades which separated this date from the creation of the Commonwealth of Australia in 1901 constitute the colonial period of Victoria's history.
2 *The Argus*, 22 October 1863.
3 Minutes of the Melbourne Ladies' Benevolent Society (27 January 1891), held at LaTrobe Library, State Library of Victoria, Melbourne.

4 *Report of the Chief Inspector of Factories on the 'Sweating System' in Connexion with the Clothing Trade in the Colony of Victoria*, VPP 1891, vol. 3, no. 138.
5 Minutes of the Melbourne Ladies' Benevolent Society (18 December 1873), held at LaTrobe Library, State Library of Victoria, Melbourne.
6 Minutes of the Melbourne Ladies' Benevolent Society (14 January 1890), held at LaTrobe Library, State Library of Victoria, Melbourne.
7 Minutes of the Melbourne Ladies' Benevolent Society (3 November 1868), held at LaTrobe Library, State Library of Victoria, Melbourne.
8 Minutes of the Melbourne Ladies' Benevolent Society (7 December 1875), held at LaTrobe Library, State Library of Victoria, Melbourne.
9 Minutes of the Melbourne Ladies' Benevolent Society (24 March 1891), held at LaTrobe Library, State Library of Victoria, Melbourne.
10 Minutes of the Melbourne Ladies' Benevolent Society (19 May 1891), held at LaTrobe Library, State Library of Victoria, Melbourne.

REFERENCES

Abberley, P. (1987) The Concept of Oppression and the Development of a Social Theory of Disability, *Disability, Handicap and Society*, 2(1), 5–20.

Archer, T. (1985 [1865]), *The Pauper, the Thief and the Convict*, New York: Garland.

Briggs, A. (1968) *Victorian Cities*, Harmondswoth: Penguin.

Brown-May, A. (1995) *The Highway of Civilisation and Common Sense: Street Regulation and the Transformation of Social Space in 19th and Early 20th Century Melbourne*, Working Paper no. 49, Urban Research Program, Australian National University.

Davison, G. (1978) *The Rise and Fall of Marvellous Melbourne*, Melbourne: Melbourne University Press.

Doray, B. (1988) *From Taylorism to Fordism: A Rational Madness*, London: Free Association.

Finkelstein, V. (1980) *Attitudes and Disabled People,* New York: World Rehabilitation Fund.

Foucault, M. (1979) *Discipline and Punish: The Birth of the Prison*, New York: Vintage.

Freeman, J. (1888) *Lights and Shadows of Melbourne Life*, London: Sampson Low, Marston Searle & Rivington.

Gleeson, B. J. (1993) 'Second Nature? The Socio-Spatial Production of Disability', unpublished PhD Thesis, Department of Geography, University of Melbourne

Gleeson, B. J. (1995) 'A Space for Women: The Case of Charity in Colonial Melbourne', *Area*, 27(3), 193–207.

Kennedy, R. (1982) 'Introduction: Against Welfare' in Kennedy, R. (ed), *Australian Welfare History: Critical Essays*, Macmillan: Melbourne.

Kennedy, R. (1985) *Charity Warfare: The Charity Organization Society in Colonial Melbourne*, Melbourne: Hyland.

Kristeva, J. (1982) *The Powers of Horror: An Essay on Abjection*, New York: Columbia University Press.

Lack, J. (1991) *A History of Footscray*, Melbourne: Hargreen/City of Footscray.

Lynn, P. (1990) 'Administrators and Change in the Penal System in Victoria, 1850–80', unpublished PhD thesis, Deakin University, Western Australia.

McCalman, J. (1984) *Struggletown: Public and Private Life in Richmond, 1900–1965*, Melbourne: Melbourne University Press.

Marx, K. (1981) *Capital: a Critique of Political Economy*, vol. 3, Harmondsworth, Middlesex: Penguin.

Mayhew, T. (1968a) *London Labour and the London Poor*, vol. 1, New York: Dover.

Mayhew, T. (1968b) *London Labour and the London Poor*, vol. 2, New York: Dover.

Mayhew, T. (1968c) *London Labour and the London Poor*, vol. 4, New York: Dover.

Oliver, M. (1990) *The Politics of Disablement*, Basingstoke: Macmillan.

Oliver, M. (1993) 'Disability and Dependency: a Creation of Industrial Societies?' in J. Swain, V. Finkelstein, S. French and M. Oliver (eds), *Disabling Barriers – Enabling Environments* London: Sage.

Rabinach, A. (1990) *The Human Motor: Energy, Fatigue, and the Origins of Modernity*, New York: Basic Books.

Ryan, J. and Thomas, F. (1987) *The Politics of Mental Handicap*, London: Free Association.

Sibley, D. (1995) *Geographies of Exclusion*, London: Routledge.

Swain, S. (1985), 'The Poor of Melbourne' in G. Davison, D. Dunstan and C. McConville (eds), *The Outcasts of Melbourne: Essays in Social History*, Sydney: Allen & Unwin.

8

HOMELESSNESS AND THE STREET: OBSERVATIONS FROM BRITAIN, CANADA AND THE UNITED STATES

Gerald Daly

•

Across the city tonight fires are stoked in rusty 55-gallon drums. Showers of sparks leap skyward from barrels serving as impromptu heaters. Gaunt figures hunch their shoulders against the bone-brittle cold, periodically adding layers of clothes or newspaper as insulation. It's wintertime in London, New York City and Toronto. Under bridges, in bus and train stations, all-night coffee shops, shopping malls, vacant warehouses, huddled under cardboard containers in the doorways of office towers and in bank machine kiosks, homeless individuals struggle to survive the icy winds and subzero temperatures. In British, American and Canadian cities this scene is repeated again and again. Life on the street is a painful experience, as they try to outwit not only the cold, snow or rain, but also police patrols, violent muggers, angry shopkeepers, and members of the public who are repelled, embarrassed or shamed by homeless people.

In addition to having political, social and cultural functions the street is a place where we confront the 'other', people who sleep rough because they have no alternative. Who are these homeless people? Are they on the street by choice? I will examine why and how they use streets and public places, how differences between 'us' and 'them' are played out, and the ways in which state control, regulation and surveillance affect people without shelter. I will explore the use by different groups of public/private spaces and the manner in which streets have come to be seen as places of deviance and contestation.[1]

WHO ARE 'THE HOMELESS'?

I feel anger toward an uncaring society . . . to be homeless means that you have no rights, but homes are a right, not a privilege.

(*Woman at a London shelter*)

While stereotypes of winos and 'shopping bag ladies' have been engraved on the public consciousness in recent years, the reality of homelessness is that it affects a heterogeneous array of people. Among these are low-income single adults (mostly men but including women as well), workers displaced by economic change, runaway youths and abused youngsters, elderly individuals on low fixed incomes, substance abusers, those who suffer from physical and mental health disabilities, people who are shelterless as a result of seasonal work, domestic strife, or personal crises; in addition, there are recent immigrants, refugees, and Natives (aboriginal people) who have migrated to the city to find work or to escape problems on the reserve, along with ex-prisoners and those recently discharged from detention or detoxification centres and mental hospitals.

Most homeless people are unemployable because of ill-health, disabilities, old age, or lack of skills; but a substantial number work at least on a part-time basis. The only available jobs, however, provide neither security nor benefits, and pay wages too low to lift them out of poverty. Moreover, if they are using shelters and soup kitchens, their employment prospects are constrained: soup kitchens operate only for a few hours during the daytime, shelters (which typically admit people only in late afternoon or early evening) do not accommodate shift work, and it is difficult to attend job interviews or to secure work when one lacks a fixed address, telephone, or access to showers and laundry facilities.

In Canada about 5 per 1,000 persons use the emergency shelter system each year; comparable figures of people using hostels in the United States are about 3–6 per 1,000, resulting in homeless populations (each year) of about 130,000–260,000 in Canada, and at least 500,000–600,000 Americans. Dennis Culhane of the University of Pennsylvania reported to Congress that, from 1990 to 1995, 3.6 per cent of New York City's and 4.8 per cent of Philadelphia's population stayed in emergency shelters on at least one occasion. Recent estimates indicate that 'as many as seven million Americans stayed in a public shelter in the first half of the 1990s' (Culhane, 1997).

Over time, the number of people who have been without homes is much higher than the level suggested by annual figures. In Britain, where households must be certified as homeless by local authorities, the rate of homelessness is 7 per 1,000 population across the country, ranging up to 25 per 1,000 in London. Each year there are well over 100,000 *households* certified as homeless in Britain, resulting in a total of several million individuals accommodated by local authorities since 1977 when the Housing (Homeless Persons) Act was signed into law (Daly, 1996: 15–16).

Although individuals without shelter have a multitude of characteristics, a few generalisations are possible. The nature and extent of difficulties encountered by homeless people are related to the length of time they have been without adequate shelter. The longer they are on the street the greater is the likelihood of their having significant problems, which in turn means that the duration of homelessness will probably extend well into the future. About one-third of homeless people are alcoholic and up to two-thirds suffer from mental illness, drug addiction or alcoholism. Many have to contend with multiple mental and physical health problems. In fact, less than one-third of homeless individuals do *not* suffer from these disabilities.

Substance abuse, like mental illness, may be the factor which precipitates loss of shelter. In a substantial number of cases, however, it is exacerbated by, and may result from, being without a roof and the social supports associated with home, family and community.[2] The distinction between these conditions and the coping mechanisms necessary to survive life on the streets is muddled. It is difficult to distinguish between precipitating and causal factors.

The gender of homeless people depends in large measure on the legislation and practices in each country. In Britain the Housing (Homeless Persons) Act of 1977 (as amended) gives housing priority to pregnant women and to families with children. As a result, 71 per cent of those certified as homeless are women and children and another 12 per cent are pregnant women without children. Because most hostels in Canada and the United States are reserved for single men, about two-thirds of shelter users are males – although this varies considerably from place to place. At least 20 per cent of the homeless population are single women or women with children (Burt, 1992: 22). Once female-headed families enter the shelter system child welfare authorities intervene and the children generally are placed in care (ie. foster homes). Often this intervention, though it may be well-intentioned, represents the first step in the creation of subsequent generations of disaffiliated and homeless people. It is difficult for families to reunite once they are separated by institutional intervention; young people on the street, for instance, often come from foster care.

In the United States, largely as a result of discrimination in employment and housing, non-whites represent about 55 per cent of the homeless population, 2.5 times their representation in American society (US Department of Housing and Urban Development, 1989: 1–2). Racial minorities in Britain are four times as likely as whites to be homeless or in very poor housing (Daly, 1996: 92). In Canada most homeless people are white but in several large cities disproportionately high numbers are Natives, who are systematically ill-treated or ignored by the government, the justice system, and the job and housing markets (Canadian Human Rights Commission, 1989).

The median age of homeless individuals is about 35 and has decreased over time. In shelters, mothers with children are, on average, about age 27. Most of these people do not have as much education as the general population and a substantial percentage are functionally illiterate, unable to deal with the job requirements of a globalised economy – though in an earlier age there would have been meaningful work for them. In Britain homeless people typically are school leavers with no marketable skills and distressingly meagre job prospects. Most are extremely poor. An American sample found that those with annual incomes below $20,000 are almost three times as likely to be homeless during their lifetimes as those with incomes of $20,000 or more (Link *et al.*, 1994: 1910).

About 180,000 children become homeless in England every year. Most of them are forced out of their homes along with their mothers, who typically are fleeing an abusive partner. Ninety per cent of these children are in poor single-parent families that lack not only financial resources but social support networks as well. Half the parents have clinically significant mental health problems; over two-thirds of the children have either a problem severe enough to require treatment, a behavioural problem, and/or significant delays in social or language development – a rate three times higher than non-homeless children from deprived inner-city neighbourhoods. Moreover, homeless youngsters' problems are aggravated by life on the streets and by virtue of their being excluded from or unable to attend school. Less than one-third of the children surveyed remained in school or nursery care on a regular basis. Even before they reach their teens some of these young people fall prey to pimps, pornographers, and drug dealers.

WHY HAS HOMELESSNESS INCREASED IN RECENT YEARS?

'Let's face it. I'm a loser. Everybody in here is a loser. You think if we weren't losers we'd be here? My father survived the Depression and the War. You think I could look him in the eye if he knew where I was?'

(Man in Toronto shelter)

Much more than a housing issue, the nature and complexity of homelessness can be more fully appreciated by examining the political economy of the state and the decisions made regarding resource distribution. In Britain, Canada, and the United States close connections are apparent among global economic changes, poverty, unemployment, welfare and housing policy, and homelessness. Those at risk are generally not well served by public policy decisions which tend to benefit individuals who are relatively well-off, vocal, and capable of exerting political leverage.

Poverty rates in these three countries fell during the 1960s; but this decline was reversed during the late 1970s and 1980s as governments reduced social spending. Shifting public policy, which diminished the staying power of the poorest households, was partially responsible for a rise in homelessness, as were changes in employment and income distribution. In the United States and Canada, for instance, the highest income group receives at least nine times as much of national income as those in the lowest income quintile – this compares to a factor of seven in Britain. The average after-tax incomes of households in all income groups, except for the richest quintile, fell in recent years.

Since 1980 the number of poor households grew in all three countries. Increases in poverty rates were recorded for specific groups: single mothers and their children, young workers, disabled persons, ethnic minorities, and inner city residents. These indicators point to substantial differences in social policies, reflected in government transfer payments to those in the lowest income quintile. The pre-transfer poverty rates (before transfer payments) are higher in Britain than in the United States; but the post-transfer payment rate in Britain is only two-thirds that of the United States. In Canada, the gap between rich and poor is diminished somewhat as a result of transfer payments and income taxes: in recent years the income share of poor families almost tripled, while the share of those in the richest quintile declined by about 5 per cent, as a consequence of public redistribution policies.

The social, economic and political conditions surrounding homelessness are similar in Britain, Canada and the United States. Responses are different, however, as they are shaped by precedent, by ideological positions, and by perceived economic, social and political constraints. All three societies are grappling with such issues as globalisation, de-industrialisation, shortages of affordable and adequate housing, and growing disparities among regions and between deprived inner-city districts and wealthy suburban areas. In the past two decades economic and social divisions became more pronounced, real wages for the working poor declined, and the gap between housing rents and income widened. In the United States, for example, many service sector employees are paid only the minimum wage. Even with two wage earners in a household, the total income received is inadequate to lift the family above the poverty line. A formerly homeless man, Terry McClintic, describes the effects of trying to make ends meet in this situation:

> 'At $5.00 an hour the take-home is around $700 per month. If the family has two children the rent will cost all of the take-home pay. . . . If we ever lose momentum, we're in trouble. Just a simple set of circumstances could make us homeless. We are treading water all the time . . . there's little room for a breather.

> (Daly, 1996: 44)

Causal factors include economic shifts, government policies, and demographic changes. Economic changes resulting from globalisation are manifested in a loss of manufacturing jobs, increased automation, and growth in involuntary part-time, low-wage, low-benefit employment in unskilled service sector jobs, along with streamlining of firms, accompanied by layoffs, redundancies and decreases in real wages, benefits and security.

Public allocation policies mirror a shift in budget priorities by administrations determined to cut spending on social programmes as they face rapidly escalating debt burdens. Government housing policies emphasise privatisation and tax subsidies for middle- and upper-income home owners. The result is an acute shortage of low-rent public housing, a depleted inventory of private rental units, and reduced shelter options for low-income groups. This situation was illustrated by a sign placed surreptitiously on a British Telecom tent (used to cover equipment) in London's Bloomsbury district (see Figure 8.1). Intended to dramatise the effects of privatisation and speculation, it advertises the tent as a designer residence, 'suitable for conversion, air-conditioned, free lighting, telephone installed, within spitting distance of all amenities, £45,000 or next offer' (*Roof*, May/June 1988).

Social policy changes are indicative of a shift away from a welfare state consensus and a resolve to shrink assistance programmes. Results are evident in increased numbers living in poverty, declining social benefits relative to the cost of living, an inability to afford available housing, and a rapid increase in evictions. In Toronto, for instance, although two-thirds of households in the metropolitan area are home-owners, over half of the city's population live in private rental units. More than one in three renters rely on social assistance, which has been severely cut during the 1990s. A 21.6 per cent reduction in social welfare benefits in late 1995 led to a 25 per cent rise in evictions. Tenants may become homeless after being intimidated

FIGURE 8.1 British Telecom tent advertised as a designer residence. Photograph by Tim Mars, courtesy of Shelter.

by landlords who wish to evict them in order to convert their buildings to condominiums. Some landlords shut off heat and water, illegally lock residents out, and arrange arson or assaults to convince sitting tenants to leave. Unaware of their rights as tenants, they may become victims of illegal eviction, harassment and violence because they are alone and lack political power.

Cities across North America report that governments and private agencies are relying on emergency shelters as dumping grounds for ex-prisoners, people from detox centres, and mental health consumer survivors. Use of emergency shelters in New York City increased by 16 per cent between 1994 and 1996; in Philadelphia, from 1992 to 1995 the rate of shelter occupancy by single adults rose by 22 per cent, and use by families grew by 58 per cent (Culhane, 1997).

Demographic shifts are apparent in the form of smaller households, many of them seniors or single mothers with young children, and in the prevalence of non-traditional households – nuclear families, though enshrined in public policy and political discourse, now represent less than 30 per cent of households in Britain. The inner city's contested terrain and the inherent contradictions of globalisation are evident in the form of gentrification in downtown districts, which led to the loss of single room occupancy units, the conversion of rentals to condominiums, and residential displacement. Changes in housing need attributable to demographic trends influenced demand for affordable shelter at the same time as supply, especially of small, low-cost units, contracted and prices rose. Not coincidentally, these shifts were accompanied by growing numbers of homeless people.

HEALTH AND THE HUMAN DIMENSIONS OF HOMELESSNESS

Why do you wait until I'm sick to give me the housing and stability I need to keep me healthy?

(A homeless man with AIDS in Boston, Massachusetts)

Ironically, within a few hundred yards of the most advanced scientific and medical establishments in London, New York City, and Toronto, thousands of homeless people suffer from a variety of health problems. In addition to economic deprivation, itself one of the most serious health hazards, these include skin diseases, gangrene, diabetes, nutritional deficiencies, sleep deprivation, hypertension, respiratory problems, tuberculosis and a number of other infectious diseases. Among young people on the street there is an extraordinarily high rate of physical and sexual abuse, unplanned pregnancy among young teenagers, and HIV infection.

Ill-health for persons without adequate shelter is attributable in part to institutional and attitudinal barriers to the provision and delivery of health care. Often the result as well as the cause of poor health, homelessness contributes to illness through a number of factors: the absence of social supports, physical and psychological stress from prolonged exposure to the elements and from living in crowded, chaotic, unhealthy environments, lack of protection from an array of bacteria and viruses, and social problems associated with poverty and the stigma of being on the streets. Wright and Weber found that 'the homeless probably harbor the largest pool of untreated disease left in American society today' (Wright and Weber, 1987: 17).

Many people lose their homes after experiencing serious illness or injury which results in hospitalisation, income loss and eviction. Over one-third are in poor health. People who lack adequate shelter, bathing and laundry facilities find it difficult to maintain personal hygiene. Their illnesses, skin diseases and nutritional deficiencies are worsened by their lifestyles.

Life on the streets is violent. Physical assaults and muggings are common, and these attacks precipitate health problems. Women and young people are extremely vulnerable. Homeless women frequently suffer from trauma and are subject to sexual assault at a rate twenty times higher than for women in general (Kelly, 1985: 75–92). Adults on the street in the United States die at a median age of about 44 – more than twenty years earlier than those in the general population. The death rate for homeless people is about three to four times greater than the rate for the total population. Almost half die violently. In American cities the murder rate among homeless people is more than twenty times higher than the US average; homicide is the leading cause of death among homeless men aged 18–24, while AIDS is the chief cause of death among homeless men and women aged 25–44 (Hwang *et al.*, 1997: 625; Wright and Weber, 1987: 126–37).

Children and teenagers also are affected by the lack of shelter. The Shout Clinic for street youth in Toronto estimates that there are from 4,000 to 12,000 homeless young people on the city's streets, depending on the season. About 70 per cent are victims of sexual and/or physical abuse. Most come through the child welfare system and are from foster homes, group homes, detention centres, or they were doubled up with relatives and friends prior to living on the street. Forty per cent are gay or bisexual, over one-third are on the street because of alcohol and drug problems in their families, 43 per cent have attempted suicide, only 32 per cent have received government assistance in the recent past, one-half are not covered by health insurance, and two-thirds have been physically abused on the street (Surbeck, 1997: 3). Wright's observations at American shelters led him to conclude that among children in homeless families 'chronic physical disorders occur at approximately twice the rate of occurrence among ambulatory children in general'. He found that poor health and chronic physical illness 'contribute to the cycle of poverty, . . . interfere with, if not preclude, normal labor force participation, and with it, the ability to lead an independent adult existence. . . . Poor health may be one mechanism by which homelessness reproduces itself in subsequent generations' (Wright, 1989: 65, 67, 72–3, 77).

Health problems affecting homeless adults are aggravated by alcohol and drug use, which also may increase the likelihood of distrust and social isolation. It is probable that alcoholism will remain prevalent among people on the street, despite the sums spent on detoxification, which typically does not deal with causal factors. People on the street drink to cope with cold weather, depression, isolation, and physical or emotional illness. Because it dulls pain, induces euphoria and fills idle time, alcohol is accepted as the drug of choice and as a means of fostering sociability among homeless men and some homeless women. There is, however, no doubt that alcohol contributes to or exacerbates health problems, hastens the death of some, and is a common factor in accidents and fatalities. Nevertheless, many experienced front-line workers feel that the key concern is behaviour, not the state of sobriety, in determining how to deal with inebriated street people. This issue represents a continuing dilemma for the operators of shelters, soup kitchens and group homes as they try to deal with individuals who are intoxicated, disruptive and represent a threat to other shelter residents, especially women with children and defenceless adults. It is relatively easy to bar entry to shelters or to require that all residents of a group home, for example, are 'clean and sober'. It may be considerably more difficult to refuse access to someone in the dead of winter if there are no alternative accommodations and he or she is likely to end up sleeping rough and, perhaps, freezing to death.

Health problems are acute for people with HIV/AIDS. Because drug use and needle-sharing are pervasive among street people, and some are gay/bisexual and/or engage in homosexual or heterosexual sex to earn money, they are susceptible to sexually transmitted venereal diseases and to HIV/AIDS. Homeless AIDS patients encounter logistical problems in dealing with their illness: they must leave shelters every morning, then spend hours waiting in line for meals, showers, and bed tickets. There is a scarcity of public bathrooms for people on the street, many of whom suffer from diarrhoea, malnourishment and chronic fatigue, and they have difficulty in taking medication regularly – AZT, for AIDS, must be taken every four hours. Because of a lack of beds or treatment facilities, there are delays between testing HIV-positive and the beginning of primary care. Housing assistance, which is essential to treatment and reduction of stress, often is provided too late. Once diagnosed as HIV-positive many also have shelter problems, including threats of violence and abuse from neighbours and landlords who learn of their illness.

Poor people in general, and homeless persons in particular, experience difficulty in obtaining access to adequate care. Normally services are provided in clinics or emergency rooms of large hospitals. Outreach is not yet common. Homeless persons find it difficult, without a fixed address or telephone, to gain entrance to the health care system. Because they lack the resources to maintain personal hygiene, and perhaps because they may appear dirty and unattractive, they find that some doctors and nurses are reluctant to provide treatment. Emergency room staff may be unreceptive, and homeless transients frequently are regarded as abusers of the benefits system. A study in St Louis found that more than 70 per cent of homeless people 'had no usual health care provider and that more than half had received no health care in the previous year' (Wright and Weber, 1987: 35). Similar concerns about inequality in health care have been expressed in Britain after the Black Report found that poor health and high morbidity rates are significantly correlated with deprivation (Black *et al.*, 1980; Townsend *et al.*, 1985: 662; Whitehead, 1987). Eligibility for benefits customarily is predicated on illness, alcoholism or drug dependency. As a result, individuals tend to adopt the behaviour patterns necessary to secure admittance. Moreover, once admitted, they must remain 'ill' or they will be forced to move on. There is little incentive to recover. Access and eligibility also are constrained because public officials typically limit programmes and funding to those with 'special needs,' such as people with severe physical or mental disabilities. In the United States it is difficult for street people to obtain Medicaid unless they are 'a parent with a dependent child or completely physically disabled' (US Congress, 1990: 38). Moreover, homeless people are discharged from hospitals very rapidly. They gravitate to the streets or to hostels where they are unable to get adequate treatment – foot lacerations, for instance, require frequent hot baths, and people with concussions need to be wakened every few hours to monitor their condition.

Mental illness may either cause or result from homelessness. Many became homeless after being de-institutionalised, when governments decided to close asylums and psychiatric institutions. Others would likely be institutionalised if adequate facilities and funding were available (Fischer and Breakey, 1991: 1121–4). Those who have been de-institutionalised are irrevocably bound to society in a dependent relationship based on illness and the dispensation of charity. Once these individuals return to the community most are not capable, initially at least, of functioning on their own. After extended periods of enforced dependency, they are cast adrift without adequate supports, facing a bewildering maze of bureaucratic hurdles. In Canada, for instance, it is necessary to pay an application fee to obtain a birth certificate, which is a prerequisite for securing a health insurance number or a social insurance card. Both are needed to get a job. Homeless individuals in all three countries rely on street-front clinics (as shown in Figure 8.2) which provide first aid, referrals to hospitals, assistance in dealing with health and welfare bureaucracies, as well as documentation in order to qualify for benefits.

For those who are disoriented, have lost touch with their families, have misplaced their papers, or had them stolen – a recurring problem on the street and in shelters – life becomes a quest for legitimacy and credibility. Immediately after being discharged they are usually unable to deal with this documentation process, which reinforces their feelings of inadequacy and frustration. Once caught in the revolving door of institutionalisation–deinstitutionalisation, escape is difficult. Frequently, discharged psychiatric patients end up in boarding houses which are unregulated regarding care standards. Residents have no security of tenure and troublesome residents are routinely sedated, confined to their rooms or strapped in wheelchairs.

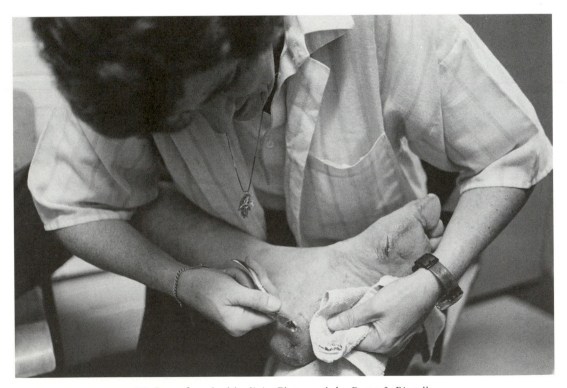

FIGURE 8.2 Street-front health clinic. Photograph by Byron J. Bignall

LIFE ON THE STREETS

I need a place to stay. My legs are badly ulcerated. . . . Not the shelters or
the flops, they are not for me. The police took me off the subway and to
the Men's Shelter a few months back, but I couldn't stay. I am 68 years old
and I can't defend myself down there.

(Coalition for the Homeless, 1983: 17)

Chronically homeless individuals, many of them elderly, disabled or malnourished, are
vulnerable to crime. When homeless persons are apprehended by police, it is usually
for victimless crimes: three out of four arrests recorded against them in Baltimore were
for disorderly conduct, trespassing, or sleeping in public parks. An additional 15 per
cent involved offences against property, including theft of food (Fischer 1988:46–51).
More than seventy American cities have criminalised activities associated with lack of
shelter. Municipalities impose fines for vagrancy, turn on lawn sprinklers in public
parks at night, place barriers across park benches, and take other actions designed to
discourage sleeping in public places. Since the mayoral election of 1991, won by a
former police chief, the City of San Francisco has cited over 4,000 homeless people and
imposed a $71 fine on each; when they are unable to pay, an arrest warrant is issued.
Justified in the name of crime prevention, this initiative was explained by Police Chief
Anthony Ribera: 'We need to stop the little crimes, the peeing in public, the public
drinking, and the big crimes will stop' (Daly, 1996: 199).

Those who are able seek alternatives to the shelters. Parks are used, if available,
as are beaches, woods, ravines, alleys, abandoned vehicles and boxcars, school buses,
old buildings, as well as storage containers and spaces under highways. Some even
live in subway or train tunnels. Single homeless individuals in particular depend on
an institutional network for survival. They move frequently but remain within a
relatively circumscribed area of the inner city, which is defined by the location of
such institutions as hostels and missions, drop-in centres, food banks and soup
kitchens, rooming houses, casual labour pools, and public baths. These areas tend to
have a concentration of retail outlets which are heavily used by street people (coffee
shops or cafés, beer and liquor stores, pawn shops, and taverns or pubs), public facilities (libraries, transit and public baths) and support services (hospitals, street health
clinics, counselling and detoxification centres, methadone clinics, welfare and employment offices). People on the street supplement their incomes by piece work, by
distributing pamphlets or selling newspapers and flowers, recycling cans or bottles,
selling plasma, making toys and artificial flowers, and, in some cases, by prostitution.
While most homeless adults do not panhandle regularly, some street youths panhandle
from time to time; they also supplement their incomes as 'squeegee kids', cleaning
car windscreens at intersections, with or without the driver's permission.

Women and men differ significantly in the nature of their networks, activity
patterns, and the ways in which they use the city. Men spend more time on the
street: 89 per cent of men are alone on the streets (though they spend some time
each day in small groups), while at least half of the women are accompanied by their
children. Typically, shelters or hostels for men require that they leave early each day

and they are not allowed to return until late afternoon. Women's refuges allow them to remain within the sheltered surroundings and permit long-term stays; they provide a number of services under one roof, while traditional hostels for men offer only space for sleeping – a few also provide meals and bathing facilities. Men range more freely throughout 'their' part of the city, staking out turf, 'hanging out' at drop-in centres or on particular street corners, making extensive use of labour pools, parks and other places where it is possible to congregate or to sleep undisturbed. Women are more restricted in their movements because they are usually accompanied by children, they are more vulnerable, and constantly must be alert to threats to their physical security. As a result, some women acquire a male friend or protector (for short or long durations) to ensure their safety and to keep aggressive men at bay. One characteristic shared by men and women on the street is their high level of distrust, which increases the longer they are sleeping rough. This is related as well to alcohol and drug abuse. A frequent observation by homeless people is that they are 'always watching their backs on the street . . . it's hard to find someone you can trust'.

Men and women are homeless for different reasons. Men usually succumb to literal homelessness because of loss of employment, leading to loss of housing and, perhaps, family connections, all of which may be aggravated by illness, injury, or substance abuse. Homeless women often are referred to as 'situationally homeless' because they are on the streets as a result of immediate economic or domestic problems, because they have been abused, or because of mental illness. Once on the streets, it is highly likely that physical and sexual abuse will continue, and perhaps worsen, unless they can gain entry to a woman's refuge. Women are younger, and (in North America) most are non-white. Despite being jobless for long spells, women (especially those accompanied by children) are on the streets for much shorter periods than are men (because they have priority for housing in many jurisdictions). They are less likely to have institutional contacts with mental health or criminal justice systems, but are in greater need than men of most social services.

A substantial percentage of women on the street are turned away from refuges and transitional housing because of a shortage of space. One in three abused women returns to an abusive relationship because they cannot locate housing (Bard, 1994: 13). Battered women usually leave home abruptly after being beaten on a number of occasions. As a result, they do not anticipate being homeless. Most see their situation as temporary and are unable directly to address their problems. Paralysed by fear, they use denial as a coping mechanism. Wherever they go, they feel unsafe, afraid that their abuser will stalk them. As a result, many homes for women do not reveal their locations and will not admit male visitors. Unfortunately, this also means that the system forces families apart.

IMAGES OF THE STREET: A CONTESTED SPACE

The problem of the vagrants, panhandlers and bums who plague our neighborhoods and commercial districts can be controlled, if there is the political will to do it . . . People do not have the right to live on our streets or in our parks. Proliferation of vagrancy is a problem of law enforcement, and should be dealt with accordingly . . . The free ride is over in San Francisco.

(Ben L. Hom, a San Francisco mayoral candidate, 1995)

The street serves a number of functions for homeless people. It is a place to socialise and to trade useful information. Bartering is widespread, enabling people to satisfy their immediate needs without having to use cash. They make arrangements for getting food (from soup kitchens and food banks), clothing (from used clothing depots), finding a place to sleep or squat, or to locate potential roommates with whom they can double up. Some sell plasma, trade in drugs or alcohol, or, perhaps, sex. Others join the queue for day labour pools, taking on whatever jobs come their way in order to earn some cash. By doing so they make themselves vulnerable to unscrupulous contractors who pay low wages, often for dangerous work, and deduct spurious charges for providing transportation and equipment or for cashing cheques. People also queue on the street to secure assistance from front-line agencies, to obtain health care, medication, psychiatric counselling, or dental treatment. Street dwellers spend a great deal of their time trying to avoid the police, constantly moving to prevent being harassed for loitering. Countless hours are devoted to walking or riding public transit, waiting for service from bureaucrats or social agencies. Life for many is a struggle to gain, reclaim, or to assert control over contested space, contending with others intent on excluding or ignoring them. In addition to the police, these include shopkeepers, punks and skinheads, drug dealers, tourists, members of the general public, as well as other homeless people.

They congregate in particular places in the city: it is common to encounter single men, along with very young people, on heating grates along Pennsylvania Avenue in Washington, DC, at the Park Street corner on Boston Common, in New York City's Thompkins Square park, on Fifth Street in Los Angeles, sleeping elbow to elbow under cardboard containers on London's South Bank, and in the centre of English market or tourist towns, like Oxford and Cambridge, where they panhandle from passers-by. The use or inhabitation of these spaces varies considerably throughout the day and night as different groups assert control over their turf.

A popular spot in Toronto is the intersection of Dundas and Sherbourne. The corner is a stop for trolleys and buses, which connect with the subway. Close by are shelters and public baths, rooming houses, soup kitchens, pawn shops, outreach health clinics, the welfare office, labour pools, beer stores, a public park and hospitals. The northwest corner of this busy intersection is occupied by prostitutes who solicit 'johns' in cars as they wait to turn at the red light; the north-east is claimed during the day by seniors and disabled individuals in wheelchairs who while away the hours watching the parade of humanity; a mini-mall on the southwest is the location of a doughnut shop frequented by police, where drug deals are transacted when the police are not present; many men hang out on the south-west corner, outside the Friendship Drop-In Centre, where they share drinks and cadge cigarettes, while waiting for soup kitchens or shelters to open. Nearby doorways or abandoned houses serve as squats for street kids or crack houses, where women are sometimes sold for drugs. In this district muggings and gratuitous violence are commonplace, typically directed against Natives, women or disabled and older men who have difficulty defending themselves. Ordinarily, women are not much in evidence at this intersection unless they are prostitutes or accompanied by a male protector. Night-time activity fluctuates, affected by the frequency of police patrols and by the presence of citizens from local residents' associations who do evening walkabouts, attempting to report 'johns', to discourage prostitution and drug dealing, and to keep homeless individuals from sleeping in the park.

It is not surpising, then, that street people expend a great deal of energy in developing appropriate coping mechanisms. While most of us seek rationality, coherence, and predictability in our lives, people on the street simply try to survive and to evade surveillance or control. We may regard some of their coping strategies as illogical or even bizarre. But mainstream judgements of this sort fail to account for the imperative of adaptation in a world that does not conform to dominant conceptions of 'reality'. Life on the streets requires adherence to a different set of organising principles, an alternate reality. Homeless people pursue immediate comfort, short-term satisfaction, and medium-term survival. Past, present and future are blurred. In an uncertain, fatalistic world where chance plays a leading role, it is difficult to distinguish between preparing for the future and living for the present.

The street and surrounding spaces consist of clearly public areas (like parks or plazas), private spaces (residences), and a host of quasi-public and quasi-private areas that are delineated by visible or invisible boundaries. These territorial demarcations are meant to organise space, to draw the line between defined districts, to regulate contact, and to protect against unwanted incursions. J.B. Jackson noted that in a well-delineated society where properties are fenced off 'we realize we are in a landscape where political identity is a matter of importance, a landscape where lawyers make a good living, and everyone knows how much land he owns' (Jackson, 1984: 15).

In such a setting, where private property is sacrosanct and boundaries are sanctified by law, those without shelter usually are spurned as non-entities, repudiated in a number of ways, rendered invisible and powerless. Space is organised in political and legal terms to express ideas about civic virtue and to differentiate between those who are deserving and those who are regarded as transient, marginal, fugitive or deviant. Even public squares or parks, intended as a manifestation of an egalitarian social order, may be declared off limits to these undesirable elements of society. Only the street and marginal spaces can be used by homeless people; but even here they may be driven off, when tent cities, for instance, or squats in parks are dismantled or shut down by state authorities who exercise surveillance in the name of social control.

The mainstream image of the homeless world is a social construction, reflecting a stereotype of the homeless individual and all that person brings to the fore in terms of anxieties, fears, and political and cultural ideologies. The streets customarily occupied by homeless people, though many are centrally located, are characterised as peripheral (in social and economic terms), because they carry the stigma of marginality and have the effect of containing or confining the unwanted 'others' of society (Shields, 1991: 3). Conversely, while the spaces inhabited by homeless people are represented as worthless, dangerous, repugnant, or nasty, they also may acquire potential value in the eyes of the dominant culture: squats may be converted to condominiums or derelict land can be reclassified and rejuvenated under an urban renewal scheme. Once again, the homeless individual is uprooted and displaced.

The desire for self-protection and its corollary, the need to exclude the odious 'other', is expressed in law, in social relations, in architecture, planning and urban design, and in language. Language is instrumental in the social construction of reality. Language is political. It is revealing of how we look at social, economic and political issues; and the ways we use language serve to convey society's messages of power, influence and authority. Homeless individuals may be silenced by such power relationships and associated control mechanisms. These processes serve not only to define social order and to set the political agenda, but also to marginalise, to devalue, and to hold at arm's length. In social, economic and political discourse the life stories of homeless people habitually are devalued, shunted aside or unconsciously limited.

Social service and health professionals, sympathetic bureaucrats and politicians, as well as many representatives of voluntary agencies, all involved in dealing with homelessness, also have constructed an image of the homeless individual – a patronising view of one who needs them to speak for him, to design his programmes, to assist him continuously, as if he were unable to help himself. A formerly homeless man in Oregon observed ruefully that 'too often people experiencing homelessness and poverty are given programmes rather than given the opportunity to design the programmes' (Daly, 1996: 250). In recent times, however, homeless individuals have begun to wrest control of the agenda that governs their lives. In Figure 8.3, occupants of Bed and Breakfasts in Bayswater, London, demonstrate against homelessness, the conditions in temporary accommodation, and the inadequaces of the state response to these problems.

Coincidental with the withdrawal of the state from many social service functions has been a growing awareness of the inappropriateness of large, bureaucratic government welfare schemes and a desire among homeless people to become the arbiters of their own fate. David Harvey notes that recent urban trends 'create all sorts of interstitial spaces in which liberatory and emancipatory possibilities can flourish' (Harvey, 1997: 69). More attention is being given, of late, to self-help housing, litigation, organised squats, sit-ins, and public education campaigns, designed to gain exposure, to educate the public and politicians, and to enhance self-esteem among homeless people who would like to leave their life on the streets behind them. Many of these individuals have considerable coping skills and resourcefulness, adapting to a difficult existence in places marginalised by society. If allowed to exercise power and control over the programmes which shape their futures, many can become advocates for themselves and active agents to improve their position. A number already have.

FIGURE 8.3 Demonstration against homelessness. Photograph by Jonathan Stearn, courtesy of Shelter

NOTES

1   Research on homelessness is difficult because it can be intrusive and may be biased by the researcher's preconceptions, values, and experience. It is further complicated by attempting to survey responses in numerous cities and in different countries. I took great pains to respect the privacy and dignity of individuals. Where possible, informal interviews were conducted with a broad array of homeless people, care givers, non-profit agency representatives, and policy-makers. I supplemented these with literature reviews, analysis of policy debates and public programmes, as well as assessments of voluntary sector efforts.

2   A former resident of a group home for recovering alcoholics observed that 'for many, they have burned their social and financial bridges beyond real repair. Relatives have long given up on these people. Without a long-term family setting, these people will not have any safety net and safe environment to restore the needed basics.' (D. Nichols, cited in Daly, 1996: 205).

BIBLIOGRAPHY

Bard, M. (1994) *Organizational Community Responses to Domestic Abuse and Homelessness*, New York: Garland.

Black, D. (1980) *Inequalities in Health*, London: HMSO.

Burt, M. (1992) *Over the Edge: The Growth of Homelessness in the 1980s*, New York and Washington, DC: Russell Sage Foundation and The Urban Institute Press.

Canadian Human Rights Commission (1989) Annual Report 1989. Ottawa: Canadian Human Rights Commission.

Coalition for the Homeless (1983) *Cruel Brinksmanship: Planning for the Homeless,* New York: Coalition for the Homeless.

Culhane, D. (1997) Testimony to the U.S. House of Representatives on H.R. 217, *Homeless Housing Programs Consolidation and Flexibility Act*, Committee on Banking and Financial Services, Subcommittee on Housing and Community Opportunities, March 5.

Daly, G. (1996) *Homeless: Policies, Strategies and Lives on the Street*, London and New York: Routledge.

Fischer, P. J. (1988) 'Criminal Activity Among the Homeless: A Study of Arrests in Baltimore,' *Hospital and Community Psychiatry* 39: 46–51.

Fischer, P. J. and Breakey, W. R. (1991) 'The Epidemiology of Alcohol, Drug, and Mental Disorders Among Homeless Persons', *American Psychologist* 46, 11: 1115–28.

Harvey, D. (1997) 'The New Urbanism and the Communitarian Trap', *Harvard Design Magazine* 1, 1 (Winter/Spring): 68–9.

Hwang, S., Orav, E., O'Connell, J., Lebow, J., and Brennan, T. (1997) 'Causes of Death in Homeless Adults in Boston', *Annals of Internal Medicine* 126: 625–8.

Jackson, J. B. (1984) *Discovering the Vernacular Landscape*, New Haven: Yale University Press.

Kelly, J. T. (1985) 'Trauma: With the Example of San Francisco's Shelter Programs', in P. W. Brickner, *et al.*, (eds) *Health Care of Homeless People*, New York: Springer, pp. 77–91.

Link, B., Susser, E., Stueve, A., Phelan, J., Moore, R. and Struening, E. (1994) 'Lifetime and Five-Year Prevalence of Homelessness in the United States', *American Journal of Public Health* 84, 12: 1907–12.

Shields, R. (1991) *Places on the Margin: Alternative Geographies of Modernity*, London and New York: Routledge.

Surbeck, M. (1997) *Shout Clinic Presentation to the City of Toronto*, January, Toronto: City of Toronto Alternative Housing Subcommittee.

Townsend, P., Simpson, D., and Tibbs, N. (1985) 'Inequalities in Health in the City of Bristol: A Preliminary Review of Statistical Evidence', *International Journal of Health Services* 15, 4: 637–63.

US Department of Housing and Urban Development (1989) 'A Report on the 1988 National Survey of Shelters for the Homeless,' Washington, DC: The Office of Policy Development and Research, Division of Policy Studies.

US Congress (1990), House Committee on Energy and Commerce, Subcommittee on Health and the Environment, 'Health Care for the Homeless: Hearing Before the Subcommittee' June 15, Washington, DC: USGPO.

Whitehead, M. (1987) *The Health Divide: Inequalities in Health in the 1980s*, London: Health Education Authority.

Wright, J. D. (1989) *Address Unknown: The Homeless in America*, New York: Aldine de Gruyter.

Wright, J. D. and Weber, E. (1987) *Homelessness and Health*, Washington, DC.: McGraw-Hill.

9

THEATRES OF CRUELTY, RIVERS OF DESIRE

THE EROTICS OF THE STREET

David Bell and Jon Binnie

•

The Pet Shop Boys' ironic coupling of the urban lament 'Where the Streets Have No Name' with the camply romantic 'I can't take my eyes off you' in a single released in 1991 could easily be played as soundtrack to this chapter, since it segues between the bleak and the bouncy, between earnestness and frivolity, between cruelty and desire. Inevitably read, as with so many of the Boys' songs, as an AIDS lament, the song also evokes a sense of being on the streets, of both wanting to seek 'shelter from poison rain' and wanting to 'let me love you, baby' – at once an erotics of looking and a desire for disappearance (also captured lyrically in some passages of Alphonso Lingis's sensuous geography, *Foreign Bodies* (1994)). It thus summarises the importance of the city streets for sexual dissidents: the flow of people, the chance both to disappear in a crowd but also to catch glances, to look and be looked at, the chance for a brief encounter – the possibilities which frame the city streets as an iconic space in the queer sexual imaginary, contributing to that 'great gay migration' to urban areas detailed in the US by Kath Weston (1995). Similarly, the PSBs' first major hit single 'West End girls' (1985) couples street fear with yearning: 'Too many shadows, whispering voices/Faces on posters, too many choices/If, when, why, what/How much have you got?/Have you got it, do you get it?/If so, how often?/Which do you choose/A hard or soft option?'.

It is this tension, between shadows and choices, hard or soft options, which we want to explore in this chapter, through a reading of Andrew Holleran's novel of 1970s gay Manhattan, *Dancer From the Dance* (1978), and Stewart Home's *No Pity* (1993), a collection of short stories based in contemporary (and future) London. From these texts we can examine the shifting narratives of urban queer sexualities and their relationship to the street as a site of both desire and dread: as a locus of violence and of sex. As Gill Valentine (1996) describes, the streets are ambivalent spaces for sexual dissidents;[1] *Dancer From the Dance* and *No Pity* signal this ambivalence through their very different strategies for writing the erotics of being on the street. In Holleran's work the streets of late 1970s Manhattan are romantically portrayed as 'rivers of desire' – utopian playgrounds for sexual experimentation, and for the realisation of an intimate public sexual citizenship. By contrast, Home's brutal, situationist eroticisation of London's streets describes a post-AIDS urban landscape of ultraviolence and rough sex. While there are some interesting similarities in their work – the fondness for ruins and dereliction, for example – there are fundamental differences, most especially those arising from the inevitable inscribing of AIDS into the sexual script. We begin by discussing the status of the street as an emblematic landscape form for sexual outlaws, then critically summarise and contextualise Holleran's and Home's work; this leads us to a discussion of their deployment of city streets as an erotic narrative trope, and finally to a reflection on the impact of AIDS on these theatres of cruelty and rivers of desire.

QUEER NARRATIVES OF THE CITY STREETS

Although historical moments in the life of a city can be isolated, the urban process never stops. Unlike works of art – or even certain buildings, which have a more determinate existence – streets are as mutable as life itself and are subject to constant alterations through design or use that foil the historian's desire to give them categorical finitude. . . . The street has been treated as a nostalgic artefact, to be restored to an ideal state, or simulated to an imaginary historical model. With both the abandonment of the public realm and the recreation of a pseudopublic realm, civic values, such as the street as a space for community, have disappeared.

Celik, Favro and Ingersoll, 1994: 1, 6

The seemingly simple, everyday act of wandering the city streets has, of course, been subject to a long history of theoretical and experiential exploration. Benjamin, Baudelaire, de Certeau, the situationists, and a host of other writers and wanderers have been followed on their *derives*, their *flâneries*, their saunters (Shields, 1996). Baudelaire's iconic figure, the *flâneur*, has been evoked in countless contexts (Tester, 1994); questions of gender and latterly of sexuality have been added in re-evaluations of this definitive urban player. Essays such as that by Elizabeth Wilson (1992), for example, have argued that *flânerie* objectifies women and denies them agency (more nuanced discussion of the multiple gendered and sexualised subjectivities in urban space can be found in Judith Walkowitz's *City of Dreadful Delight* (1992)). This criticism has been supplemented by work challenging feminist gaze theory and raising questions of queer agency, queer subjectivity and queer spectatorship (see for example Dyer, 1992, 1993; Evans and Gamman, 1995; Fraser and Boffin, 1991; Gamman and Markinen, 1994). Inserting queer agency, Paul Hallam's (1993) 'circuit walk' through Sodom texts and London streets, and Sally Munt's (1995) essay on 'The lesbian flaneur', both open up the possibilities for queer appropriations of street life;

both also signal the dangers and constraints which mark urban public space as homo-phobic and exclusionary.

Tracing the *flâneur*'s own sexual history, Peter Horne suggests:

> It is possible to see the flaneur as having evolved out of and reconverged with another of Baudelaire's modern types, the 'dandy', whose sense of self became the self-consciousness of the figure determined to stand out.
>
> (Horne, 1996: 134)

He goes on to argue:

> One may see the aesthete's adoption of this role as an active reconstitution of the flaneur, whose original fancy was that the city was a visual text for a modern allegory of visual desires. For the aesthete, the city becomes instead the site for a play in which roles can be adopted as masquerade, as a way of being transgressive in a modern style.
>
> (Horne 1996: 135)

While the practice of *flânerie* has been criticised for an obsession with surface details and impressions, Horne notes that Oscar Wilde was punished for exactly such sexual aesthetics; the dandy 'became a transgressor[, a]nd the revenge of the new hegemonic order fell upon Wilde for his sexual escapades and class transgressions' (Horne, 1996: 151). Wilde's crime – and his punishment – echoes through the city streets to this day; the pleasures of transgression always carry a cost (Bell, 1995; Califia, 1994).

WALKING AND WATCHING

Michel de Certeau (1984: 97) famously wrote that '[t]he act of walking is to the urban system what the speech act is to language or to the statements uttered. At the most elementary level, it has a[n] . . . '"enunciative" function: it is a process of *appropriation* of the topographical system on the part of the pedestrian.' Appropri-ating the street, walking, looking and being looked at; these are fundamental aspects of the formation of queer consciousness (Binnie, 1995). Queering the streets, then, is an enunciative act, a moment of transgression, when the pseudo-public realm gets reinscribed as a site of possibility (Bell, Binnie, Cream and Valentine, 1994); this possibility has been seized upon by activists intent on destabilizing the assumed heteronormativity of urban public space with theatrical displays of queer affection, desire and rage (Geltmaker, 1992).[2]

The streets also serve as a metaphor for dissident sexual cultures, with wandering, mobility, arrival and departure serving to embody the tracing of sex lives across urban space:

> This is how he came to us:
>
> Boy was walking down the street. Our street, though he didn't know that yet. And his head was spinning from walking so far; he walked everywhere, and though he stopped to eat every day when he was on one of these journeys

of his he did not I expect eat especially properly. He was worn out. Worn out with his own personal brand of window shopping; all that staring and never buying anything, all those shop windows, all those men to stare at and not dare follow, as if there was indeed a sheet of plate glass between him and them. And worn out with all that thinking because there was no one to talk to. No one whose advice he could ask. Some days he would follow a man, a man he'd just seen in the street, for minutes or for hours. Thinking he would go up to him and ask him if he knew the way. I can remember doing that in my time. Thinking that maybe this man was the right man, that maybe it was him I could ask for directions, him who would take me home or wherever it was I was trying to get to. Boy was like that, he was hoping that somebody would take him to the place where everybody else was. Or at least tell him how to get there, or give him the money to get in when he did get there, or at least lend him a map with a cross marked on it, or give him an address. But he never did ask any man for directions; he walked and he walked. In fact when Boy first came to us he was at the point of exhaustion. This is partly what made us seem like a destination to him; he was in that simple sense ready to arrive, ready to get somewhere and rest there for the night.

(Bartlett, 1990: 18–19)

It must be remembered here that particular streets have special resonances for sexual outlaws, as yellow-brick roads on the path to sexual self-realisation. Far from being streets with no name, these are streets whose names are known across the queer world, and whose pavements are wandered by a multitude of pilgrims and adventurers: Old Compton Street in London, (C)anal (S)treet in Manchester, Christopher Street in New York, Santa Monica Boulevard in Los Angeles. . . . These are streets whose mythologies are intimately interwoven with the histories of sexual dissident communities and activities; streets where to walk is to enunciate queerness itself. Such are the streets walked (and cruised) by the characters in *Dancer From the Dance*.

'NIGHTS IN THE CITY IN GAY NEW YORK'

Dancer From the Dance has played a pivotal role in the self-definition of many urban gay men since its publication in the 1970s. The novel is a celebration of the erotic potential of the city. As Edmund White says on the dust jacket of the paperback edition, '*Dancer From the Dance* accomplished for the 1970s what *The Great Gatsby* achieved for the 1920s . . . the glamorization of a decade and a culture.' Indeed, the novel's iconic status translated readily across the Atlantic, with the British gay scene taking much from Holleran's description of US clones, clubs and communities (a very different kind of cross-Atlantic translation can be traced for Home's work; see below). Central to the eroticization of the city and the streets in *Dancer From the Dance* is the notion that these sexualized spaces are profoundly democratic – they are open to all. In this description of the experience of clubbers, for instance, there is a sense of a queer version of the American Dream:

They lived only to bathe in the music, and in each other's desire, in a strange democracy whose only ticket of admission was physical beauty – and not even that sometimes. All else was strictly classless: the boy passed out on the sofa from an overdose of Tuinols was a Puerto Rican who washed dishes in the employees' cafeteria at CBS, but the doctor bending over him had treated presidents. It was a democracy such as the world – with its rewards and penalties, its competition, its snobbery – never permits, but which flourished in this little room on the twelfth floor of a factory building on West Thirty-third Street. Because its central principle was the most anarchic of all: erotic love.

(Holleran, 1978: 41)

Gay identity is here wrought out of the derelict buildings and ruins of the city; the inhabitants of the scene live for the night, the dance floor, the street – spaces which frame their experience of the city:

They were the most romantic creatures in the city. . . . If their days were spent in banks and office buildings, no matter: their true lives began when they walked through this door – and were baptized into a deeper faith, as if brought to life by miraculous immersion. They lived only for the night.

(Holleran, 1978: 43)

The novel, then, occupies a key position as a memorial, an elegiac landmark to a foregone era, a forgotten city. Its hero, Malone, sees the emerging gay community space of the Lower East Side as a theatrical backdrop for the realisation of his desires; it is only here that he can be himself, where he does not feel 'out of place':

Malone regressed when he came to live alone on the Lower East Side: he went back to the dreams of adolescence, became the girl on her prom night, dreaming of clothes, of love, of the handsome stranger, of being desired. He'd wanted to live a life like this, of self-indulgence, long days, gossip and love affairs, and in this shabby room on the Lower East Side he was completely free to love this timeless existence.

For there is no sense of time passing when you live in that part of town. . . .
He'd finally found a place whose streets he could roam, where time passed and one wasn't conscious of it, no one cared. He was a literal prisoner of love.

(Holleran, 1978: 125–6; our emphasis)

In his critical essay on Holleran, Joseph Cady (1993) suggests that the novel reproduces homophobic psychological stereotyping of gay men; Holleran's description of Malone 'regressing' here certainly echoes such infantalising discourses of stunted development. Cady also argues that Holleran writes with an 'antiseriousness', which acts as a 'rule and necessity in the novel's entire world' (1993: 259); like the Pet Shop Boys' ridiculing of the earnest posturing of rock band U2 ('Where the streets have no name' is a U2 song), which reveals a camply ironic antiseriousness, Holleran's novel of dancing queens gives frivolity significance in the face of surrounding seriousness. We find a similar stance – a similar antiseriousness – in the work of Stewart Home.

DEFIANT POSING

It is important when introducing Stewart Home's work to discuss the legacy of an earlier writer, Richard Allen. Home's stories are blatant parodies of the pulp splatter fiction Allen (real name James Moffatt) churned out under assorted pen-names for pulp publishing houses in the 1970s. His series of *Skinhead* novels, highly popular in 1970s Britain, are trashy, violent and lurid (they have also all been recently reprinted). They have also been subject to sustained queer appropriation, fuelling many fantasies and enjoying renewed celebrity among the queer skinhead culture of the 1990s (Healy, 1996). As Healy notes, the circuit of queer culture between the UK and the USA can be charted by examining this 'queer skin' scene on both sides of the Atlantic, with the 'Queercore' movement especially embodying a radical, politicised enactment of subcultural identities often more closely associated with London (punks, skinheads and so on; on Queercore see, for example, LaBruce, 1995).

Home extends Allen's textual practice, steeping it in irony; where Allen's protagonist, Joe Hawkins, is unambiguously heterosexual and relentlessly sexist, racist and homophobic, the characters inhabiting Home's world are polymorphous, outrageous and libertarian: here are the queers who bash back. And while an ultraviolent street aesthetic pervades both Allen's and Home's work, Home is more willing to satisfy his one-handed readers; the thought of queer reading would have horrified Allen:

> The anarchist London of Home's novels, the queer activities of his skin heroes, and a culture where such appropriations are not so much feasible as inevitable, hardly constitute the kind of tradition Allen would have wanted the conservative political vision of his books to inspire.
>
> (Healy, 1996: 100)

In an interview in *i-D* magazine in 1993, Home comments on one of his stories as 'playing the skinhead theme against the gay theme'. He goes on to say that 'Gay skinheads are great because they totally upset people's notions of what a skinhead should be. We also challenge the ridiculous idea that skinhead equals racism, which it clearly doesn't' (Cornwell, 1993). This extract from *Defiant Pose* exemplifies Home's take on skinhead culture:

> Terry was a boot-boy, a skinhead, a working-class warrior who was ready and able to take on those who would oppress him with their determinist doctrines. . . . As a skinhead, Terry was justly proud of his severe appearance. To the bourgeoisie, his cropped hair made him featureless and thus less than an individual. In a sense they rendered him anonymous and invisible. . . . Terry's appearance aroused moral indignation, which led innumerable trendies to fantasize about possessing him. They wanted to fuck, and be fucked by someone who rejected plastic individualism.
>
> (Home, 1991: 47–9)

Contrast that with this description of Allen's protagonist Joe Hawkins:

> Basically Joe Hawkins has a 'feeling' for violence. Regardless of what the do-gooders and the socialists and psychiatrists claimed, some people have an instinct bent on creating havoc and resorting to jungle savagery . . . Joe was one of these.
>
> (Allen, 1992: 101)

Home's heroes are also prone to shouting revolutionary aphorisms while meting out violent sex or sexy violence. In the *No Pity* story 'Pusher', for example, Adam is engaged in a S/M scene with Reggie, a 'wealthy young socialite' picked up at a leather club. As the violence of the scene escalates, Adam's proclamations include the following:

> 'Capitalism is pornographic because it turns individuals into ciphers, representations of the human potential it inhibits.'
>
> 'There's more to be learnt from wearing a dress for a day, than wearing a suit for a year.'
>
> (Home, 1993: 122)

These situationist-inspired slogans, which seem radically out of place amid a scene of violent (eventually murderous) sex, heighten Home's comic irony, giving a twist to the pulp genre which is rarely seen elsewhere.

QUEER STREETS

Obviously, then, the ways the streets are used in Holleran's and Home's work couldn't be more different; in *Dancer From the Dance*, the streets are to be cruised, in search of love or sex. In *No Pity* the streets are the site of incredible violence, grubby sex and acts of revolutionary agitation (frequently all at the same time). In the story 'New Britain', for instance, the protagonist, anarchist small press publisher Hamish McKane, goes from assaulting a security guard (at a toll booth on a privatised London street) to rough sex with a stranger, to leading an attack on a large publishing house ('Crud and Crumb') with jeers of 'Old farts out! . . . We want books full of sex, violence and anarcho-sadism, not boring literary crap in which nothing happens!' – and that's just in the first two pages (Home, 1993: 137–8). The streets are still intensely erotic, but this is a brutal eroticism of polymorphous perversity, or 'post-modern sex in postindustrial spaces' (Binnie, 1992). The privatised city, the rule of large corporations, right-wing political morality – this is the backdrop against which Home's characters fight and fuck for freedom. In *Dancer From the Dance*, desire flows free, and the streets are utopian and democratic; in *No Pity* they are dystopian and authoritarian.

In contemporary sexual practice, as well as in erotic fiction, we get a sense of the 'double-edginess' (Pile, 1996: 236) of the experience of being on the city streets. There is always the Holleranesque moment of possibility – for glances (or stares: 'I can't take my eyes off you') and encounters – but the threat of (homophobic) violence remains ever-present. It is precisely this double-edginess, of course, which is eroticised in certain public sex and S/M scenes, which theatricalise (and maybe even radicalise) both danger and pleasure (Golding, 1993). Dystopian, authoritarian urban space, in these cases, forms the perfect setting for sexualised transgression (in much the same way as the mournful griminess of dereliction can provoke excitement and arousal); as Viegener (1993: 130) writes, in the context of the queercore movie *No Skin Off My Ass*, we have in these scenes, on these streets, 'a model of cultural resistance [that works] by fragmenting the pervasive imagery of dystopia into utopian change through local pleasure'.

QUEER MORTALITY

> The Street and the Road are 'Theatres of Cruelty' par excellence. Places where extreme adventures, experiments at the limits, are performed, places of fights, fantasies. . . . End of the century? Absolutely. Standing in the middle of the streets, backs to the wall, amidst rubbish and ruins, watching old dreams crumbling. But standing fierce, ready for more. Who knows . . .
>
> (Marsault, 1990: 7)

For both Holleran and Home, the streets are more than sites of action; like the iconic streetscapes listed earlier, they are symbols of an attitude to life. But the attitudes described in their streets are radically different; as we have already noted, the impact of AIDS on this sexual cityscape must be placed centre-stage in our analysis, for while *Dancer From the Dance* predates the AIDS crisis and its remapping of desire, *No Pity* bears the trace of AIDS in a similar way to the writings of American author Dennis Cooper. As Earl Jackson, Jnr (1994: 143) writes: 'The AIDS epidemic is . . . a non-explicit horizon of Cooper's writing – a terrible historical accident that imposes an unanticipated literalness upon the risks to the body and the self that sex constitutes in much of his work.' Cooper himself has written that AIDS has 'ruined death', radically changing the erotic aesthetic (Cooper, 1988: 5). While death and desire hang heavy in both Holleran's and Home's texts, AIDS is only a 'non-explicit horizon' in *No Pity*; Holleran would tackle the epidemic explicitly in a later work, *Ground Zero* (1988).[3]

There is a second important distinction: where Holleran's protagonists are blessed with a firm, affirmed gay identity (the novel is thus a mirror of the post-Stonewall, pre-AIDS generation also fabulously captured in Edmund White's 1980 gay USA travelogue *States of Desire*), Home presents a more polymorphous perversity. His characters – Terry Blake, Fellatio Jones and others – are sexual in a raw way, fucking anyone and everyone. The brutal, nihilistic promiscuity of Home's characters might be read as a pornographic supplement to safer sex, their spurts of 'liquid DNA' or 'liquid genetics' (as Home is fond of calling semen) an antidote to the 'sex without secretions' provoked by AIDS panic (Kroker, Kroker and Cook, 1989).

The shift from the utopian landscapes of *Dancer From the Dance* to the dystopia of queer urban sex after AIDS, is also traced in James Miller's (1993) essay in the collection *Writing AIDS*. He focuses on Holleran's use of Fire Island as an iconic space for sexual citizenship:

> If this was indeed the vision of heaven entertained by such men before the days of HIV-antibody tests, it has been mercilessly extinguished by the AIDS crisis. Now Fire Island is commonly imagined as a zone of apocalyptic despair, a nuked prospect of desolate dunes, windswept beachgrass, and bare ruined pines where late the sweet boys cruised.
>
> (Miller, 1993: 268–9)

Home's work is always based around such zones of apocalyptic despair – visceral spaces where sex (often quite literally) has a 'sacrificial' quality (Jackson, 1994: 158). Of course, the reader of *Dancer From the Dance* in today's AIDS-infected world inscribes a nostalgic melancholia into Holleran's text (as Miller's essay makes clear); in *No Pity* such sentiments are erased in, to use the title of one of Home's stories, a 'Frenzy of the Flesh' (which is also a frenzy of the streets). That 'non-explicit horizon' which Jackson (1994) writes of in Dennis Cooper's work is, we see with hindsight, just around the corner for Holleran's dancing queens, but it surrounds and encloses Home's characters. It has redefined their relation to sex, violence, and the city. As Kobena Mercer (1993: 321) writes: 'In the wake of the wholly contingent and arbitrary events brought about by the AIDS health crisis, the metropolitan spaces of gay male subculture have been changed irrevocably.'

CONCLUSION

In looking at the fictional depictions of the erotics of the street in Holleran's and Home's work, we have tried to explore a shift in both queer representations of and queer experiences of urban public space. We have not traced other aspects of the 'double-edginess' of the streets explicitly (the threat and presence of homophobic violence especially), but have chosen to examine the impact of AIDS on writing the erotics of the street. Given the centrality of the urban to queer self-definition (Bell and Valentine, 1995) – and the iconic status of the streets within that – the shifts from Holleran's romantic democracy to Home's nihilistic brutalism can be read as signalling an important renegotiation of the relationship between queer sex and the streetscape in the time of AIDS. Of course, the streets retain some of the democratic potential explored in *Dancer From the Dance* (as sites for flirting and cruising, spaces of public sex, and for acts of political resistance and rage (Golding, 1993)), but AIDS has at once radicalised such spatial practices and recast the sexual landscape as a 'zone of apocalyptic despair'. The task – which Home takes up in one sense, and which countless other sexual outlaws are also engaged in – is to find new erotic (and political) possibilities in this zone, on these streets.

ACKNOWLEDGEMENTS

This chapter draws on material from Jon Binnie's PhD thesis (Binnie, 1996); an earlier version of the chapter was presented at the City Limits conference, Staffordshire University, September 1996.

NOTES

1   While it is not possible to discuss homophobic street violence within this chapter, we must note its overbearing significance in sexual outlaws' experience of being on the street. The erotic potentials we read from Holleran and Home are too often overwritten by the threat of violence in reality (though this is also used erotically by Home, where taking possession of the streets is always a powerful strategy, and where it is the homophobes who often suffer the violence. In this sense, at least, we can see something utopian in Home's work (although it is a very dystopian utopia!)). For scholarly work on homophobic violence, see for example Comstock (1991), Herek and Berrill (1992), and Valentine (1996).

2   The use of the term 'pseudo-public space' here begins to signal the erosion of the public sphere by forces of privatisation (used as a motif by Home in stories such as 'New Britain'). In the particular contexts of sexual dissidents, it has been argued that the distinction between the public and the private, so essential to legal definitions and restrictions of sexual behaviour (as used, for example, in the Wolfenden Act), has been completely collapsed as law has sought to both render the private public, and simultaneously to privatise the public, punishing those caught in the interstices (see Bell 1995 on Operation Spanner and the regulation of same-sex sadomasochism).

3   It's worth remembering here that AIDS itself has a geography (so far very poorly attended to by geographers; see Brown, 1995) – and that an important part of that is its *cultural* geography (especially in terms of resistance and protest, with groups like ACT-UP never really establishing themselves in the UK). There is clearly a subtext of this AIDS geography that could be read into our discussion.

REFERENCES

Allen, R. (1992) *The Complete Richard Allen*, vol. 1, Dunoon: ST Publishing.

Neil Bartlett (1990) *Ready to Catch Him Should He Fall*, London: Serpent's Tail.

Bell, D. (1995) 'Perverse dynamics, sexual citizenship and the transformation of intimacy', in D. Bell and G. Valentine (eds) *Mapping Desire: Geographies of Sexualities*, pp. 304–17, London: Routledge.

Bell, D. and Valentine, G. (1995) 'Introduction: orientations', in D. Bell and G. Valentine (eds) *Mapping Desire: Geographies of Sexualities*, pp. 1–27, London: Routledge.

Bell, D., Binnie, J., Cream, J. and Valentine, G. (1994) 'All hyped up and no place to go', *Gender, Place and Culture* 1: 31–47.

Binnie, J. (1992) 'Fucking among the ruins', paper presented at the Sexuality and Space Network conference, London, September.

Binnie, J. (1995) 'Trading places: consumption, sexuality and the production of queer space', in D. Bell and G. Valentine (eds) *Mapping Desire: Geographies of Sexualities*, pp. 182–99, London: Routledge.

Binnie, J. (1996) 'A geography of urban desires: sexual culture in the city', unpublished PhD thesis, University of London.

Brown, M. (1995) 'Ironies of distance: an ongoing critique of the geographies of AIDS', *Environment and Planning D: Society and Space* 13: 159–83.

Cady, J. (1993) 'Immersive and counter-immersive writing about AIDS: the achievement of Paul Monette's *Love Alone*', in T. Murphy and S. Poirier (eds) *Writing AIDS: Gay Literature, Language, and Analysis*, pp. 244–64, New York: Columbia University Press.

Califia, P. (1994) *Public Sex: The Culture of Radical Sex*, Pittsburgh: Cleis Press.

Celik, Z., Favro, D. and Ingersoll, R. (1994) 'Streets and the urban process: a tribute to Spiro Kostof', in Z. Celik, D. Favro and R. Ingersoll (eds) *Streets: Critical Perspectives on Public Space*, pp. 2–19, Berkeley, CA: University of California Press.

de Certeau, M. (1984) *The Practice of Everyday Life*, Berkeley: University of California Press.

Comstock, G. (1991) *Violence Against Lesbians and Gay Men*, New York: Columbia University Press.

Cooper, D. (1988) 'Dear secret diary', in R. Hawkins and D. Cooper (eds) *Against Nature: A Group Show of Work by Homosexual Men*, pp. 5–7, Los Angeles: LACE.

Cornwell, J. (1993) 'Carry on class war', *i-D* 122: 105–6.

Dyer, R. (1992) *Only Entertainment*, London: Routledge.

Dyer, R. (1993) *The Matter of Images: Essays on Representation*, London: Routledge.

Evans, C. and Gamman, L. (1995) 'The gaze revisited, or reviewing queer viewing', in P. Burston and C. Richardson (eds) *A Queer Romance: Lesbians, Gay Men and Popular Culture*, pp. 13–56, London: Routledge.

Fraser, N. and Boffin, T. (1991) *Stolen Glances: Lesbians Take Photographs*, London: Gay Men's Press.

Gamman, L. and Makinen, M. (1994) *Female Fetishism: A New Look*, London: Lawrence and Wishart.

Geltmaker, T. (1992) 'The queer nation acts up: health care, politics, and sexual diversity in the county of angels', *Environment and Planning D: Society and Space* 10: 609–50.

Golding, S. (1993) 'The excess: an added remark on sex, rubber, ethics, and other impurities', *New Formations* 19: 23–8.

Hallam, P. (1993) *The Book of Sodom*, London: Verso.

Healy, M. (1996) *Gay Skins: Class, Masculinity and Queer Appropriation*, London: Cassell.

Herek, G. and Berrill, K. (eds) (1992) *Hate Crimes: Confronting Violence against Lesbians and Gay Men*, London: Sage.

Holleran, A. (1978) *Dancer From the Dance*, New York: William Morrow (Penguin edition 1990).

Holleran, A. (1988) *Ground Zero*, New York: New American Library.

Home, S. (1991) *Defiant Pose*, London: Peter Owen.

Home, S. (1993) *No Pity*, Edinburgh: AK Press.

Horne, P. (1996) 'Sodomy to Salome: camp revisions of modernism, modernity and masquerade', in M. Nava and A. O'Shea (eds) *Modern Times: Reflections on a Century of English Modernity*, pp. 129–60, London: Routledge.

Jackson, Jnr., E. (1994) 'Death drives across pornotopia: Dennis Cooper on the extermities of being', *GLQ* 1: 143–62.

Kroker, M., Kroker, A. and Cook, D. (1989) *Panic Encyclopedia*, Basingstoke: Macmillan.

LaBruce, B. (1995) 'The wild, wild world of fanzines: notes from a reluctant pornographer', in P. Burston and C. Richardson (eds) *A Queer Romance: lesbians, gay men and popular culture*, pp. 186–98, London: Routledge.

Lingis, A. (1994) *Foreign Bodies*, New York: Routledge.

Marsault, R. (1990) 'Foreword', in 25/34 Photographes *Fin de siècle*, 7, Paris: Les Pirates Associes.

Mercer, K. (1993) 'Reading racial fetishism: the photographs of Robert Mapplethorpe', in E. Apter and W. Pietz (eds) *Fetishism as Cultural Discourse*, pp. 307–29, Ithaca, NY: Cornell University Press.

Miller, J. (1993) 'Dante on Fire Island: reinventing Heaven in the AIDS elegy', in T. Murphy and S. Poirier (eds) *Writing AIDS: Gay Literature, Language, and Analysis*, pp. 265–305, New York: Columbia University Press.

Munt, S. (1995) 'The lesbian flaneur', in D. Bell and G. Valentine (eds) *Mapping Desire: Geographies of Sexualities*, pp. 114–25, London: Routledge.

Pile, S. (1996) *The Body and the City: Psychoanalysis, Space and Subjectivity*, London: Routledge.

Shields, R. (1996) 'A guide to urban representation and what to do about it: alternative traditions of urban theory', in A. King (ed.) *Re-presenting the City: Ethnicity, Capital and Culture in the 21st-Century Metropolis*, 227–52, Basingstoke: Macmillan.

Tester, K. (ed.) (1994) *The Flaneur*, London: Routledge.

Valentine, G. (1996) '(Re)negotiating the 'heterosexual street': lesbian productions of space', in N. Duncan (ed.) *BodySpace: Destabilizing Geographies of Gender and Sexuality*, pp. 146–55, London: Routledge.

Viegener, M. (1993) ''The only haircut that makes sense anymore': queer subculture and gay resistance', in M. Gever, J. Greyson and P. Parmar (eds) *Queer Looks: Perspectives on Lesbian and Gay Film and Video*, New York: Routledge.

Walkowitz, J. (1992) *City of Dreadful Delight: Narratives of Sexual Danger in Late-Victorian London*, pp. 116–33, London: Virago.

Weston, K. (1995) 'Get thee to a big city: sexual imaginary and the great gay migration', *GLQ* 2: 253–78.

White, E. (1980) *States of Desire: Travels in Gay America*, London: Andre Deutsch.

Wilson, E. (1992) 'The invisible flaneur', *New Left Review* 195: 90–110.

10

OF HEROES, FOOLS AND FISHER KINGS

CINEMATIC REPRESENTATIONS OF STREET MYTHS AND HYSTERICAL MALES

Stuart C. Aitken and Chris Lukinbeal

•

On the first day, the aliens arrive. On the second day, the streets of New York, Los Angeles and Washington DC are torn apart in a series of graphic conflagrations. On the Eastern Seaboard, the Empire State Building and the White House explode below 15-mile-wide spaceships that spit out green rays; in Los Angeles, ground zero is the First Interstate World Center, the tallest building west of the Mississippi River. Fireballs erupting from these icons sear down familiar streetscapes. Grid-locked streets and sidewalks filled with panic stricken citizens are ripped up by the energy of the fireballs. Cars, trucks and buses career off exploding buildings. Faces of citizens are locked in terror as they watch the numbing spectacle only seconds before they too become part of it. In *Independence Day* (1996), one of the biggest grossing movies ever, the impressive destruction of familiar city streets is the best amongst a series of spectacular special effects. From these impressive displays of combusting streetscapes, we argue, arise mythic male archetypes that are contrived by the obsessive repetition of the *mise-en-abyme*. The *mise-en-abyme* is a fractal geography that Diane Elam (1994: 27–8) refers to as a 'spiral of infinite deferral . . . [where] representation can never come to an end, since greater accuracy and detail only allows us to see even more of the same representation'.[1] Dean Devlin, who produced *Independence Day* and wrote the screenplay with director Roland Emmerich, proclaims that they 'intentionally spiced a dark stew' where 'the human spirit is part of the mix . . . [surfacing after] the barrage of explosions and fires' (*LA Times,* 7 July 1996: A27). Male heroes in the movie arise phoenix-like from the rubble of the streets in an infinite regression, but from the obsessive repetition of images nothing more is revealed about maleness and masculinity. Side-stepping the possibility that male archetypes are complicated and unstable social constructions, the *mise-en-abyme* inscribes them with essential and immutable qualities.[2]

We begin this chapter with *Independence Day* in part because the destruction of familiar streetscapes in this movie is followed by such an auspicious rise of mythic heroes. Does this movie suggest that without streets there is no restraint upon what can be done by masculinity? In a mythic and metaphoric sense, the hero encompasses Dean Devlin's 'human spirit' as he rises phoenix-like from the ashes of the streets. The streets are metaphoric places where masculinity's public and private constitution meet, where acts of virility, heroism and hysteria are played out, and where the 'boundaries of conventional and aberrant behavior are frequently drawn' (Anderson, 1978: 1). With this chapter, we investigate the conflation of street myths, masculinities and representations of hysteria with a specific focus on the work of American film director Terry Gilliam. Without recourse to obsessive repetition, we argue that Gilliam's artistic ambitions rise above the stultified Hollywood focus on formula spectacles. Gilliam's work is not 'thrilling' in the sense that watching the destruction of familiar streetscapes in *Independence Day* thrills (see Robins, 1996). Rather, through the fantasy of the vaguely familiar, Gilliam focuses on the complex relationships, hysterias and political identities that undergird urban streetscapes. None the less, as a director of films that are often on the margins of mainstream Hollywood acclaim, he draws upon the mythic qualities of male heroes, fools and fisher kings in ways that are not necessarily unproblematic. Despite this, we suggest that Gilliam's work seems to avoid the mundane prison of spectacle and *mise-en-abyme* to play upon fantastic and disturbing images at the edges of hegemonic masculinity. Our central argument is that Gilliam's work contrives a *mise-en-scène* that is a 'continuous space . . . a positioning and positioned movement' (Aitken and Zonn, 1994: 17) for multiple male masculinities.

We begin by looking closely at how 'the mobilized gaze' inscribes gendered power relations in cinematic street space and suggest that transgressive sites can be created wherein gender power relations are negotiated. In so doing, we elaborate on how the notion of the hysterical male upsets preconceived notions of hegemonic masculinity. We then investigate the relations between fragmented myths, archetypes and male hysteria and show how these themes are woven through Gilliam's *Brazil* (1984), *The Fisher King* (1991), and *12 Monkeys* (1995).

MYTHIC STREETSCAPES AND THE *MISE-EN-ABYME* OF CINEMATIC SPACE

New premises about the production of space suggest that we should conceptualise streetscapes as multi-layered social moraines filled with competing powers that are both metaphoric and mythic. We use the term myth in the sense that Roland Barthes (1972) meant when he suggested a process whereby a concept becomes naturalised and, in effect, depoliticised:

> For the very end of myths is to immobilize the world: they must suggest and mimic a universal order which has fixated once and for all the hierarchy of possessions. Thus, everyday and everywhere, man is stopped by myths, referred

by them to this motionless prototype which lives in his place, stifles him in the manner of a huge internal parasite and assigns to his activity the narrow limits within which he is allowed to suffer without upsetting the world.

(Barthes, 1972: 255–6)

Barthes warns that a myth's power lies in its ability to naturalise specific ways of action and growth. We argue that streets are often imbued with mythic meaning and, as a consequence, a fairly precise cognitive mapping of streetscapes is reproduced and structured through popular films. In the American social imaginary these maps reveal a relatively unambiguous response to urban streets as the visual signifiers for the loss of innocence (Holtan, 1971) and the alientation of city life (Lukinbeal and Kennedy, 1993). City streets are the social space that contextualises films about gangsters, violence and crime, inner urban slums, homelessness and urban disinvestment (Gold, 1984; Ford, 1994; Natter and Jones, 1993). In addition, the 'mean streets' of the city are often portrayed in stark contrast to the escapism of life on the 'open road'. In road movies, for example, the open road metaphor often highlights a juxtaposition between place-based sedentarism (urban streets) and disengaged mobility (on the road). The juxtaposition of these metaphors is used to suggest a disassociation between self and object, and a re-affirmation of political identity through mobility. We use this argument elsewhere to suggest that road movies often romanticize non-urban streets as symbols of hope, signifying a 'geographic cure' to the hegemonic subjugation and violence encountered on the mean streets (Aitken and Lukinbeal, 1997). For example, in Gus Van Sant's *My Own Private Idaho* (1991), the life of two rent boys on the streets of Portland and Seattle is juxtaposed against the search of one of them, Mike, for self and family on rural roads that begin and end in Idaho. Van Sant uses Mike's narcolepsy to muddle time, space and place in a feeling that leaves us diverted from his neurosis and concerned about his character, his relationships and his spatial mobility. For example, in a narrative that is slippery because of geographic jump-cuts, we experience in one moment Mike's bucolic fantasies of roads, Idaho and mother love and, in the next, his climax at the hand of an unseen client in a Portland hotel. Importantly, it is spatial symbolism and the mobility of the road that focuses Van Sant's narrative conclusion: In the last scene of the movie Mike is back on his Idaho road where he collapses and folds into narcolepsy. The camera remains focused from a distance on the road and Mike's soporific form. After a while, a pickup truck stops and the occupants jump out to steal Mike's shoes and duffel bag. A little later, another car stops and an unidentified man gets out quickly to inspect Mike whom he then picks up and gently lays out on his back seat. He drives off in a remarkably inconclusive but none the less satisfying ending that establishes the multiple possibilities of liberatory masculinities through Mike's innocent and poetic mobility:

I always know where I am by the way the road looks. Like I just know that I've been here before. I just know that I've been stuck here. Like this one fucking time before . . . There is not another road that looks like this road. It is one kind of place. One of a kind.

We discuss Van Sant's portrayal of 'disassociated masculinities' more fully elsewhere (Aitken and Lukinbeal, 1997) but raise it here to highlight that Gilliam's movies are intriguingly different because his narrative resolves issues of male political identity through performances of male hysteria in place on, and under, urban streets. It can be argued that his characters are trapped by, and absorbed into, the street with little prospect of escape to the 'open road'. Although their political and sexual identities are constructed out of the street, in a very real but non-essentialist sense the streets are an embodiment of Gilliam's male hysteria. These complex associations between mobility/sedentarism, hysteria, male political identity and the filmic construction of streetscapes are difficult to untangle, but we believe that a reconceptualisation of Laura Mulvey's (1975) notion of 'the gaze' is a useful way to begin describing male subjectivities, delimiting how they bound the street, and suggesting possibilities for their reconstruction.

Mythic boundaries, male subjectivities and geographies of the gaze

Vision is key to the ways that filmic streetscapes are constituted. As we gaze we capture the world, but when gazed upon we are placed in the world. If 'the gaze' orders internal and external space in this way, then it can be argued that it also bounds the self. The distinctions and boundaries between internal and external space, and the policing of those boundaries, helps construct the power of the street. We argue that these are mythic, ideological controlled boundaries that are maintained to enforce a naturalized patriarchal normalcy. As a possible 'in-road' to understanding male hysteria, it may be fruitful to unravel the complex spaces of the gaze and consider their usefulness for reconstructing the politics and sexuality of street representations.

Kathleen Kirby (1996: 125–7) suggests that two divergent models of the gaze purportedly underlie much feminist writing. The first constructs the gaze as a projectile propelled from one subject to another and then deflected back. This gaze has some properties that are detachable from already determined subjects, who are complete and differentiated outside of the gaze's trajectory. The spatial structure of a gaze formulated in this way is one of immobility because women cannot empower themselves through the gaze. In Kirby's second formulation the gaze is a plastic medium constituting a field of power differences between subjects. This is perhaps more interesting to radical feminism because the 'direction and quality of the gaze determines the gender positionality of the participating subjects. In this case the gaze would be reversible; whoever wields the gaze has power, and thereby turns the gender positionality of the participating *subjects*' (Kirby, 1996: 126) This configuration resonates with a spatial plasticity of the gaze wherein power relations are very much predicated upon instruments of surveillance, control and visual pleasure.[3] Rather than prescribing the one-way trajectory of a rigid projectile, the space of the gaze is considered to be infinitely plastic and, thus, reversible by those who take on the power to control the space. The problem with this configuration is that those who take control are implicitly masculine. Kirby goes on to speculate that reversing spatial valences in order to contrive a feminine gaze may not be culturally possible. If the concept of the gaze is to be useful and liberatory, she asserts, then we must break the continuities between the gaze, power and masculinity. She suggests a third model of the gaze that positions it as part of a field with flexible geographic boundaries delimited by continual gender play:

> To my mind, the gaze denotes not an instrument of the already powerful, nor a field open to infinite play, but instead a graphic illustration of what happens when the genders come together. It is an illustration of the complementary shapes drawn for their subjectivities by culture, an indication of the mutability of gender shaping, and a place to test, manipulate and attempt to reform both the form of the genders and the relation between them. In this depiction, the gaze appears as the middle area of a tripartite space, a flexible boundary taking shape in response to the pressures exerted by either side. The gaze indicates the contours they already possess, but also can be a medium for one side to effect changes in the shape of the other.
>
> (Kirby, 1996: 127)

Kirby's reconfiguration of the gaze with flexible and indifferent boundaries has important implications for understanding streetscapes and male hysteria. In a moment, we will show how important the conceptualisation of subject boundaries is to understanding male hysteria, but for now we need to elaborate more fully on relations between the mobile gaze, 'streetwalking' and cinematic space.

Streetscapes and the mobile gaze

Guilliana Bruno (1993) argues that women struggle for a mobile gaze in attempts to create indifferent transitory boundaries. None the less, she asserts that the *flâneur* is traditionally and etymologically a male loiterer and gazer:

> A female equivalent was made impossible by the division of sexual realms that restricted female mobility and confined women to the private space. As a result, the 'peripatetic' gaze of the *flâneur* is a position that a woman has had to struggle to acquire and to liberate from its connotations of social ostracism and danger.
>
> (Bruno, 1993: 50)

The myth of the urban street is informed, for the most part, by the masculine power inherent in Kirby's characterization of projectile and reversible gazes. The mean streets are expressions of the masculinities that perpetuate the urban wastelands of our postmodern cityscapes: Gang violence, drug deals, homelessness, graffiti, rap and metal music reign in a conflicted hegemonic masculinity. Within this configuration of the street, women are the objects of the *flâneur's* gaze. If they roam the streets, they are often typecast as streetwalkers, prostitutes and objects of consumption similar to objects viewed in arcade windows. The respectable *flâneuse* was not possible until she was able to roam the city on her own and this did not occur until the first mobilization of the gaze when shopping became a socially acceptable leisure activity for bourgeois women.[4] Film also helped to liberate the female gaze by providing a public space in which she could 'renegotiate, on the new terrain of intersubjectivity, the configuration of private/public [by experience of the] erotics of darkness and [urban] wandering denied to the female subject' (Bruno, 1993: 51). Bruno argues that a gaze mobilized through cinema accommodates difference and liberates sexuality through the practice of transgressive, third spaces. Transgressive spaces do more than allow avenues for the contestation of the gaze: To borrow Kirby's arguments, these spaces offer indifferent boundaries between external and internal space where genders come together and change one another which, in turn, allows the gaze to be reconstituted as a flexible medium.

Although political identity may be constituted from a gaze that can liberate through the practice of transgressive spaces, representations often are girded by powerful narrative devices that find form in myths whose purpose, according to Barthes, is to translate histories and geographies into something 'natural'. These stories are naturalized to the extent that we suspend our disbelief of the context within which they are experienced. At this point, we need to address the ways in which mythic archetypes naturalise and culturally inscribe prototypical masculinities.

Fragmented myths and male hysteria

We argued earlier that archetypes naturalise men's actions through the obsessive repetition of the *mise-en-abyme*. We now take this further by suggesting that archetypes, in an important sense, are institutionalised myths which defer male hysteria by naturalising it as heroic. This proposition needs some justification that we think it finds in Diane Elam's (1994) fleshing out of the relations between subjects and objects in the *mise-en-abyme*. Elam argues that within the *mise-en-abyme*, the relationship between the part (the specific actions that make an archetype) and the whole (what constitutes an archetype) is inverted: On the one hand, the object cannot be grasped by the subject and, on the other hand, it produces a parallel regression in the subject or viewer of the *mise-en-abyme*. According to Elam, the

> mise-en-abyme . . . opens a spiral of infinite regression in representation. . . . The subject and object infinitely change places within the mise-en-abyme; there is no set sender or receiver of the representation. The infinitely receding object in the mise-en-abyme closes down the possibility of a stable subject/object relation. On the one hand, the object cannot be grasped by the subject; it slips away into infinity.
>
> (Elam, 1994: 27–8)

If naturalisation through obsessive repetition is a large part of the role of the mythic hero, then we need to consider more fully why this role is needed.[5] In so doing, we suggest that archetypes are useful to patriarchal hegemony because they defer male hysteria.

If we assume for a moment that hysteria is a socially constructed phenomenon that is 'seen' as deviant from a norm rather than found through physiological or psychological diagnoses, then it is not difficult to argue that the gaze controls the form taken by femininities and masculinities. Femininities and masculinities are defined in relation to a mythic hegemonic masculinity and when women or men refuse to accept these definitions they may be denounced as 'hysterical' (Moore, 1988). This argument positions hysteria as a bodily expression and reaction to oppression by patriarchal society and the social roles assigned to individuals (Showalter, 1993). We argue that institutionalised forms of male hysteria also open doors for hegemonic masculinity to express itself without condemnation. Institutionalised male hysteria takes on the form of territorial games such as sports, wars (including the 'war on drugs' or the 'war on poverty'), hostile corporate takeovers, gang turf battles and graffiti. James McBride (1995) suggests that these forms of hysteria are

> the product of unresolved desire for and aggression against the mother – conflicts that are too dangerous to express in the childhood home. . . . Because war and sports seem to be common enterprise to most men, indeed, a defining moment in what it means to be a man in our culture, participation in territorial games is not looked upon as being aberrant. In a patriarchal social order the very opposite is assumed true! The refusal to partake of institutionalized male hysteria, with its attendant emotionality, violence, sentimentality, and romanticism, is taken as evidence of effeminate, that is, deviant, behavior.
>
> (McBride, 1995: 179)

Institutional forms of hysteria often maintain and allow forums of naturalised and acceptable expressions of the desire for the phallic mother and against castration anxiety.

Etymologically, hysteria derives from the Greek *hystera* which means 'uterus' (McBride, 1995). Egyptians and Greeks first diagnosed hysteria as the flight of the uterus away from the female body. Alternatively, male hysteria is the flight of the phallus. Male hysteria is about emasculation in the sociological sense because men become separated from patriarchal gender roles. It is a physical and psychological expression of a man's sexual and social repression by patriarchy and hegemonic masculinity. To a large extent, then, male hysteria is part of the Oedipal crisis because it comprises rivalry with the father (hegemonic masculinity and patriarchy) and fear of castration (emasculation and removal from the social interactions that constitute and construct masculinity). This crisis is also over the fear/desire of the phallic mother who must be controlled, dominated and marked with signifiers of patriarchy.

Hegemonic masculinity is encompassed within male myths and thus defers the liberatory potential of male hysteria. Part of the cinematic genius of Gilliam is his ability to caricature male hysteria within mythologies while at the same time constituting transgressive spaces wherein the male gaze is challenged. We argue that he does this in part through the *mise-en-scène* of the street.

OF HEROES, FOOLS, AND FISHER KINGS

Homelessness, mental illness, intellectual curiosity, foolishness and loneliness live side by side on Gilliam's streets but none of his films are about intellectuals, the homeless or the mentally ill, nor does he propose a political solution to the social ills he so graphically depicts. Rather, Gilliam's films are about our need first to embrace the hysteria of the streets and, second, to respond with love and selflessness to those around us. In this sense, his work is liberatory in a humanistic way, but we argue that it may also be trapped for the most part in a hegemonic masculine myth that remains mute on the potential of masculinities beyond the gaze of patriarchy's heterosexual norm.

The street images in *Brazil* (1986) are a fantasy of 1920s and 1930s fascist architecture (see also Chapter 2 of this volume), post-Fordist consumerism and late industrial environmental degradation. The characters are also fantasies that are complexly woven into the day-dreaming of the central protagonist, Sam Lowry. In the opening sequence, dreamlike clouds fill the screen and the soundtrack plays a few bars of the title track 'Brazil'. This sequence is part of Sam's day-dream of flying through the clouds as a winged warrior-hero in silver armour on a quest to save a helpless maiden. Sam is a clerk in an Orwellian Ministry of Information doomed to an unheroic existence in the crazed and chaotic bureaucratic consumer capitalism represented in *Brazil*. Sam's ambitionlessness is more than made up for by his wealthy dowager mother whose two main concerns in life are surgically altering her looks to appear younger and advancing her son's career (Glass, 1986). Life's everyday tedium is transgressed when Sam meets the first of many father-figures in the form of Harry Tuttle, a heating duct repairman and 'freelance subversive'. Tuttle monitors Sam's hysterical call for help when his air-conditioner breaks down. The chaotic heating ducts that are ubiquitous in *Brazil* symbolise an apparent connection between individuals and an oppressive and disorganised society; when Tuttle opens a panel in Sam's apartment the duct system moves and sounds like some leviathan respiratory system. Tuttle starts working on the needed repair while explaining the clandestine nature of his work: 'Listen, this old system of yours could be on fire and I couldn't even turn on the kitchen tap without filling out a 27b/6 . . . bloody paperwork!' When Sam asks Tuttle who he is and what he is about, Tuttle starts humming the tune 'Brazil', giving the first indication of the movie's liberatory signifier. In a stunning exit that presages Tim Burton's *Batman* (1989), Tuttle hooks up an aerial runway and slides to the streets far below Sam's apartment.

When Sam sees Jill Klayton, the apparent object of his day-dreaming, and she turns out to be a suspected terrorist, he decides to save her at his own peril. Jill is trying to find out what happened to a neighbour, Mr Buttle, who was 'deleted, nonoperationalised, excised' by the forces of Information Retrieval. Informational Retrieval is a secret police force of the Ministry of Information that terminated Mr Buttle by mistake (they were meant to go after Tuttle). To save his paranoid and scheming supervisor's reputation, Sam takes a reimbursement check to Mrs Buttle. After seeing Jill, Sam discovers that her file is in the hands of 'the boys from Information Retrieval'. Sam uses his mother's influence to get a new job that will give him the clearance he needs to get the file. He eventually 'saves' Jill by erasing her file from the archaic computer in the Ministry of Information but, in so doing, he has to pay the price of his heroism. The ensuing narrative of Sam's fall from grace with this Orwellian society is interspersed with his flights of heroic fancy.

Sam's fantasy begins with him flying, Icarus-like, over a bucolic pastoral landscape. Jill is in the clouds and he swoops over for a gentle kiss. Later, giant skyscraper-like monoliths thrust up from the ground to fragment the rural landscape. In the interstices of the monoliths, a dark nightmarish streetscape evolves. A giant robotic samurai warrior bars Sam's way to an opportunity to rescue Jill who is now in an (Weberian?) iron cage suspended above the street. The samurai bears a close resemblance in the way it appears and imposes to the red knight in Gilliam's *The Fisher King*. The knight and samurai are representations of absent fathers in a scrambled but recognisable Oedipal battle. The samurai slices Sam's wings off with his sword thus ending his ability to soar. Unable to fly, Sam grabs one of the ropes tethered to Jill's cage as it floats off but giant hands and the cobble-stone face of his supervisor emerge from the street to grab his lower body; the street literally holds him down. In order to fulfil his quest, Sam the warrior-hero is neither defeated by the samurai nor betrayed by the maiden/mother-figure, but rather he is emasculated by the street upon which he stands. For Gilliam, the streets are a mythic embodiment of an emasculating corporate bureaucracy and, in conjunction with the samurai, male patriarchal dominance. When Sam lifts up the helmet of the samurai he sees his own face and, by so doing, he discovers his own participation in this Oedipal drama.

Brazil's oppressive street space in conjunction with the melding of past, present and future technologies provides a clue to deciphering how Gilliam's Oedipal drama is played out. The workplaces and homes are parodies of new technology (everything is automated but nothing works properly), old-fashioned heating ducts and pneumatic communication tubes in an impossible syncretism. The car Sam drives to deliver Mrs. Buttle's refund check is built from parts of a Second World War Messerschmitt. The slogans and posters – 'Suspicion Breeds Confidence' and 'Don't Suspect Your Neighbor – Report Him' are fashioned after British wartime originals. In front of the Ministry of Information, a teacher slaps a schoolboy in front of a statue of an iron eagle embossed at its base with the CIA slogan 'The Truth Shall Make You Free'.

In an act of resistance against this oppressive landscape, Sam destroys Jill's file and in a wonderful Oedipal reversal tells her that he has killed Jill Klayton in the office of the deputy prime minister to which she replies, 'Care for a little necrophilia? ... Hmmm?' The morning after, their lovemaking is disturbed by the police of Information Retrieval who burst through the walls and ceiling to seize Sam. During the ensuing interrogation (the samurai figure is replaced by a torturer father-figure wearing a baby-faced mask), Sam is rescued by Tuttle and his fellow terrorists.[6] Sam escapes with the terrorists and then witnesses Tuttle and what remains of the rescue team killing his torturer and blowing up the Ministry of Information. As Sam follows Tuttle's flight through a consumerist streetscape of glaring lights and arcades, Tuttle is overwhelmed by flying memos and pieces of paper. Sam scrambles to rescue

Tuttle from the chaos of paper but by the time he tears his way through, Tuttle has disappeared, apparently devoured by the amassed memos and documentation. With the demise of his inspirational father-figure, Sam's torturer father-figure will soon reappear but first Sam must deal with the maiden/mother-figure.

Sam's mother's plastic surgery ultimately regresses her into a young woman and, at a party for her departed best friend whose similar surgical treatments ended with tragic complications, she is surrounded by adoring young men. In a penultimate fantasy (although by this time fantasy and reality are difficult to disentangle), Sam runs hysterically to her shouting 'Mother!' and as she turns we see the face of Jill. The forces of Information Retrieval burst into the funeral parlour, Sam's hysterics escalate and he topples over the coffin of his mother's friend. A glutinous mass of bones and gore slips out of the coffin and onto the floor as Sam looses his balance and falls through the coffin/womb. He lands in the street where he fought the samurai. This time Sam is chased by death creatures (wearing baby-faced masks) over a pile of heating ducts to a blank wall where he finds a door into a prefabricated house that turns out to be the same one Jill was delivering on her truck earlier in the movie. At that earlier time, while pursued by police from Information Retrieval, Jill and Sam drove the truck down a street lined with billboards that are enlarged to the extent that it is almost impossible to see the devastated industrial wasteland beyond. This street is symbolic of Sam's fantasies that myopically lead him on to seeming liberation. We will return to the importance of this streetscape in a moment, but for now we need to take Sam's story to its conclusion. Jill starts the engine and drives off with Sam in the house on the back of her trailer. In the next shot it is morning and Jill's rig is parked in a bucolic countryside; the prefabricated house is now in place with a white picket fence. With this suggestion that Sam's heroic fantasy has come to its Oedipal conclusion, the image flips to the torturer/father-figure peering into Sam's bulging eyes: 'He got away from us.' The daring rescue and escape with Tuttle is another fantasy: Sam's tormentor and his assistant are alive while Sam escapes to a state of vegetative catalepsy.

The world created in *Brazil* is interpreted by Fred Glass (1986) to be a conjoining of the Frankfurt School's classic analysis of fascism and late capitalism with Freud's fears about human potential in history. He notes that the film unites an understanding of the crisis of capitalism around bureaucratization, unemployment and post-Fordist consumerism on the one hand, and the psychological responses called into being by these larger structural forces:

> Either the possibilities to change things, both for the individual and society, are nil; or – and I think the preponderant weight of *Brazil*'s evidence rests in this direction – change is possible only if it is attempted outside the ideological terms dictated by oppressive social structures. This means, with Lowry's fate the alternative, steering clear of the ever-present temptation to fantasize one's way toward a solution. ... The last scene, lingering on through the credits, sums up the argument of the film. The visual presentation of Sam dwarfed by the immense torture chamber, the room filling

with the clouds of his . . . dreams, graphically charts the individual's capit-
ulation to the torture chamber of society, even – especially – in the very
moment of attempting a private escape through fantasy.

(Glass, 1986: 22)

Gilliam draws from the heroic myths of Daedalus, Icarus, King Laius and
Oedipus but his genius highlights the creation of political identities by mapping
fascist imagery and state icons over the myths. Sam's silver armour and angelic wings
are the images of Nazi propaganda. The point here is that fantasy is no escape from
oppression when it is tightly constrained by the myths that are shaped by that oppression. Nonetheless, *Brazil's* sombre worldview is countered with Tuttle's humming
and the opening and closing soundtrack where we find the only references to Brazil.
The festival suggested by the song 'Brazil' opens a tenuous conduit to everyday
carnival that cannot be superimposed by mythic patriarchal structures. Sam's resistance to the social norms of the Orwellian landscapes in *Brazil* is well-intended but
wrongly directed through mythic fantasies that are as myopic as the billboarded
street that Jill and Sam use to escape the Information Retrieval police. Sam is blinded
to the moral and environmental decay beyond the billboards and the path of the
street is directed by the mechanisms of state oppression. As Fred Glass points out,
the utopian promise of festival implied by the song 'Brazil' is completely absent
from *Brazil* because, in many ways, the movie defines the 'other' by caricaturing
what it is not.[7] By so doing, Gilliam highlights how myths – and the moribund
fantasies that draw their strength from them – offer few possibilities for resisting
the political and cultural oppression of late capitalism.

The song that begins Gilliam's *The Fisher King* (1991) – 'Hit the Road Jack' –
is a spatial metaphor for the journey that Jack Lucas undertakes on the streets of
New York City. Jack is a call-in radio star, a Howard Stern-like shock jock whose
'taunting rant on the airwaves is like the city's soul sick mantra' (Rainer, 1991: 1).
Sitting in the bathtub of his luxury apartment far above Manhattan's streets Jack
calls out 'forgive me' in a contrived ritual of self-aggrandisement. Ironically, a search
for forgiveness will embody his grail quest on the mean streets below. At the beginning of the movie, Jack is the fisher king drunk on visions of power and glory.[8]
Jack's journey begins after a caller, John Babbit, takes Jack's anti-yuppie tirade literally and executes a gruesome mass murder in a Manhattan restaurant. The shock of
the incident is the wound through which Jack looses his creative edge: It throws
him out of his high-priced luxury sarcasm and into lower street life cynicism.[9] Three
years later, Jack is a jobless self-absorbed drunk living with Anne, a jaded and needy
owner of a video shop. His attempt at suicide by jumping into the watery womb of
the East River begins the trials that he must go through to return to hegemonic
masculinity. As he gets ready to jump, two street thugs arrive to defend their neighbourhood's integrity and, labelling Jack as a homosexual street bum, soak him in
gasoline in preparation for sacrifice. As they begin their ritualised foreplay, Parry
appears and cautions the boys to yield or 'feel the sting of my shaft'.[10] Without
waiting for a reply, Parry's bow-strung suction-cupped arrow hits one of the youths

between the legs. Parry sends the fear of castration anxiety into the heart of their ritual; he destabilises the institutional hysteria, creating a transgressive space in which hegemonic and effeminate hysteria perform a street battle. Parry's poetic banter and dishevelled knightly garb help set the tone for the superimposition of a medieval street topography. Surrounded by the citadel architecture of bridge pylons, the street space becomes enlivened with the hysteria of homelessness as Parry and his friends rescue Jack and take him back to their serf-like village. Jack burns his hand and, by so doing, contextualises Gilliam's version of the fisher king story. The villagers clap in delight – he has fallen and is wounded as they have fallen and are wounded – and as a ritual of his acceptance to the street he is baptised with a brown-bag-bottle-of-booze bath. The streets are filled with madness, but it is a naturalised reality that thrives in medieval grunge and reflects back the crisis of modern-day homelessness. Almost exactly the same kind of landscapes are created in Gilliam's *12 Monkeys* (1995), where homelessness and medieval grunge are combined in a street festival that includes children and a soap-box prophet. With these kinds of representations, Gilliam plays hysteria as 'a historical body in trouble with its representation of itself and its historical place' (Corrigan, 1991: 143). Later, Jack awakens in the basement of an apartment block in Parry's home/boiler room (his wound is so deep that he lives under the streets) that bears close resemblance to the heating duct-bedraggled apartments in *Brazil*. In the future society of *12 Monkeys,* everybody lives in these kinds of environments because a deadly virus has made the streets unsafe for humans. The quest of James Cole, the hero in *12 Monkeys,* is to return to 1996 to discover how the virus was let loose. Similarly, Parry's quest is to get back something that God lost: the holy grail which conveniently resides in a 'castle' on Fifth Avenue belonging to a financial entrepreneur, London Carmichael.

When Jack flees from Parry's hysterical abode, he runs into the landlord of the apartment who explains that Parry's psychological wounds were caused by the death of his wife at the hands of Babbit. A little later Jack tells Anne, 'I feel cursed. Like a magnet, I attract shit. I wish there was some way I could just pay the fine and go home.' Jack's holy grail quest becomes a struggle to help Parry return to some form of sanity. Back on the streets, Jack finds Parry sitting on top of a car in the financial district. Parry gazes up at a clock on a high-rise awaiting the arrival of the object of his unrequited love, Lydia. Parry, outfitted with a monkish robe and red ear-flapped hat, is in all appearance the errant medieval knight at home in the towering 'vision of New York as medieval fortress' (Rainer, 1991: 1). Jack tries to 'pay the fine' to Parry so that he can go home, but Parry gives the money to another mad street hysteric. Parry now imposes upon Jack his own quest: to retrieve the grail from the castle on Fifth Avenue.

By joining the two fisher kings' quests, Gilliam resolves the myth of male bonding in a problematic buddy movie formulation (see Aitken and Zonn, 1993; Fuchs, 1993; Silverman, 1988). Gilliam fixes the gaze in a contrived and monolithic patriarchal space: Lydia is the love inspired in the hero/fool Parry; Anne is needy of Jack's love and attention. The potential for latent homosexuality in the relationship between Jack and Parry is foiled by the movie's emphasis on heterosexuality. Occasional asides to a transgressive homosexual space are made with nude 'cloud-busting moons' in central park and a wonderfully hysteric segment when Jack and Parry get a schizophrenic drag queen to deliver an Ethel Merman style singing telegram to Lydia.[11] When Parry eventually ends up catatonic and back in the mental hospital, he cannot be awakened by Lydia's kiss. Rather, Jack must consummate their quest by retrieving the grail and, by so doing, he revives Parry to marry Lydia. At the same time, Jack is freed to confess his love to Anne. Highlighting its pagan origins, seeking the grail is not only a search for redemption but also a quest for the mythic safety of the womb.

Parry's fantasies take on the hysteria of Sam Lowry in the latter half of *Brazil*. His street-bound nightmare is constituted by a red knight breathing fire and sitting astride a ribbon adorned charger. Flashbacks to the murder of Parry's wife draw connections between this medieval phantasm and the graphic massacre in the restaurant. When Parry says goodnight to Lydia after a date prearranged by Jack, the red knight is waiting for him on the street. Parry runs hysterically, but he cannot escape the clutches of this wound/womb. The chase ends under the bridge on the East River where Jack's journey began; the image of the red knight is superseded by the same street thugs who tried to torch Jack. Parry, now marked as an effeminate hysteric, is beaten into a catatonic stupor.

In the mental hospital, Jack, now resigned to stealing the grail, whispers to Parry, 'If I do this I want you to know that it wouldn't be because I felt I had to, because I felt cursed or guilty or responsible or anything . . . just because I want to do this for you.' Jack's healing of the wound – transcendence through the womb – must be through his fears and through a completely selfless act. Dressed as the errant knight, Jack scales the fortress walls of the Fifth Avenue castle. He enters a tower window and begins descending down a set of spiral stairs beside walls draped in red velvet. Light from a blazing fire fills the room at the bottom of the tower. The recovery of the grail is a ritualised decent into the womb to recover manhood. Ironically, when he finds the grail he also finds its owner, London Carmichael, prostrate on a chair. As a symbol of the height of materialistic and patriarchal capitalism, Carmichael now reifies the crisis of hegemonic masculinity in an attempted suicide. If the recover of the grail by the fisher king is strong enough to heal Jack and Parry, it also is strong enough to heal hegemonic masculinity: Jack purposefully trips the burglar alarm on his way out thus saving Carmichael. This finale may reinforce the need for masculinity to accept its own institutionalised hysteria as madness but if anything it highlights hegemonic masculinity rather than opening up transgressive spaces for liberatory masculinities.

Where *Brazil* is about the uselessness of fantasy and myth in an oppressive society, *The Fisher King* seems to reify those myths as the basis of masculine redemption. Alternatively Pat Dowell (1991: 47) suggests that *The Fisher King* 'points to a possible trajectory for sorting out the war between men and women in its most personal phase, suggesting that heterosexuality may not succeed in relationships until each gender helps itself to find conscience and courage.' Although this may be so, we are troubled by the way that Gilliam seems to endorse in *The Fisher King* the myths he exposes as problematic in *Brazil*. Ultimately, the complex characters in *The*

Fisher King are reduced to caricatured outcomes from a chivalric code. None the less, Gilliam spares nothing of what he sees of the chaos and sorrow in the streets of contemporary Manhattan. He constructs a *mise-en-scène* that highlights the crisis of masculinity embodied in hysteria. *12 Monkeys* takes this setting for male hysteria further by never reconciling whether James Cole is actually from the future or whether he is completely psychotic. In one of the best sequences of the movie, Cole's doctor tries to convince him that he is from the future while Cole, who has now 'become addicted to that dying world', prefers to stay with her and be mad rather than return to the subterranean world of the future. The future in *12 Monkeys* is not only below the streets but it also resembles the mishmash of old and new technologies that Gilliam created in *Brazil*. One of Cole's returns to the future world is opened with his revival from time-travel and a doctor peering into his face in a sequence that bears a strong resemblance to the torturing of Sam Lowry in *Brazil*. None the less, the conflation of mythic fantasy and overbearing society are played out at street level to the extent that several myths of masculinity can be interrogated through all three of Gilliam's movies. Like the interrogation of Sam Lowry, Gilliam's narrative conclusions on masculinity may be more twisted and convoluted in *The Fisher King* and *12 Monkeys* than they first appear. In part, the energy of maleness is violated and deadened by an overbearing use of fantasy but, with this, Gilliam's real genius is revealed through his ironic distancing of myth not only from his narrative, but also from his representation of streetscapes and his character development.

CONCLUSION

Encompassed within hysteria are many forms of political identity and so we use hysteria in this essay not just as a single psychoanalytic category but also as an expression of multiple identities. We argue that these identities are imagined well, but not unproblematically, in the gazes and streetscapes created by Gilliam. Whether represented as fascist monuments (*Brazil*) or medieval citadels (*The Fisher King, 12 Monkeys*), Gilliam's streetscapes are integral plot devices, maybe even actors, in a Brechtian/ Wagnerian apotheosis of male hysteria. It is ambiguity, parody and ironic distance that work best for Gilliam's representations of the cultural logic of late capitalism. In the sense that he creates a *mise-en-scène* for male hysteria, Gilliam's representations of the street construct transgressive spaces that offer no political solutions but simply highlight an interpersonal space of selflessness, gentleness and passion. Institutionalised male hysteria, on the other hand, is represented as fear and a power that requires masculinity and femininity to be immutable, monolithic and coded in the mythic sexuality that creates heroes.

The liberatory masculinity in Gilliam's work is not based on sexual identity *per se* but rather, perhaps like the work of Gus Van Sant, it focuses on how the gentleness of interpersonal space can usurp the hard edge of hegemonic space. Like Mike's escape into narcolepsy in Van Sant's *My Own Private Idaho*, Sam Lowry's rural, bucolic fantasies and Parry's fantasy of the red knight are about the search for mother love. But this is not a simple Freudian scramble out from under repression. Van Sant's penchant for representing liberatory sexual identities in movies like *My Own Private Idaho* is through a subtle engineering of political identity, but he also focuses on relationships and it is here that he and Gilliam are able to expose a transgressive space.

Returning as we close to *Independence Day,* it seems clear that this blockbuster movie highlights mythic constructions of institutionalised male hysteria while deferring multiple meanings of maleness in favour of a monolithic patriarchal male political identity. The aliens are not battered drug addicts, homeless people, homosexuals, prostitutes or 'illegal aliens', the normal subjects of institutionalised male hysteria; rather they are slimy, smart, thoroughly evil and worthy of our wrath. By deferring to an acceptable institutionalised hysteria, *Independence Day* enables a grandiose display of heroic male agitation: an African American topgun pilot turned down by NASA kicks alien butt; an alien abductee turned alcoholic gives up his self-pity and fear to fly against his tormentors; the President leads the final cavalry charge piloting an F-14. Male hysteria is made mythic and depoliticised by the fractal geometries created through obsessive repetition: We embrace the heroes as if they were a natural part of our culture. *Independence Day* is ultimately upbeat and optimistic. It ends with all the human survivors cheering and hugging as the mighty alien spaceships tumble to earth. No-one seems to be affected by the fact that most people they knew are dead. The sad face on the son of the alcoholic who sacrificed himself is transformed by a much too big smile when he is told that his father is a hero. The heroes and their actions are unbelievable caricatures and yet we are willing to suspend our disbelief in part because the movie's plot offers a safe foray into the myths that darken our spirit and highlight our own cultural hysteria. This narrative solution is disturbing to us because it maintains a silence about the streets that bare these myths. If the streets are places where hegemonic masculinity is inscribed and formed then we need a forum through which it can be challenged.

NOTES

1 Elam illustrates the *mise-en-abyme* by using the example of the man on the Quaker Oats box who holds an identical box of Quaker Oats. The representation folds into itself ad infinitum in the same way fractal clusters generate exactly the same patterns at higher levels of resolution.

2 Audrey Kobayashi (1994: 77) notes that the power of naturalised and essentialised categories stems in part from an ability to build upon phenotypical and genotypical attributes as if they were simple, natural and unviolable. Our focus on masculinity in this essay is not narrowly defined by the male body but rather recognises different masculinities and the power relations and alliances that comprise these masculinities.

3 This model derives from the work of Michel Foucault who, in *Discipline and Punish* (1979), invoked Jeremy Benthan's panopticon device (1791) to suggest that the power of surveillance transcends the power of spectacle. The Panopticon was a twelve-sided eighteenth century prison made of iron and glass. The jailer was positioned in a central tower where he had a panoramic view of separate peripheral cells. This view granted the jailer visual power and domination over the inmate who suffered a sense of disciplined surveillance.

4 To make this argument, critical theorists such as Anne Friedberg (1993) trace the sense of voyeurism embodied by the panopticon to other visual devices of the eighteenth and nineteenth centuries such as the panorama, the diorama, and phantasmagorias. These devices allowed people either to walk through or past spectacles wherein all time and space is lost and 'the real' is presented. The origins of the (male) gaze are found in these apparatuses but the initial mobilisation of the gaze, Friedberg argues, is derived from the creation of *flânerie* (see Aitken, 1997).

5 A current renaissance of men seeking the roots of masculinity has lead some 'feminist activists and academics [to] confess a nervousness and even a visceral sense of danger about the resurgent androcentric consciousness in a patriarchal social order' (McBride, 1995: 210). The base of much mythopoetic writing, for example, suggests that we live in an age without myths to guide men towards their manhood and, consequently, men need to colonise primitive rituals from other cultures in the hopes of bringing the mythic male back to modern society (Betcher and Pollack, 1993; Bly, 1990; Keen, 1991). By conceiving our society as bereft of mythological thinking and attempting to reappropriate male arche-types through ritual, mythopoetic writing often constructs either a largely womanless past or it falls into the trap of blaming *the* mother (see Aitken, 1998).

6 The attacks by Tuttle's forces and those of the Information Retrieval that occur throughout the movie bear a strong resemblance to those of the contemporary British SAS as well as parodying the central element (terrifying scenes of police bursting through doors) of Ben Shahn's Second World War propaganda poster 'This is Nazi Brutality' for the British Office of War Information (Glass, 1986; Rubenstein, 1986).

7 In Jack Matthew's book *The Battle of Brazil* (1987) Gilliam explains where the title stemmed from, while he was in Port Talbot, Wales: 'Port Talbot is a steel town, where everything is covered with gray iron ore dust. Even the beach is completely littered with dust, it's just black. The sun was setting, and it was quite beautiful. The contrast was extraordinary, I had this image of a guy sitting there on this dingy beach with a portable radio, tuning in these strange Latin escapist songs like "Brazil". The music transported him somehow and made his world less gray.' (1987: vii–ix).

8 There are several different variations of the Arthurial legend and the fisher king story, some involving the holy grail and others not. All speak of a young king receiving a terrible wound that cannot heal. Gilliam's version is told by Parry in the middle of the movie:

> It begins with the king as a boy, having to spend the night in the forest to prove that he is a king. And while he is spending the night alone he is visited by a sacred vision. Out of the fire appears the holy grail, the symbol of God's divine grace. And the voice said to the boy, 'You shall be keeper of the grail so you can heal the hearts of men.' But the boy was blinded by a greater vision of power and glory and beauty. And in this state of radical amazement he felt for a brief moment not like a boy but invincible, like God. So he reached in the fire to take the grail, and the grail vanished. Leaving him with his hand in the fire to be terribly wounded. Now as this boy grew older his wound grew deeper. Until one day, life for him lost its reason. He had no faith in any men, not even himself. He couldn't love, or feel loved. He was sick with experience. He began to die. One day a fool wandered in the castle and found the king alone. And being a fool he was simple minded, he didn't see a king, he only saw a man alone and in pain. And he asked the king, 'What ails you friend?' The king replied, 'I'm thirsty, I need some water to cool my throat.' So the fool took a cup from beside his bed and filled it with water and handed it to the king. And as the king began to drink he realised that his would was healed. And he looked in his hands and there was the holy grail, that which he sought all his life. He turned to the fool and said with amazement, 'How could you

find that which my brightest and bravest could not?' And the fool replied, 'I don't know, I only knew that you were thirsty.'

Prior to its appropriation by Christendom in the twelfth century, the grail (in the movie, Anne calls it 'Jesus' juice glass') was thought to be a sacred cauldron symbolising the womb. In some grail stories the wounded king's original sin is rape (Dowell, 1991: 46).

9 According to most legends, the fisher king's would is in the male, generative, creative part of a man's being (Johnson, 1993).

10 Parry is an alias for Parsifal which is phallic in origin meaning 'spearman' or 'piercer of the valley'. Parsifal was a knight at Arthur's round table and a failed grail seeker. The gradual transformation of his character through Christian influences in the twelfth and thirteenth centuries completes a transition of the grail from pagan and female to Christian and male. Pat Dowell (1991: 46) points out that Parsifal helps achieve the masculine usurpation of feminine religious power.

11 Gilliam also uses a singing telegram in *Brazil* to invite Sam to his mother's party to celebrate the completion of her recent plastic surgery.

REFERENCES

Aitken, Stuart. (1998) *Family Fantasies and Community Space*. Rutgers: Rutgers University Press.

Aitken, Stuart. (1997) Contesting the Mobilized Virtual Gaze: A Review of Anne Friedberg's Window Shopping. *Environment and Planning D: Society and Space* 15, 1: 113–14.

Aitken, Stuart C. and Leo Zonn. (1993) Weir(d) Sex: Representation of Gender–Environment Relations in Peter Weir's *Picnic at Hanging Rock* and *Gallipoli*. *Environment and Planning D: Society and Space* 11: 191–212.

Aitken, Stuart C. and Leo Zonn. (1994) Re-Presenting the Place Pastiche. In *Place, Power, Situation and Spectacle: A Geography of Film*, eds Stuart Aitken, and Leo Zonn, pp. 1–18. Totowa, NJ: Rowan and Littlefield.

Aitken, Stuart C. and Chris Lukinbeal. (1997) Disassociated Masculinities and Geographies of the Road. In *The Road Movie Book*, eds Inar Rae Hark and Steve Cohan, pp. 349–70. London: Routledge.

Anderson, Stanford. (1978) People in the Physical Environment: The Urban Ecology of Streets. In *On Streets*, ed. Stanford Anderson, pp. 1–12. London: MIT Press.

Barthes, Roland. (1972) *Mythologies*. New York: Hill and Wang.

Betcher, William and William S. Pollack. (1993) *In a Time of Fallen Heroes: The Re-Creation of Masculinity*. New York: Guilford Press.

Billson, Anne. (1991) *The Fisher King*, a film review. *New Statesman and Society*, 8 November: 30.

Bjørnerud, Andreas. (1996) Beckett's Model of Masculinity: Male Hysteria. In *Not I*, eds Lois Oppenheim and Marius Buning, pp. 27–35. Madison: Fairleigh Dickinson University Press.

Bly, Robert. (1990) *Iron John: A Book About Men*. New York: Random House.

Bruno, Giuliana. (1993) *Street Walking on a Ruined Map: Cultural Theory and the City Films of Elvira Notari*. Princeton: Princeton University Press.

Caput, Mary. (1994) *Voluptuous Yearnings: A Feminist Theory of the Obscene*. Lanham, MD: Rowan and Littlefield.

Corrigan, Timothy. (1991) *A Cinema Without Walls: Movies and Culture After Vietnam*. New Brunswick, NJ: Rutger University Press.

Denby, David. (1989) *Drugstore Cowboy*, a movie review. *New York Times*, 10 October.

Denby, David. (1991) *My Own Private Idaho*, a movie review. *New York Times*, 10 July, 79.

Dowell, Pat. (1991) *The Fisher King*, a film review. *Cineaste* 18(4): 46.

Driscoll, Mark. (1994) James Bond and Immanuel Kant's War on Drugs: A Nosography and Nosegrammatics of Male Hysteria. In *Body Politics: Disease, Desire, and the Family*, eds Michael Ryan and Avery Gordon, pp. 104–14. Boulder, Colo.: Westview Press.

Elam, Diane. (1994) *Feminism and Deconstructionism: Ms. En Abyme*. New York: Routledge.

Ford, Larry. (1994) Sunshine And Shadow: Lighting and Color in the Depiction of Cities on Film. In *Place, Power, Situation and Spectacle: A Geography of Film*, eds Stuart C. Aitken and Leo Zonn, pp. 119–36. Totowa, NJ: Rowan and Littlefield.

Foucault, Michel. (1965) *Madness and Civilization: A History of Insanity in the Age of Reason*. New York: Random House.

Foucault, Michel. (1979) *Discipline and Punish*. New York: Vintage.

Friedberg, Anne. (1993) *Window Shopping: Cinema and the Postmodern*. Berkeley: University of California Press.

Fuchs, Cynthia. (1993) The Buddy Politic In *Screening the Male: Exploring Masculinities in Hollywood Cinema*, eds Steven Cohan and Ina Rae Hark. London: Routledge.

Glass, Fred. (1986) Review of Brazil. *Film Quarterly* 39(4): 22.

Gold, John R. (1984) The City in Film: A Bibliography. *Architecture Series: Bibliography*, August: 1–12. Monticello, IL: Vance Bibliographies.

Greenberg, Harvey R. (1992) Review of *My Own Private Idaho*. *Film Quarterly,* Fall: 23.

Holtan, O. (1971) Individualism, Alienation and the Search for Community: Urban Imagery in Recent American Films. *Journal of Popular Culture* 4, 933–42.

Johnson, Robert A. (1993) *The Fisher King and the Handless Maiden : Understanding the Wounded Feeling Function in Masculine and Feminine Psychology,* San Francisco: Harper.

Keen, Sam. (1991) *Fire in the Belly: On Being a Man*. New York: Bantom.

Kirby, Kathleen. (1996) *Indifferent Boundaries: Spatial Concepts of Human Subjectivity*. New York: The Guilford Press.

Kirby, Lynne. (1988) Male Hysteria and Early Cinema. *Camera Obscura* 17: 112–31.

Kobayashi, Audrey. (1994) Coloring the Field: Gender, 'Race,' and the Politics of Fieldwork. *The Professional Geographer* 46(1), 73–60.

Lefebvre, H. (1991) *The Production of Space*. Oxford: Blackwell.

Lukinbeal, Christopher and Christina Kennedy. (1993) Dick Tracy's Cityscape. *Association of Pacific Coast Geographers Yearbook* 55: 76–96.

Mason, M.S. (1991) *The Fisher King*: film review. *Christian Science Monitor*, 27 September.

Matthews, Jack (1987) *The Battle of Brazil*. New York: Crown Publishing.

McBride, James. (1995) War, Battering, and Other Sports: The Gulf Between American Men and Woman. New Jersey: Humanities Press.

Moore, Suzanne. (1988) Getting a Bit of the Other – the Pimps of Postmodernism. In *Male Order: Unwrapping Masculinity*, eds Rowena Chapman and Jonathan Rutherford, pp. 165–92. London: Lawrence & Wishart.

Mulvey, Laura. (1975) Visual Pleasure and Narrative Cinema. *Screen* 16(3): 6–18.

Natter, Wolfgang, and John Paul Jones. (1993) Pets or meat: Class, Ideology, and Space in *Roger and Me*. *Antipode* 25: 140–58.

O'Pray, Michael. (1989) *Drugstore Cowboy*, a movie review. Monthly Film Bulletin, December: 362.

Rainer, Peter. (1991) *The Fisher King*, a film review. *Los Angeles Times*, 20 September: Calendar, p. 1.

Robins, Kevin. (1996) *Into the Image: Culture and Politics in the Field of Vision*. London: Routledge

Rubenstein, Lenny. (1986). Review of *Brazil*. In *Cineaste*, 14(4): 48.

Saco, D. (1992) Masculinity as Signs: Poststructuralist Feminist Approaches to the Study of Gender. In *Men, Masculinity, and the Media*, ed. S. Craig. pp. 23–39. Newbury Park: Sage.

Showalter, Elaine. (1993) Hysteria, Feminism, and Gender. In *Hysteria Beyond Freud*, eds Sander Gilman, Helen King, Roy Porter, G.S. Rousseau and Elaine Showalter, pp. 286–344. Berkeley: University of California Press.

Silverman, Kaja. (1988) *The Acoustic Mirror: The Female Voice in Psychoanalysis and Cinema*. Bloomington: Indiana University Press.

Smith, Neil. (1993) Homeless/Global: Scaling Places. In *Mapping the Futures: Local Cultures, Global Change*, eds J. Bird, B. Curtis, T. Putman, G. Robertson and L. Tickner, pp. 87–119. London: Routledge.

Vineburg, Steve. (1990) *Drugstore Cowboy*, a movie review. *Film Quarterly*, Spring: 27.

11

THE STREET IN THE MAKING OF POPULAR GEOGRAPHICAL KNOWLEDGE

David Crouch

●

INTRODUCTION

The Street is an everyday site of geographical knowledge and leisure practice. In recent geographical literature, there is the impression that the street is spectacle, display and epitomised in the gaze, the *flâneur*. The mall has come to be seen as archetype of the turn-of-this-century street (Shields, 1992; Jackson and Thrift, 1995). These are important dimensions of what we know of the street, but perhaps they overstate particular dimensions of its imagery in terms of making sense of contemporary cultural practice. It is argued in this chapter that streets are themselves sites of cultural practices, and part of our knowledge of the city because they link sites of activity, of cultural practice, make escape possible and are a step to somewhere else, and someone else. They connect all sorts of sites of everyday cultural life, both spectacular and humble. People meet in the street, and they can avoid engagement in the street.

Cultural practices spill out onto the street from diverse corners of ordinary life; numerous sites of cultural practices are visible from the street, and are part of the image of the street. The limits of the street, similarly, merge into the spaces around them. Each of these practices and places contribute to the images of the street. These images emerge that are both embedded in cultural practice, and composed of dreams, fantasies, imaginations, departure and longing, and these themselves are included in cultural practices. Representations of such an everyday geographical knowledge include maps that depict, and themselves make, images of local space, where streets provide an imaginative force (Crouch and Matless, 1996).

In this chapter we explore distinctive practices in an effort to articulate ways of contemporary practice that make streets important in everyday lives. In particular, we explore ways in which streets need a reappraisal: a reappraisal of their position in contemporary popular culture; in the production–consumption debate; in terms of the production and transfer of representations in an especially commodified world – and of course not all of it is. An especial concern in this chapter is to see how people at large make sense of their lives by way of the street. In this, we examine practices and meanings in the empirical world, everyday life at street level. We look at ordinary, folk or popular geographical knowledges. It is necessary to unlock the political issues that surround the street, not simply as site of the mass, the demonstration and the carnival, or the solitary individual, all of which include the use of the street in expressing an interest, making claims on life, space, and issues, values. We discover the street in everyday life to be a site of claims, shared expression, of alternative values and practices in subtler ways, too. Our practices are happening in and along streets, in my and your neighbourhood every day and in ways often unobserved in street writing.

The places where these practices happen are bits of streets, themselves bits of the city, a focus of Sharon Zukin's attention in her recent analysis (Zukin, 1995). These practices and these bits develop an interpretation of images of the street that unsettles the Gaze and its kind of image-ination. We consider parallel discourses and make an effort towards a greater inclusivity of geographical knowledge in cultural practice. These discourses seek to situate the street in practice; to relate the gaze to other dimensions of being, with a glance at the body, so riskily, perhaps heretically, confronting Foucault with Merleau-Ponty (Foucault, 1981; Merleau-Ponty, 1962; Crossley, 1996). We go to one side of the consumer–producer relationship to discover people in the street; we refigure the making of geographical knowledge amongst the signs people make themselves, in their own everyday practice, and where they fit the signs of commodification. There is a profound issue of power, empowerment and contestation that surrounds these images.

There is a determined effort in each of these ways to attend to the empirical world, walking with discourses and stopping by at places people inhabit. In this turning to, and tuning into, the empirical world we make a persistent effort to check where everyday cultural practice may prompt our theoretical work. In each of these discourses, we map a contingent and fragmented geographical knowledge but one that is identifiably situated in people's own everyday lives as well as interpenetrated with significations that are found to be at once important, but perhaps semi-detached from the lives of people in the street. In this we step alongside the familiar iconographies of the street-as-consumption, street-as-gaze, surveilled and surveilling; street-as-display, street-as-detachment, to unearth a fragmented, but participatory iconography. In many different ways we discover the street carries abstracted images.

These concerns are worked in this chapter through five main sections: 'Practices and representations', which introduces the possibility of people themselves representing their streets and how they come to know them; 'Street practices, rituals of the street', which situates leisure as a useful informer of geographical knowledge and thus of images of the street through rituals and relationships made in leisure practices; 'Embodying the street', which seeks to unlock the intimacy of movement and the multi-sensuality of the body in street space; 'Decommodifying the street', which encounters other means of knowing; and 'semi-detached images of the street', the unstable articulation that occurs in the working of images. Final conclusions include attention to empowerment in the images of the street.

PRACTICE AND REPRESENTATIONS

The street is a series, a fragmented and uneven broken series of bits. Each is a site of something else, of someone else, of somewhere else. There are houses, vacant sites and car parks; the edge of the town, schools, shops, malls, workplaces. There are parks, corners, homes, playgrounds, vacant lots, pubs, allotments, football grounds and meeting rooms. In many months in the year, many streets are an outdoor play-room or meeting place. Streets are used in numerous and diverse ways, and we may anticipate numerous and diverse meanings to attend them. Streets may be commod-ified in places but only in places; actually, most streets are found not to be explicitly commodified, nor do they contain obvious or intended commodification. How much does the commodification of the street change what we do there, what it means to us? The sites just listed – pub, club, tennis court – prompt the idea of leisure. In this chapter we take leisure practices to be one means to unpack the signs of the street. Indeed, if shopping is a leisure activity, then we need to reconsider what is happening in making a trip along to the shops; it is much more than just a moment of consumerism (Shields, 1992).

An insight into the possibilities of the street in *popular* geographical knowledge may be attempted by considering the way people have sought to articulate what they value on their 'home' ground – at least, in the immediate physical space where they live. Akin to Bourdieu's idea of the ragged and changing yet structuring *habitus,* people have represented their loose, unstable, contested and contesting knowledge of a familiar place in Parish Maps, stimulated by ideas of popular geography of the environmental campaigning group, Common Ground (Crouch and Matless, 1996). Streets emerge as very visible features of these maps, suggesting that the street can be a familiar organiser of local knowledge (Bourdieu, 1984; Crouch and Matless, 1996). In these 'maps' are people who live in a village, estate or corner of the city. These maps are full of anecdote, memory, hopes and evasions; notes of celebration and irony, all depicted as writing or sketches inscribed across or around a usually central shape of the streets of the place.

Around and across these streets people depict all sorts of events, memories and feelings, shop signs, house prices, about places and people, and practices, inscribed on the map surface in an effort to represent what their locality means to them, and many of these are leisure practices of just the kind included in the narrative of this chapter. Often baffling, perhaps even locally nationalistic and exclusive, sometimes fecund, in these representations of the street we discover a chaos of marks, striations and inscriptions of memory, actions, places, emotions and love, fear, care, inconve-nience, opposition, celebrated and worried over the map surface (Crouch and Matless, 1996). This frequently includes leisure practices, and the particular sites where these happen are presented as embodied with this knowledge of practice. 'The life of consciousness – cognitive life, the life of desire or perceptual life – is subtended by an intellectual one' (Merleau-Ponty, 1962: 136).

These are, then, documents that people produce themselves; produced through the shared experience of everyday life. These are maps as popular culture; and they *represent* cultural practices. These amateur maps of popular-expert knowledge suggest how streets may be significant in 'ordinary' lived experience. Streets trigger memories; they are a site where events happen, cut across, depart from, spill onto. In just one of the 2,000 maps made like this, one made from notes handed in on one day of a local fair in a Bristol suburb, there are these inscriptions: 'food for when the fridge is empty' (a Chinese shop); lose yourself in wilderness' (an allotment); 'slip in doggy do' (Redland Green). In other maps, depictions of clubs, sports fields, views from bedroom windows, playgrounds, schools and shops, local services and environmental features are interwoven with remarks, objections (for example to gravel working, or road noise), and sites where 'strange' things happen. These are notes along a street about everyday events and interactions *and* local myth that constitute a kaleidoscope of events and people. These notes can be stiflingly parochial, but often merge with wider references, to places at the other end of the street out from the map; to people who have called through. These local events are rendered as representations in the map itself. The map, in turn, triggers recall, excitement and concern as these features arrive in a new life as representations. This is knowledge that is contextualised and contextualising, people making claims on their street and on their lives. These representations of what locality means is made partly in terms of leisure practices.

Leisure is considered here as cultural practice; an act of engagement, empowerment, enjoyment and identity; contest and transgression; and of consumption. Contemporary leisure practices may be significant in being detached from other components of life, happening in hyperreal space (Rojek, 1994). However, leisure often happens in particular places, and the street is just one site. We pay attention to ways in which leisure practices may engage the street and themselves provide significations for the street, and similarly the street, significations for the leisure practice. The practices, the human social interplay, and the bits of the street themselves, contribute to this geographical knowledge in a complex commingling of metaphor and materiality in ways that we will explore in this chapter.

The bits of the street are both along the street space itself, huddled around it; just hidden from its view, and very clearly the wider setting of the streetway itself. In this way, we see the street less as a way of focused presentation, and instead discover its overhang, its lack of borders and boundaries, and fluidity. Our different knowledges may render the same street to end in different places, overhung by different memories or cut back, revealing all; in our own lives, streets may have different limits at different times. So sites around the street and practices that happen in these sites may be more or less prominent in the signs around us. To understand images of the street, it is necessary to attend to its margins, to what happens in the corners, what is seemingly off the street, and where it connects, leads, anticipates.

The examples of sites and of practices used to explore those spaces in different ways include the following: Football crowds walking to and from a match, along the streetway itself; youths using the street to meet in; people in the park; people working allotments; and young people using pieces of 'vacant' ground. Parks are seen from the street; people approach them from the street, and they provide life to the street (Willats 1982; Crouch, 1989, 1994; Willis, 1990; Nielsen, 1995). What happens in the park can change the meaning of a street to those who use it, or who just walk by. Allotments, whilst being very private places of cultural practice, as people represent their lives and 'play' with space, are usually very visible, too, and contribute a critical dimension to the image a street aquires. A piece of vacant land may be used in a way that demands secrecy, but the site and the way it is used, and simply its 'availability' may dramatically alter the image of the street for those to whom it is important in their lives. We explore how these meanings are contextualised and disrupted through cultural practices, and how their signs emerge through an active process of geographical knowledge.

STREET PRACTICES: RITUALS OF THE STREET

Whether leisure practices happen alongside the street, in it, behind it or just off its main frontage, the street is a significant part of what is happening; the street informs, and its image is informed by, these practices. Youth culture finds the street valuable for its own practices. These include display, and gauging and swapping representations visually and in conversation. However, it is also a culture where practices include talking, walking, meeting, making friends and fighting; swapping bootleg tapes and escaping home, going somewhere else. These streets can be central, across the park, in the narrow walls of the estate, or through the leafy roads of the suburbs. Paul Willis's exploration of youth culture tracked what young people did and the street was a focus and a conveyor of their activities (Willis, 1990). In each of these cases, the street is not only an actual material place, but embodied with all sorts of meanings and metaphors; of escape, discovery, of home, too. The street offers an opportunity, a place to *be* (as well as to be seen). The street becomes a place of sharing; of conflict, confrontation between different members; of ownership, especially after dark or outside time when other age groups may be there.

In a second example, our consideration turns to that part of spectator leisure of attending a sports event where people are on the move. Walking to a football match or amongst the streets with other youth follows de Certau's 'long poem' of walking that 'manipulates spatial organisations, no matter how panoptic they may be: it is neither foreign to them (it can take place only within them) nor in conformity with them (it does not receive its identity from them)' (de Certau, 1984: 101). Through the practice the street's image is reconstructed. Walking like this renders the street 'no longer a diffuse and placeless space of competition . . . but a space of united action or solidarity' (Nielsen, 1995: 32–3). By their presence, as groups or even relatively unconnected people, with a shared goal and probably shared concern, celebration and event, they transform the image the street holds for them. In both our examples, people are finding themselves together with a degree of shared practice; people are getting together and getting somewhere else.

Parks are typical of those spaces that make the edge of the street ambiguous, that extend the space the street signifies. They are situated alongside many streets, and what happens there informs the image of the street; but also, people approaching the park view the street, engage with the same anticipation or wariness that the football crowd shares. Approaching and leaving the park by the street, the street becomes an imaginary space of anticipation and memory; often shared with people with whom the park is enjoyed, sharing moments of reflection and a range of emotions.

Parks are places where people do many different things; where spaces may be marked out, imaginatively, for different events and engagements, memories. Many activities in the park are self-generated, unaware of commercial signification and uncharted by management. In this people discover an often magical space of possibility (Crouch, 1994; Comedia 1995).

In these mundane everyday practices, we discover something distinctive. There are elements of shared, co-operative practice, where in sometimes uneven ways there are shared values, attitudes; the practices involve friendships, as well as alienations, but for many these are not experiences of isolated individualism held tougher only by the marketplace. Many of these practices are hardly recognisable in terms of decentred, disembedded leisure and hypereal sites, where the images are 'constituted by references which stretch over vast expanses of geographical space' (Rojek, 1995: 147). In this familiar scenario, space, the street, is 'decontextualised'. This is the oversliding of global and real space, of fictional and fantasy space – 'cyberspace'. The information world 'heightens interdependence and multiplies encoding and decoding'; we are 'offered' only representational purchase as signs' (Rojek, 1995: 147) The mall, Las Vegas, supplies the familiar referents and reference, alongside Zukin's commodified cities, the gentrified streets of the city downtown (Zukin, 1995).

However, rather than these claims, our examples suggest ways in which the images of the street are situated in leisure practices that are very different. It is valuable to turn to recent counter-interpretations of making sense of spaces in leisure, of how places are used in making sense of leisure practices. Moorhouse observed in his fascinating study of hotrodding in the USA, that leisure can be marked by making community, friends, engaging one another's enthusiasms (Moorhouse, 1992). Instead of distanciation and de-differentiation, instead of unanchored and uprooted leisure, we discover something else. Moreover, young people and football supporters are using the street not as individuals alone, but doing so together. This does not happen only in loose, transient and uncommitted sociations (Shields, 1992), although the engagements may be temporary, intermittent; the boundaries of practice and knowledge may be diffuse.

The leisure practices across and around the street resemble Gorz's dimensions of communal participation, of 'communication, giving, creating and aesthetic enjoyment, the production and reproduction of life, tenderness the realisation of physical, sensuous and intellectual capacities, the creation of non-commodity values' (Gorz, 1985). The youth fighting in the street may not always live up to Gorz's more peaceable values, but many of them are likely to apply. In particular, there is the assertion of non-commodity values in these street practices, this 'long poem'. What is being read is not only the signs of the mall, but what people are doing, together and in relation to each other. There is a considerable social dimension to many of these leisure practices that become important in making the images of the street. Our attention focused upon the gaze of the *flâneur* begins to demand reappraisal, but does not efface, as we will suggest later, the position of these commodified signs on the street.

As we move from identifying the practices and the sites along the street to articulate their construction as imagery, we note the importance of memory. Memory is important in the practices, informed of a past and informing a future, even if the geographical knowledge accumulates in a half-understood, half-chaotic way.

> Memories are part of culture and depend, in various ways, upon the physical setting and how people remember the course of events leading up to the present. . . . Artefacts and the fabricated environment are also there as a tangible expression of the basis from which one remembers, the material aspect of the setting which justifies the memories so constructed.
>
> (Radley, 1990: 49)

Memories are of course contingent, unstable, and constantly reworked in further practices. They inform us today in a different way from the way they may inform a practice tomorrow. These practices are rituals which are inscribed on the knowledge of the street, as the bits of the street where they happen, or lead to, and from, are inscribed in the rituals themselves.

EMBODYING THE STREET

In the many dimensions of these practices, we discover that the gaze, the visual, is accompanied by more complex knowledge-making. Here we turn attention to the body in making geographical knowledge and informing images of the street. There are two ways in which we can understand 'embodying the street' in our images. The practices themselves, in their own materiality and metaphor, are embodied in what the street means to us. Furthermore, the body itself is important in the way we make the images of the street.

Allotments are in many towns in the UK a significant part of the street, amongst houses, near the park. For many they are a part of the observed street, and for those who leave the street to go into an allotment site, an important image of anticipation and then engagement. Having an allotment can provide a distinctive geographical knowledge which contains a complex play of significations, many of which are social, and a very bodily experience of space, and making artefacts very directly; concerning land, nature, control and empowerment, as well as more directly sensual meanings, often in pursuit of ideas about using the environment.

> Working outdoors feels much better for your body somehow . . . more vigorous than day to day housework . . . much more variety and stimulus.

The air is always different and alerts the skin . . . unexpected scents brought by breezes . . . only when on your hands and knees do you notice insects and other small wonders.

(Crouch and Ward, 1994: 157)

Whilst for many people who pass an allotment site along the road the image is one of anachronism, there are others who identify with an image of the site that is about human labour, a creative landscape and a direct engagement with land; they see intense human activity, in practice or in the marks it leaves, inscribed on the edges of the street. For this plot holder the image is situated in practice; in a practice that is very bodily; of immediate and close movement, sensuous and sensual; intimate in scale.

Referring back to Nielsen's depiction of the crowd on the way to a football match, he connects the socially contextualised knowledge of the street with the body: 'During the walk in the street, the bodies themselves transform the body space, which becomes no longer a diffuse and placeless space of competition, with uniformed bodies heading for some coincidental spot, but the space of united action or solidarity' (Nielsen, 1995: 32–3). It is the presence and movement of bodies that assists the knowledge of a shared purpose and participation; and of a meaning that transforms the materiality of space itself. This situates knowledge both in shared action, and through the work of the body itself. In their presence and in their movement, the bodies of self and of others, identified with similar purpose, feeling and direction, are dissolved into an intention and an identity that overwhelms any other image the street may hold; shops, houses, traffic, as these become culturally deafened by the ritual of occupation.

Repeated regularly en route to the home ground, and repeated in similar spaces across the country on away fixtures, the street becomes a place in a ritual; with an especial meaning being inscribed on the streets and buildings. We may apply a similar idea of solidarity, embedded in the street, amongst other supporters on their way by car, club scarves seen waving from the window. In each case, the street becomes an identity, an image of shared interest; their memories of the match will include the image of the street, where people are moving knowing they have the same pulse, to a shared spot.

There are similarities in process of geographical knowledge in the allotment, with an identity of shared practice and in this case artefacts too, that situate images of the street. Allotments spill over into the surrounding streets both visibly and in arrival and departure. In a comparable way, children use vacant sites for play, especially where their home may leave little physical or cultural space for play, discovery, identity and empowerment. Children interviewed living on one estate in north-west London use a piece of 'vacant' ground only a short walk from their home, a tower block of typically run-down flats. The vacant ground, hidden just behind the road, is transformed, and for them being 'vacant' may mean available for new inscription. 'We sit around and talk about mums and dads; where you'd hidden your packet of fags at home . . . we just went there to get away from everything and forgot about everything . . . this place is another world really. We never thought about (the flats) over there' (Willatts, 1984).

These children demonstrate the construction of image through a shared practice, where they are able to swap stories, and take all sorts of artefacts with them from home to make both connections, and separations real; these artefacts are re-imagined, reworked in their new places. They take old prams, a stool, clothes, fragments of identity from 'home' that can be used to imprint identity in a place they feel is their own; making their own memories. They leave these artefacts at the site, to return to later, to build and inform their practice and meaning. Their presence and their activity transforms the site; they move around the site, simply spend time there; on ground where they know there is a sensuality of play otherwise unavailable. The street on the way is transformed as it bears the hope, the escape and the identity of somewhere else; the partial concealment of the site from the street enhances its 'ownership' and their identity. The street becomes more ambiguous, embedded with the journey in both directions, towards the flats, and the vacant space; ambiguous in ownership, control, identity and transgression.

One of the complexities of this geographical knowledge that informs the images of the street emerging from these examples is the importance of both body as gazed at and gazing, with the power of the gaze – and the importance of the body as a site of action, agency, empowerment. The body becomes important in reading the multidimensionality of events, their intensity, uncertainty. The body reads the space and the surrounding action of events, making sense of where it is, in an encompassing sense of culture, social relations and space, movement and artefacts; imagination and metaphor. To interpret these practices in this way is to acknowledge both Foucault and Merleau Ponty (Foucault, 1981; Merleau Ponty, 1962; Crossley, 1996).

Our bodies provide us with an important means of access to the world, not only its gaze: 'my surroundings . . . a collection of possible points upon which this bodily action may operate.' (Merleau Ponty, 1962: 100–3). For him, bodily actions are much more than just seeing, or even within that, standing to stare. This makes us wary of privileging the gaze and making simple suppositions of how we construct images of the street. Not only is there multisensuality, but an awareness of the presence of other people in our examples of the street becomes crucial. They are not important as visual elements, but because they are known to be there, to be acting in partic-ular ways, and in some ways that resonate with our own; some practices, indeed, we may be doing, explicitly, together. Thus, the information, emotions, struck by the body, ignite our imagination and move beyond the material. This ranges from the depth of meaning, feelings, values and memory to other material geographies in other places and at other times. Although Merleau-Ponty is looking away from the gaze, he is arguing for less 'universal constituting consciousness' and more attention to the uncertainties but manifest 'union of essence and existence' that the body knows.

There is a strong sense of physical awareness; the body moves on, reads messages in the movement, reshapes one version of observed material into another, more complex form; whether simply out of sight, or in terms of a complexity of surface, vibration, shape, form. Merleau-Ponty seeks to remind us of this complexity as 'fortuitously agglomerated contents' (1962: 147). Merleau-Ponty is turning to the empirical to con-sider the variety of experience, the element of 'senselessness in it, the contingency of its contents'. The examples in this chapter argue that we engage the material and the metaphorical; reflecting on both practice and meaning; space and knowledge; the cultural practices that we consider in this chapter as they inform the images of the street.

People are not found in motionless movement of the anonymity of a Lowry painting. They are a series of bodies in spaces, moving, active agents in their lives and in our own. In the street on the way to the match, along the side of the road at the park, we do know something about the people around us, their shared anticipations, loyalties perhaps. Their active bodies enliven the space around us. Across the park, the people are going about their lives, presenting themselves as ambiguous images to us, of reflection and play, in each case agents in their, and our, lives. The bodies are themselves about a whole complexity of existence, to us, movement, bodily form, practice, memory. We discover all of this in our vision, imagination, other senses, and through our own bodily movement, fast or imperceptible, and knowledge, imagination and prejudice; around and across the space itself. Our body informs in different ways, and reads the surroundings through our own emotions of the time. We engage people we do not know; they become part of the artefacts we piece together in our image of the street, and we in theirs.

The images are formed and reformed in complex and unstable ways; images become incomplete fusions of imagination and reality; these include a reality and are not detached from reality. Unironically, we discover that we experience the world at a very human scale; the image of the global is drawn down – or up – to that dimension. Despite the recent fascination for the posited 'global space' of the airport lounge and similar sites (Augé, 1995), most people spend a great deal of their time in 'real space' however much in our knowledge that is worked and reworked, contested, perhaps confused, by other sources of coding and meaning.

Many of these practices are distinguished less by the lone individual, *flâneur*, but in a social practice (Shields, 1991: 53). In our examples, the individual is making knowledge through taking up 'fragments . . . to actualise them. . . . He [sic] dooms certain sites to inertia or to decay, and from others he forms "rare" or illegal spatial "shapes" . . . the walker, in relation to his [sic] position, creates a near and a far, a here and a there . . . indicative of an actual appropriation of space by an "I", . . . thereby establishing a conjunctive and disjunctive articulation of places' (de Certau, 1984: 130–3). However, this does not contend with the collective form that inhabits many contemporary practices, as we have observed in this section. It is not only 'I', but with others. Being with others, and being in the presence of others with shared intent, values, excitements, is more than the isolated consuming individual. This knowledge reworks our images of the street

DECOMMODIFYING THE STREET

Our knowledge of 'real space' that includes the metaphor we bring to bear upon it does include the images of commodification. However, in our examples, there is a different but interesting play between commodification and other means of human relationship from the one which dominates much geographical thinking (Jackson and

Thrift, 1995). In the *flâneur*'s 'play' with signs along the street there is a vacuum of alternative contexts of this playfulness. Our examples begin to offer some of these. Warren's argument, that too much earlier work on place and representation ignored the commodified, was right (Warren, 1992), but perhaps this has leant too far the other way in more recent work. Or rather, we need to attend to how those commodified images are negotiated in our wider experience. The crucial issue with regard commodification concerns how it is used in the way images of the street are constructed by agentive human beings (Giddens, 1984).

Focusing on working-class youth in the late eighties UK, otherwise time of the rising mall and its attendant commodification of spaces, Willis found that images of consumption, in terms of commodified landscapes, were important, indeed significant in informing practice. Indeed, the images of commodification are used in asserting street credibility. The lived street is found in these signs; the buildings, the shops, amongst the material young people play with in making sense of their lives. Television and commercial music were crucial in providing patterns for cultural activity amongst the young, and used in triggering their own practice. The images of consumption were used as a trigger to make sense of what was going on in their own world: 'they can be used as symbolic material to make sense of that incoherent but exciting experience which in the heat of the conflict seems to be without its own meaningful signs and symbols' (Willis, 1990; Crouch and Tomlinson, 1994).

These images were interfaced with plenty of mutual support, confrontation, on terms that they make their own. For youth in working-class UK, the street is the very site of survival, or struggle. The street can be made to hold its own images for young people, and Willis gives the example of music: 'Music is used to create and mark off physical and cultural space as young people's space, be it in the bedroom, . . . the shopping precinct, street, park or concert hall' (Willis, 1990: 71). This provides an example of what Willis calls grounded aesthetics, where symbols and practices are mixed and selected again, 'recomposed to resonate further appropriation and particularised meanings' (21). The image of consumption can be used as a trigger to make sense of what is going on in their own world: 'they can be used as symbolic material to make sense of that incoherent but exciting experience which in the heat of the conflict seems to be without its own meaningful signs and symbols' (106). The images of their streets are much more complex than the images of the commodified landscapes, because of relationships, escapes and fantasy that are also related elsewhere.

The recent changes in city parks has made for changes in their imagery (Zukin, 1995). Their gentrification, linked with tourism and business investment, has rendered some very obviously as objects of consumption, to visit in order to experience a provided signification. However, many parks have not been included in this programme of change.

Parks, like the vacant land sites used by children, can be sites of imagination and self-realisation; of play on one's own terms, and frequently in shared practice. People using parks build their own grounded aesthetic, in numerous different bodily knowledges and practices, relationships, memories and values like those remarked by Gorz. The commodification of the park brings additional source material, as well, sometimes, as making everyday enjoyment of a park greater, shifting fear to safety (Crouch, 1994: Comedia, 1995: Zukin, 1995: 285).

Of course, we cannot now imagine the street without images from commercial culture, but these interpenetrate, contest and reaffirm or disrupt our own knowledges, make contingency more fragile. However, to privilege TV or film and other outward sources of imagery is as incomplete as regarding them as supplanting knowledge that includes lived cultural practice. We may watch *EastEnders* or *Coronation Street* in a pub; but that watching reverberates with our own arrival and departure at the pub, today, before, and its likelihood in the future.

As the street has become an icon of late twentieth-century postmodernist gaze and *flânerie*, we rediscover that it is as likely that the street is the site of numerous interactions and events that engage bodily, socially, in ritual and outside the contextuality of commodification. The street remains an embodied place, with lives and relationships inscribed across and along its surface as we observed in the examples of Parish Maps. Once we engage the empirical world it is difficult to sustain the claims of the street as a space only inscribed, or prescribed, by anyone but ourselves; we are not only '*flâneurs*'. Liberated from only a controlling gaze, in our last section of this chapter, we turn these four themes towards a new way of envisioning the street in contemporary, popular geographical knowledge.

SEMI-DETACHED IMAGES OF THE STREET

Rather than replace reality with sign, or situated signification with commercial, or vice versa, the challenge for geography is to unravel how these all interpenetrate. People made their representations of their streets in Parish Maps; they represented practices, memories, values, and they interconnected with images from a wider world, not simply narrowly focused. Although their streets figure in their intimacy, they connect with messages from the consumer worked and an almost universal iconography of the housing market, of places to go to away from the immediate streets. Their streets include metaphors of this wider, sometimes global, reality. But they are not dominated by it.

It is surely right that the iconography of the street is assembled from popular culture. As Fiske argues, that popular culture is always in process, and is something that people themselves do, make, construct (Fiske, 1989), and we have observed, the street is profoundly a site of popular culture. So popular culture is a process, with no 'dominant popular culture' (ibid.), but where people make use of commodified and other materials and metaphor, significations, as we have observed in process through geographical knowledge in leisure. The acts and practices of that everyday popular culture, as we do it ourselves, works erratically and unevenly with the signs we can draw upon, and from this interpenetration we make our own significations – our own images of the street. However, of course, this does not avoid the contestation, appropriation and possibilities for empowerment that are of the street, and informed by images of the street.

Commodified images, global images – these are only semi-detached, or semi-attached, in our geographical knowledge. In one way, the images of the street that emerge from the commodified representations along the street, or seen in the mall-bag label someone else is carrying are interpenetrated with what we are doing socially, bodily, making that process of popular culture in geographical knowledge. Like the young people Willis interviewed, for many of us those commodified signs we may gaze upon, or simply catch out of the corner of the eye, we use as one of numerous resources in making sense of where – and who – we are, and the place, street, we move through.

We may contest the apparent localism of these examples, but the crucial issue is neither signs wrought elsewhere or made only, insularly, 'at home'. The football stadium in view, the streets we walk through may be just channels to get some-where else, but our presence makes more than this; the street is not limited to its immediate physicality, but its boundaries broken to the League, the European Cup, and beyond. There is another way in which this mutuality of images is at work. The images/meanings of the commodified are semi-detached from the ways we inscribe our lives on the street, and how we insert the street into our lives. The images at which we gaze as *flâneur* are semi-detached in relation to the images we make through the practices themselves, ourselves, our friends, our bodies and actions. These both gathered and fragmented images are in turn only moments semi-detached from/to knowledges and experiences and imaginations that occur elsewhere, in other places, in other parts of our lives. Looking over the allotment fence makes the *flâneur* gaze, but in looking, we may observe, judge, reject, connect, reflect on much wider issues, and also the intimacy of human activity. The youth at the vacant site or in the estate street may feel intense disempowerment in their flat, but empowerment outside; the need to escape may contain its own oppression; swapping bootlegs they can penetrate the world of fashion and feel isolated. The street bears all of these images, and much negotiation and unease.

Our actions, relationships, human engagements and values, as well as imagina-tions, are juggled and played alongside representations. 'Rather than representations being the primary focus of understanding, they are islands in the sea of our unfor-mulated practical grasp of the world' (Taylor, quoted in Pile and Thrift, 1995). Yet representations are constructed from those actions, memories and knowledges, in our own imaginings, and in ways that may be shared and contested.

Transgression has rightly fascinated geographers recently (Cresswell, 1993, 1996), yet empowerment is something that features more widely in geographical knowledge, as the examples of leisure practices considered in this chapter show; knowledges cleave along lines of class, ethnicity and other social divisions, and are marked out by power (Zukin, 1995: 274; May 1996). Part of this is to transgress commodification. Ignoring the culture of commodities is not confined to the kids in the mall who won't buy (Goss, 1993), but is germane to most leisure practice. They may be simply not very important. Although we have focused on practices that may be considered relatively self-generated (produced, contextualised, participated in by individuals and groups of people themselves) and market cool (Crouch and Tomlinson, 1994), this should be extended by renewed efforts to understand leisure practices anywhere. If shopping is a leisure activity, it becomes necessary to stretch the investigations beyond the commercial symbolism and mall-planning that has confined the available work, leaving only the young meeting in the mall for 'fags' on Sunday. People shopping partake, and participate, in a wide range of popular culture, about meeting each other, making freedoms, being out anywhere with the family – or away from it. The key experience of the World's Fair in Vancouver was a great day out for the family, not some embedding with symbolism of nation, global existence or commodification (Ley and Olds, 1988).

There are other important issues of empowerment in images of the street. Zukin (1995) documents the appropriation of sites of shopping in a deeply wrought narrative on New York, including its parks. In reaction to the appropriation of familiar shopping streets by unfamiliar business cultures or simply privatised, as they find their very identity fractured, people shift elsewhere to something which reverberates more closely with their own familiar knowledges even of shopping streets (Zukin, 1995: 251ff.). Yet in the process there is a real contest and competition for meaning and place; place becomes appropriated and its memory and meaning effaced. The street can become a site of loss and disempowerment.

In all our cases, there is the prospect of appropriation and re-appropriation: parks gentrified away from local use and knowledges; allotments appropriated for 'profitable' (sic) leisure (Crouch 1994); vacant land kept enclosed and policed; the street gentrified from working-class youth; the football ground out of town and too expensive. Commodification can appropriate memories, knowledges, ways in which images have been embodied. Although we can pick over the materials of commodification in our urban tactics, those we know are disrupted, and this can happen in a way that is very unequal, making access and empowerment very different between class, ethnicity and gender, and age. We observed the importance in many of our examples of sites on the street holding 'ownership' – not in a financial or legal way, but as knowledge and use; of security and identity, play. These are important images of the street that can be reason enough for contestation, and sense of loss, although Zukin observes the potential for increased access for women to parks that have become reduced of fear following enhanced policing presence, and their shared sense of 'ownership' increased: they could go there (Zukin, 1995: 275).

The young people in the vacant site considered earlier are transgressing in a move to establish an identity and a separation. But more aspects are important than that in their geographical knowledge, and that includes sharing, together producing and reproducing human relationships, love, care, and those other crucial qualities and human, non-commodity values that Gorz acknowledged (Gorz, 1985). The street is simply not a series of marketable images; even 'the contemporary flaneur might still speculate that the consumer-centred postmodern city is based on unstable foundations', so contesting the power of the market to outline what the image of the street is in our lives, in popular geographical knowledge (Savage, 1995: 214), as Finnegan challenged the power of earlier planning (Finnegan, 1989). Rather than focus 'marketing, packaging and appearance' crucially informing our knowledge of the street (Rojek, 1995: 169), we have discovered other things. 'Ideologies, or as Foucault puts it, discursive practices, are created in specific spaces. These spaces then provide the pictures in our minds when we conceive our identities' (Zukin, 1995: 293). She points to the ongoing production of urban space, its elevated recommodification, which importance we cannot ignore, and which is at the core of reason for the questions of unstable empowerment we note above. However, there is a continued insistence towards our own images of the street.

We observe a series of different contexts of geographical knowledge, some of which may be in shared practice. In terms of the body, this is more than simply adding to the visual more senses to the list. It is identifying the ways in which the body informs, and interprets, engaging space reflexively. The meanings are embodied with practice; their practices embodied, contingently, with fragments of the street and other people; of conflict and of people sharing space, people sharing practice. We struggle, as de Certau argues, in the gaps, lapses and allusions of other people's structures (de Certau, 1984: 101–2), to make sense of the complexity of contingent images, to put together our own grounded aesthetic, and leisure provides one means of doing so. Leisure practices are laden with metaphor, and place is complicit in this process. The street is embodied in the practice and in its memory. The street triggers recall of a contingent knowledge of values, actions, relationships and anticipations; the street is an image of solidarity, loss, and shared practices, too. This is more than a singular story of empowerment and free play, but one of contestation and challenge, as the materials – both concrete and metaphorical – are themselves the object of power, and the exercise of power. This chapter has tried to shift the narrative from a bourgeois privileged position and explore a wider world of knowledge, and images of the street.

The claim to attend to the multidimensional human subject challenges the notion that there is no subject because there can be no transcendental subject, as Giddens challenged nearly two decades ago (Giddens, 1979: 47, 69–70). Buttimer has criticised the breaking of engagement between thought and reality as simply unnecessary (Buttimer, 1993; Grimes, 1996). We may argue the relativity of the way we 'know' reality, its contextuality and especially contextuality in its making sense. Enlightened from an inclusive bourgeois gaze on voyeurism, we find another street, another city, one even more transparent in its contestation, freed from a compulsive celebration of commodification. We can still gaze *and* feel, be surveilled, know and act, if we are permitted to be inclusive. The images of the street emerge from a very complex human process of geographical knowledge.

REFERENCES

Augé, M. (1995) *Non-Places: An Introduction to an Anthropology of Supermodernity*, London: Verso.

Bourdieu, P. (1984) *Distinction: A Social Critique of the Judgement of Taste*, London, Routledge.

Buttimer, A. (1993) *Geography and the Human Spirit*, Baltimore, Johns Hopkins University Press.

Comedia (1995) *Parklife*, London and Stroud, Demos-Comedia.

Cresswell, T. (1993) Mobility as Resistance: A Geographical Reading of Kerouac's *On The Road. Transactions of the Institute of British Geographers* 18: 249–62.

Cresswell, T. (1996) *In Place/Out of Place*, Minnesota, Minnesota University Press.

Crossley, N. (1996) Body-Subject/Body-Power: Agency, Inscription and Control in Foucault and Merleau-Ponty. *Body and Society*, 2(2): 99–116.

Crouch, D. (1994) *The Popular Culture of City Parks*. London and Stroud, Demos-Comedia.

Crouch, D. (1989) Patterns of Co-operation in Outdoor Leisure. *Leisure Studies* 8(3): 189–99.

Crouch, D. and Matless, D. (1996) Refiguring Geography: the Parish Maps of Common Ground, *Transactions of the Institute of British Geographers*, 21(1): 236–55.

Crouch, D. and Tomlinson, A. (1994) Collective, Self-generated Consumption: Leisure, Space and Cultural Identity. In Ian Henry (ed.) *Modernity, Late Modernity and Lifestyle*. Brighton, Leisure Studies Association.

Crouch, D. and Ward, C. (1994) *The Allotment: Its Landscape and Culture*. Nottingham, Mushroom Books.

de Certau, M. (1984) *The Practice of Everyday Life*, California, University of California Press.

Fiske, J. (1989) *Understanding Popular Culture*. London, Routledge.

Foucault, M. (1981) *The History of Sexuality*, vol. 1. Harmondsworth, Penguin.

Finnegan, R. (1990) *The Hidden Musicians*, Milton Keynes, Open University Press.

Giddens, A. (1979) *Central Problems in Social Theory*. London, Hutchinson.

Giddens, A. (1984) *The Constitution of Society*. Cambridge, Polity Press.

Goss, J. (1992) 'The Magic of the Mall': An Analysis of Form, Function, and Meaning in the Contemporary Retail Built Environment. *Annals of the Association of American Geographers* 83(1): 18–47.

Gorz, A. (1985) *Paths to Paradise*. London, Pluto.

Grimes, S. (1996) Geography and Metaphysics. Paper presented to the *Conference of Irish Geographers*, Galway.

Jackson, P. and Thrift, N. (1995) Geographies of Consumption. In D. Miller (ed.) *Acknowledging Consumption*. London, Routledge.

Ley, D. and Olds, K. (1988) Landscape as Spectacle: World's Fairs as Heroic Consumption. *Environment and Planning D: Society and Space* 6: 191–212.

May, J. (1996) Globalization and the Politics of Place: Place and Identity in an Inner London Neighbourhood, *Transactions of the Institute of British Geographers*, 21(1): 194–215.

Merleau-Ponty, M. (1962) *The Phenomenology of Perception*. London, Routledge and Kegan Paul.

Moorhouse, B. (1992) *Driving Ambitions*. Manchester, Manchester University Press.

Nielsen, N.K. (1995) The Stadium and the City: A Modern Story. In J. Bale (ed.) *The Stadium and the City*. Keele, Keele University Press.

Pile, S. and Thrift, N. (1995) *Mapping the Subject*. London, Routledge.

Rojek, C. (1995) *Decentring Leisure*. London, Sage.

Savage, M. (1995) Walter Benjamin's Urban Thought: A Critical Analysis. *Environment and Planning D: Society and Space*. 13: 201–16.

Shields, R. (1991) *Places on the Margin*. London, Routledge.

Shields, R. (1992) *Lifestyle Shopping*. London, Routledge.

Warren, S. (1992) 'This Heaven Gives Me Migraines': The Problems and Promises of Landscapes of Leisure. In D. Ley and J. Duncan (eds) *Place/Culture/Representation*. London. Routledge.

Willatts, S. (1982) *The New Reality*. Derry, The Orchard Gallery.

Willis, P. (1990) *Common Culture*. Milton Keynes, Open University Press.

Zukin, S. (1995) *The Cultures of Cities*. Oxford. Blackwell.

12

DOMESTICATING THE STREET

THE CONTESTED SPACES OF THE HIGH STREET AND THE MALL

Peter Jackson

●

INTRODUCTION: THE POLITICS OF THE STREETS

At least since the nineteenth century, if not before (Harrington and Crysler, 1995; Nead, 1997), streets have been regarded as a lively and contested public domain, the site of popular protest and political struggle. For Marshall Berman (1988), the politicisation of the streets was a key component of the 'experience of modernity' as this public domain became subject to increasing regulation and control. Berman traces this process through Haussmann's uncompromising 'modernisation' of the streets of Paris through Le Corbusier's vision of the street as a 'machine for traffic' to Robert Moses's formidable plans for metropolitan redevelopment in New York. More recently, the debate has moved to Los Angeles where Mike Davis warns of the destruction of public space and the spread of private surveillance in 'Fortress LA' with its panoptican malls and carceral spaces (Davis, 1990).

In lamenting the privatisation of public space in the modern city, some observers have tended to romanticise its history, celebrating the openness and accessibility of the streets. Such spaces were, of course, never entirely free and democratic, nor were they ever equally available to all. Various social groups – the elderly and the young, women and members of sexual and ethnic minorities – have, in different times and places, been excluded from public places or subject to political and moral censure. In nineteenth-century New York, for example, Christine Stansell shows how women, along with their 'delinquent' children, who ventured unchaperoned into public places were subject to summary arrest and institutionalisation under the new vagrancy and truancy laws (Stansell, 1986). Similarly, in late Victorian London, the streets were experienced simultaneously as a place of sexual danger and erotic delight according to one's social position (Walkowitz, 1992). So, too, in the contemporary city, there is a 'geography of fear', structured by gender, class and racialised fears (compare Smith, 1987; Valentine, 1989). The streets remain places of desire and dread, pleasure and pain, fantasy and fear, their history is full of such paradoxes and tensions.

From this perspective, Richard Sennett's lament for 'the fall of public man' (Sennett, 1977) seems over-drawn and overly nostalgic. Contrasting the modern city with other times and places, he comments elsewhere on the failure of contemporary public spaces to fulfil our human potential: 'the shopping mall, the parking lot, the apartment house elevator do not suggest in their form the complexities of how people might live' (Sennett, 1993: xi). Compared to 'the ancients' who used their eyes 'to think about political, religious, and erotic experiences' in the city, Sennett argues, modern culture reduces and trivialises the city by producing only carefully orches- trated spaces of consumption and tourism. He writes wistfully of 'the modern fear of exposure', contrasted with 'streets full of life' and 'places full of time'.[1] In his earlier work, Sennett traced the demise of the public city and the growth of socially segregated suburbs where the privatisation of space had allowed middle-class fami- lies to turn their backs on the perceived ills of city life. He documented this process in *Families against the city* (1970), a study of Union Park in late nineteenth-century industrial Chicago where the intensity of family life was the accepted reward for a limited degree of social mobility.

The relationship between a romanticised view of family life and a pervasive fear of social difference remains strong in Britain at the end of the twentieth century where 'law and order' have returned to the top of the political agenda. Once again, the fear of crime and insecurity has focused on the public places of the city and on the fear of various racialised Others. The relocation of retail space from central cities and town centres to suburban malls and 'out-of-town' shopping centres offers a useful window on this process as will be illustrated here from recent research on consumption and identity in north London.[2] Drawing on this research, I will argue that shopping malls and other planned retail developments involve the domestication of public space, reducing the risks of unplanned social encounters and promoting the familiarity of privatised places. As with other exercises in the 'purification of space' (Sibley, 1988), however, such developments also have their costs in terms of social exclusion and increased inequality.

DOMESTICATING THE HIGH STREET

As is now well known, the history of 'out-of-town' shopping centres in Britain and North America resulted from a wide range of economic, political and social forces including the growth of post-war consumer affluence, increased private car ownership and other technological changes, the growth of female labour force participation, and, notably in the US, a process of 'white flight' as middle-class residents fled from the cities following the disturbances of the civil rights era. The built form of the shopping mall reflects all of these trends including 'a modernist nostalgia for authentic community, perceived to exist only in past and distant places' (Goss, 1993: 22). This can be observed in the many evocations of other times and places in the contemporary

North American shopping mall, from nostalgic re-creations of the historic Middle-American Main Street to Parisian boulevards, Mexican paseos, Arabic souks and casbahs. Such places, Goss claims,

> reclaim, for the middle-class imagination, 'The Street' – an idealized social space free, by virtue of private property, planning, and strict control, from the inconvenience of the weather and the danger and pollution of the automobile, but most important from the terror of crime associated with today's urban environment.
>
> (Goss, 1993: 24)

Jeff Hopkins (1990) follows a similar logic in his account of the West Edmonton Mall, describing it as a 'landscape of myths and elsewhereness' though he makes less of the connection advanced by Goss between idealised suburb and demonised inner city.[3]

As Goss and others have shown, the 'magic of the mall' is carefully contrived to differentiate it from the perceived ills of the contemporary city and the alleged disorder of its fearful streets. In these accounts, the social space of the mall is represented as a kind of metaphorical or idealised version of the city street. But whereas the idealised form of the public street is a relatively open and democratic space, the shopping centre offers only a parody of participation: where 'credit card citizenship' allows the consumer to purchase an identity, engaging however vicariously in their chosen lifestyle (compare Shields, 1992). In such accounts, the mall is designed to protect middle-class patrons from the moral confusion that might result from an unmediated confrontation with social difference. It is in this sense that one can write about the 'domestication' of city streets in the contemporary retail environment.[4]

While these trends may be less evident in Britain, similar processes can be discerned. Here, too, the contrived spaces of the shopping mall are a direct response to the perceived incivility of the city streets. The links between the high street and the mall are often very close with many observers blaming the rise of large-scale, 'out-of-town' shopping centres for the demise of high-street shopping in traditional city centres. As centres moved out of town (or, more accurately, to the suburban fringes of metropolitan centres), over 40,000 shops (around a third of the total) closed between 1976 and 1987 (Raven *et al.*, 1995: 37).

Responding to these developments, a House of Commons Environment Select Committee was set up to inquire into *Shopping centres and their future* (1994), leading to a series of revisions to the Department of the Environment's Planning Policy Guidance, seeking to impose tighter restrictions on the building of new out-of-town shopping centres. Earlier guidance on *Major retail developments* (DoE, 1988) had spoken glowingly of such developments in terms of the extension of choice, the improvement of efficiency, better service to the public and more pleasant shopping. In Parliamentary Questions on 5 July 1985, the Secretary of State argued that it was 'not the function of the planning system to inhibit competition among retailers' and that the possible effects of major retail developments on existing retailers was 'not a relevant factor in deciding planning applications and appeals' (DoE, 1988: 2). When revised guidance on *Town centres and retail developments* was issued (DoE, 1993), emphasis was placed on the important contribution that retail activity can make to securing the vitality and viability of town centres and on the need for a 'suitable balance' between town-centre and out-of-town facilities. Moreover, the scale, type and location of out-of-town developments 'should not be such as to undermine the vitality and viability of those town centres that would otherwise serve the community well' (DoE, 1993: 5). Accompanying these changes, the Department of the Environment commissioned a special report on *Vital and viable town centres* (URBED, 1994). Invoking the idea of the 24-hour city or town centre, with people living above businesses, a more vibrant night life, better street lighting and safer car parks, Secretary of State John Gummer claimed that the new guidelines 'hail the revival of the British High Street and the demise of the out-of-town shopping centre' (*The Times*, 8 February 1994). That such a view may be exaggerated is suggested by the fact that no new restrictions were placed on existing centres or on those which had already received planning permission. These include large-scale developments at Dartford (Blue Water Park), Leeds (White Rose Centre), Bristol (Cribbs Causeway) and Manchester (Trafford Centre), and over 300 others, amounting to an additional 9 million square feet of retail space.[5]

Objections to such developments have tended to focus on their aesthetic, social and environmental costs, uniting commentators from the right and the left. In a snap survey published in the *Daily Telegraph* (1 January 1994), for example, readers from Hampshire to Hertfordshire, Surrey to the West Midlands, Gloucestershire to Essex all claimed to prefer 'small, nice shops' in 'little old welcoming towns' rather than the dubious pleasures of 'hatchback shopping'. Similar objections to out-of-town shopping were voiced by Patrick Wright in the *Guardian* (9 February 1995), describing such developments as 'a collection of vast metal sheds along the bypass'. The social costs include the loss of civil liberties, such as the right of free assembly or free speech, which are routinely surrendered when the public enter the privatised spaces of the shopping mall. Drawing on Canadian evidence, Jeff Hopkins (1994) describes the conditions of access and laws of trespass that govern entry to these commodified spaces. Failure to leave on request of the management is punishable by fine ($100 in Alberta, $1000 in Ontario), failure to pay resulting in the threat of imprisonment.[6] Environmental costs include the loss of agricultural land for large-scale retail developments and the energy costs of lengthy car journeys to say nothing of the development implications for supplier economies of the increasing domination of a handful of supermarket chains (see Jackson and Thrift, 1995).[7] As Sarah Whatmore notes, consumer experiences of food have become profoundly distanced from the social and economic organisation of agriculture and from the increasingly globalised process of food production (Whatmore, 1995: 36).

There is no doubting the profitability of shopping centres as a form of prime investment for pension funds and other major corporations. But the demand side should not be underestimated, as many consumers have lost their faith in town-centre shopping, preferring the comfort and security of the controlled environment of the suburban mall or out-of-town shopping centre. Much of the demand for shopping in such sanitised spaces is fuelled by a fear of crime and incivility in town and city centres. A recent survey on 'managing the risk to safe shopping' reported in the *Guardian* (8 November 1995) suggests that 'Shoppers are deserting the high street for shopping malls because they fear crime and feel threatened by beggars, drunks and vagrants'. Based on a survey of 622 shoppers in six British towns and cities, the authors of the report claim that six times as much crime was suffered by shoppers in town centres as in shopping centres (Beck and Willis, 1995). Several notorious cases of abduction or child murder have contributed to these fears such that 'Only half the population now dares go out after dark, fewer than a third of children are allowed to walk to school, and public fear of strangers regularly erupts after such public murders as Jamie Bulger and Rachel Nickell' (The *Guardian*, 17 November 1994).

Based on our own research in north London, we suggest that the popularity of shopping malls and other planned shopping centres can be attributed to their success in managing diversity, reducing the risks of social difference and promoting the virtues of familiarity. Through the privatisation of public space, the city streets have been 'domesticated' and shopping made safe for the majority of middle-class consumers.

FAMILIARITY AND DIFFERENCE: EXPERIENCING THE HIGH STREET AND THE MALL

The project from which this chapter is drawn involved survey, focus group and ethnographic research in two north London shopping centres. Brent Cross was Britain's first purpose-built, out-of-town shopping centre, opened in 1976, at the southern end of the M1 motorway at its junction with the North Circular Road in the suburbs of north-west London. It was a private initiative (developed by the Hammerson Property and Investment Trust and managed by the agents, Donaldsons) with over seventy retail units, including several department stores and other retail fashion outlets (John Lewis, Fenwick, Hennes, Laura Ashley, Miss Selfridge, The Gap). The management company controls the access roads, car parks, and circulation space within the centre (regarded by many consumers as 'public' space, though in practice it is privately owned and managed), while the individual stores are leased or rented from Donaldson's. Opening in 1981, Wood Green Shopping City is a more recent 'mixed-use' development, including fifty retail units, a library and public housing complex. The development was undertaken with the financial support of the local authority and the Greater London Council. Located a couple of miles inside the North Circular Road, between Wood Green and Turnpike Lane underground stations, it has more of an 'inner city' clientele with a higher proportion of working-class and ethnic minorities among its customer base (for further details, see Holbrook and Jackson, 1996a). Besides the indoor 'shopping city', Wood Green also has traditional high-street shops and both indoor and outdoor markets providing a variety of retail environments within close proximity. This makes the configuration of public and private space more complex in Wood Green than in Brent Cross. Here, the streets and pavements outside the Shopping City are publicly owned (as is the library and post office) while the interior spaces of the Shopping City are privately owned and managed.

Property rights have become increasingly blurred with the sale of council housing into private ownership and with the proposed privatisation of former public services such as the post office. While Brent Cross is more clearly bounded space (because of its 'out-of-town' location), the boundaries of public and private space are blurred in Wood Green as one moves from the Shopping City to the high street, or from the library to the post office (see Figures 12.1 and 12.2).

Despite the privatised character of 'public' space in both centres (in terms of management and control), people use them as quasi-public places (in terms of access) for socialising and other non-commercial activities. Almost two-thirds of shoppers at both centres had been browsing or window shopping as well as or instead of making purchases. More than one in ten had met up with friends or family members or had recognised someone they knew. About a quarter had visited a café or restaurant and a similar proportion had sat down while in the centre (see Table 12.1). Focus group and ethnographic evidence confirm that the pleasures of shopping are tempered by widespread fears, focused on particular places and on certain social groups. The risks of shopping at the centres are perceived in relation to the degree of (un)familiarity with those who shop there, including 'foreigners' and others who are considered 'out of place'. As one US shopping centre consultant has argued, to maintain the 'family appeal' of such places: 'You have to create a safe, secure feeling and make sure it's not intimidating to anyone' (quoted in Goss, 1993: 28). Degrees of familiarity are associated with particular boundaries of inclusion, with codes of social exclusion working along lines of gender and generation, and especially along lines of 'race' and class.

We undertook two phases of six focus group discussions in youth and community centres in and around Brent Cross and Wood Green (see Table 12.2).[8] In these discussions, Brent Cross shoppers often tried to dissociate themselves from the perceived incivilities of Wood Green: 'Wood Green isn't part of this area', 'It's a bit scary, Wood Green', 'more violent', 'some of the teenagers round here are a little frightened of Wood Green' (various members of the Finchley & Whetstone branch of the National Women's Register group). Conversely, these same participants were thoroughly aware of the 'middle class' character of Brent Cross:

> I think Brent Cross is middle-class . . . you tend to see a lot of middle-class people there, not a lot of working-class . . .
>
> Yeah, Boot's are terribly middle class. I mean, if you wanted, you'd go to, you'd want Superdrug [a discount chemist], if you wanted, you know, a lower price shop.
>
> (National Women's Register)

The fact that Boot's have a branch in Wood Green as well as in Brent Cross and that the actual class differences in the clientele of the two centres were not as marked as these observations suggest (Holbrook and Jackson, 1996a) only confirms the extent to which the two centres (as opposed to their individual shops) served as symbolic markers of class and other social differences (compare Bourdieu, 1984). Participants in all of the Brent Cross groups were highly conscious of such distinctions. Aware

TABLE 12.1 Social activities at Brent Cross and Wood Green shopping centres

Percentage of respondents who had:	Brent Cross	Wood Green
been browsing	62.2	65.7
window shopping	52.1	65.0
talked to shop employee/security guard	38.7	35.4
looked for bargains	30.2	39.9
sat down while in centre	29.8	21.0
visited café/restaurant	27.6	23.4
watched passers-by	16.8	13.3
talked to someone	14.6	14.7
met with family/friends	14.0	10.8
recognised someone	12.4	12.6
returned faulty goods	4.4	2.1

Source: survey results

TABLE 12.2 Focus groups in Brent Cross and Wood Green

Round 1: Wood Green area (January–February 1994)	Round 2: Brent Cross area (July–September 1994)
Tottenham Mothers and Toddlers' Group	Unitarian Church Coffee Club
Woodside Senior Citizens' Centre	Brent Cross Residents' Association
Woodside Luncheon Club	National Women's Register
Wood Green Area Youth Project	Canada Villas Youth Centre
Greek-Cypriot Youth Centre	St Andrew's Prampushers
English-as-Second-Language Job Club	Jewish Women's Network

that other centres are 'much more down-market', they were 'very aware of the different sorts of people who shop there' (National Women's Register).

Judging by our survey results and an extensive review of local newspapers, Wood Green has a worse reputation for street crime and burglary. While some focus group participants complained that Brent Cross 'attracts a very big criminal element' (Brent Cross Residents' Association), others found the level of security there very reassuring: 'Compared with an open air shopping parade, I would think I'd feel much happier at Brent Cross 'cos it's lit and there are security guards, there's always plenty of people' (Unitarian Church group). Brent Cross has a reputation for motor vehicle crimes including car theft, but this gave rise to fewer concerns for public safety than the perceived level of street crime in Wood Green. Whereas 42 per cent of Brent Cross respondents described the centre as 'safe', only 25 per cent did so in Wood Green. Both Brent Cross and Wood Green have experienced major security problems

FIGURE 12.1 Brent Cross: an interior 'street', privately owned and managed, but used as quasi-public space by consumers

FIGURE 12.2 Wood Green: with no clear boundaries between the public space of the high street and the private space of the indoor Shopping City

in the relatively recent past, including bomb explosions at Brent Cross in December 1991 and at Wood Green in December 1992 ('Fire bomb blitz', *Hendon and Finchley Times*, 19 December 1991; 'Police and shoppers injured in bomb blasts', *Independent*, 11 December 1992). But shoppers rarely spoke of the fear of terrorism, focusing instead on the fear of other forms of interpersonal violence. As an Indian woman in the English-as-a-Second-Language job club told us: 'I don't like to walk at nights in Wood Green Shopping City or anywhere because I'm scared ... I always go by car.' The local press (whose significance in the dissemination of fear of crime is discussed by Smith, 1985) is full of stories about crimes of violence in the Wood Green area: 'Beast boy faces a life sentence', *Haringey and Wood Green Independent* (23 September 1994); 'Cashpoint man acts to beat off muggers', *Haringey and Wood Green Independent* (23 September 1994); 'Thief KO's shop boss in £13,000 raid', *Haringey and Wood Green Independent* (30 September 1994); 'Mad arsonist strikes twice', *Haringey Advertiser* (13 July 1994); 'Armed burglars raid home', *Haringey and Wood Green Independent* (7 October 1994); 'Gang kicks victim to ground', *Haringey and Wood Green Independent* (8 July 1994). Despite a declining level of reported crime at the time of our focus group research, shoppers at both centres generally supported the introduction of closed-circuit television and other forms of customer surveillance (see also chapter 17 of this volume).[9] A significant exception were those groups (particularly ethnic minority teenagers) who felt they were the target of such surveillance. Focus group participants from the Wood Green Greek-Cypriot youth club clearly expressed such anxieties:

> I was followed by two security people just because of the clothes I was wearing. If I went in like this [smartly dressed] after work in a suit or whatever, they wouldn't even give me a second look. But I was young, I was in a tracksuit ... and automatically, you know, it's 'shoplifter'.

Another member of the same group agreed:

> It scares me, all this security. ... It's like they're trying to pry into all aspects of your life. You can't even go two miles over the speed limit and you're going to get booked. It's going to become like 1984, George Orwell, you know, like Big Brother and everyone spying on you. Next we're going to have a camera in the corner over there to listen, just in case we say anything against them ... it's terrible.

Members of the Wood Green Area Youth Project had had similar experiences:

> I walked into W.H.Smith and the security guard followed me the whole way round the music store ...
>
> It's intimidating 'cos they're watching you; it makes you feel a bit dodgy.

A young woman in this group reported similar feelings:

> Before Christmas, yeah, the security guard comes up to us. We were standing there, we weren't doing nothing wrong, weren't causing no trouble. We

wasn't doing anything to nobody and the security guard says to us, 'Can you move along, please?'

At Brent Cross, too, members of the Canada Villas Youth Club had also been watched and followed round:

You get kicked out [by the security staff] for doing nothing, for doing nothing whatsoever, just standing there looking over a balcony. All of a sudden, these guards start crowding round us and chasing us round Brent Cross, and they kicked us out.

While people in both centres welcomed the 'convenience', 'familiarity', 'safety' and 'easiness' of shopping in a secure and climate-controlled environment, a smaller number described the centres as 'artificial', 'soulless', 'intimidating' or 'alienating' (see Table 12.3). Participants in the focus group discussions voiced similar views. According to one member of the Brent Cross Residents' Association,

I mean it's artificial, everything – artificial air pumped in, air conditioning, artificial light – I mean it's not like being out in the open. Personally, I'd rather sort of take a chance and shop in an open-air market or open-air street.

Other participants said they preferred high street shopping because

There's more fresh air. I think that's what bothers me, the fact that there isn't any fresh air in some shopping centres. I don't like the false atmosphere.
(National Women's Register)

Focus group participants frequently complained of claustrophobia because of the crowds and lack of fresh air and natural light. One person described the Brent Cross fountain as 'very refreshing . . . like an oasis really' (Brent Cross Residents' Association) while others complained, 'I find Brent Cross dreadfully claustrophobic . . . if you could see the sky it would be lovely' (Unitarian Church). Members of this group also objected to the artificial music ('this ghastly music that comes out of each shop . . . it just assaults the ears as you go along'; 'the sort of background sickly muzak, wall-to-wall unnecessary noise'); to a sense of general disorientation ('nobody realises how far you're walking in a shopping centre . . . there's no clocks'); and even to the lack of seasonality ('I don't like eating things out of season . . . now it's Christmas all year round'). Shopping centre managers are well aware of these issues, hence their eagerness to incorporate features like trees and shrubs, fountains and waterfalls, all of which serve to 'naturalise' what might otherwise seem an excessively sterile environment. The recent refurbishment of Brent Cross (a joint venture between Hammerson plc and the Standard Life Assurance Co.) includes several such features, opening up the centre to 'natural light' and incorporating a much-enlarged fountain.

Concern for the 'artificiality' of the built environment also appeared to be connected with people's feelings regarding the increasing artificiality of the social relations of consumption: 'we've become more mechanised'; 'we're going to become

TABLE 12.3 Words used to describe Brent Cross and Wood Green shopping centres

	Brent Cross	Wood Green
convenient	69.5	66.8
easy	45.1	44.8
modern	44.1	45.1
local	43.5	43.7
attractive	42.9	32.5
safe	41.6	24.8
familiar	38.4	30.8
enjoyable	38.1	32.2
expensive	21.0	8.8
lively	14.0	17.5
artificial	9.5	11.9
soulless	5.1	7.7
alienating	2.2	2.4
intimidating	1.6	2.8
unsafe	0.0	3.8

isolated, somehow we're never going to really know Chris who sells stamps at the post office any more'; 'it's human contact you want, isn't it' (all from the Unitarian Church group). The transformation of nature into the commodity form has generated a long-standing critique (Haug, 1986), now widely reflected in popular disquiet about conspicuous consumption.[10] For members of the National Women's Register group, shopping for 'luxuries' often provoked feelings of guilt, compared to the virtues of economy and thrift: 'I feel guilty spending money on clothes and things like that', 'I don't often splash out on luxuries [because] I feel guilty'. Guilty feelings often emerged in describing personal indulgence compared to self-sacrifice for the good of the family ('I have to go, really for my children's sake'). One teenager in the Wood Green Area Youth Project compared the apparent selfishness of some parents with her own mother's commendable level of self-sacrifice:

> My mum's not like that, she don't think of herself, she thinks of us two [children] first. It's like if we go shopping, she'll think of us two first and she'll only get herself something every two months, like, get herself a really nice outfit, like kids are first, you get what I mean?

While a young mother reflected ironically on her apparent inability to indulge herself:

> Yeah, but don't you find if you go shopping – this is my problem always – I go out shopping for myself and I say, right, this is it, I'm going out today, I'm going out to get something for myself and I will come back, guaranteed, with something for everyone else at home except what I want.
> (Tottenham Mothers' and Toddlers' Group)

Not all such feelings were expressed in familial terms. Members of the Unitarian Church focus group objected to the sheer scale of Brent Cross: 'It's enormous,

enormous, it's disgusting ... going round the shops and goggling at them'. The impersonality of contemporary shopping was frequently contrasted with a previous generation's investment in 'personal service', the decline of which was widely lamented particularly by the elderly: 'You had personal contact, which you haven't got these days', 'There's a lot of couldn't care less attitude', 'I mean, you got service, didn't you, in those days?' (various members of the Woodside Women's Group).

As we have argued elsewhere (Jackson and Holbrook, 1995), these concerns about the changing nature of contemporary consumption are frequently associated with racialised fears about the nature of neighbourhood change:

It used to be beautiful, lovely shops ...

When there were ordinary shops and that sort of thing down there, the shopkeepers got to known you and you got to know them ...

If there's any corner shops left, either Indians, Pakis or what have you, sweep it up. You see ... Haringey is completely spoilt ... they admit 80 per cent ethnics in Wood Green and Haringey. When we've worked all our lives and now we're strangers because they've taken over.

(Various members of the Woodside Bowling Club)

Concerns for the purity of 'real linen' and 'nice quality shops' are contrasted with a vocabulary of dirt and pollution, where prices are 'dirt cheap', based on imported 'rubbish' and other 'foreign muck'. Racialised fears of dirt and pollution are contrasted with an exaggerated respect for the cleanliness of established stores, like Marks and Spencer's, now considered archetypally 'English' despite its immigrant Jewish roots:

As standards go, you can't get much higher than Marks & Spencer's. I mean, you know, they carry very, very high standards – more than people realise – they are so meticulously clean, it's incredible.

(Brent Cross Residents' Association)

For others, the perceived decline of 'community' is associated with specific memories of the high street or idealised images of village life:

It's not like, you know, the little street where women ... went to school with the kiddies, when they went shopping together, it's definitely not like that at all.

(Brent Cross Residents' Association)

Or, as another participant in this group concluded, 'there's something different about high street shopping from a shopping centre', the former being 'more like a village thing'. While the shopping mall represents an idealised, domesticated version of the high street, free from the risks of uncontrolled public spaces, high street shopping is nostalgically recalled as more spontaneous and less artificial than the mall. While many people expressed a verbal preference for high street shopping, their actual shopping practices suggest that, however sanitised the experience has become, the mall is now generally considered more attractive than the high street.

CONCLUSION

In Marshall Berman's critique of urban modernism he argues that, for most of our century, 'urban spaces have been systematically designed and organized to ensure that collisions and confrontations will not take place there' (Berman, 1988: 165). Rather than simply deploring the loss of public space in city streets and lamenting its trans-formation into privatised commodity spaces, this chapter has argued for a more nuanced social geography of the high street and the mall, sensitive to the socially differentiated nature of people's experience of these highly contested spaces. The meanings of the high street and the mall are not dictated by technology or deter-mined by changes in the built environment. The meanings that people attach to particular places are culturally mediated through experience and use, and vary according to differences of ethnicity and class, gender and generation. The semi-public spaces of the shopping mall are experienced by some people (including many elderly, white, working-class women) as fearful precisely because they are enjoyed by others (male, ethnic minority youth) as desirable places to 'hang out'. Likewise some (white, middle-class) people perceive surveillance cameras and other security measures as a reassuring presence increasing their safety and improving the civility of urban life ('I think it's nice to see them, it's reassuring'), while they are regarded much more ambiguously by other (ethnic minority, working-class) people as a potential threat to their security and an invasion of their privacy.

The findings presented in this chapter might be seen as substantiating the fears of those who are concerned about the democractic consequences of the increasingly blurred boundaries between public and private space. While Brent Cross and Wood Green are some way short of the kind of 'Fortress City' described by Mike Davis (1990) in Los Angeles, questions of access, exclusion, surveillance and control are all too present. Reviewing these issues, Susan Christopherson (1994) has argued that an emphasis on playfulness, spontaneity and interaction within the contemporary urban environment may actually disguise more sinister issues of control, manipulation and isolation. She warns of the 'end of public space', the 'illusion of diversity' and the substitution of 'consumer citizenship' for more open models of genuine democracy. My own findings would suggest that notions of 'consumer citizenship' need to be carefully situated and socially differentiated (by class, race, gender and generation). Rather than assuming that commodification and privatisation are inherently unde-mocratic and reactionary social processes, as Christopherson implies, for example, in her distinction between 'genuine ethnic culture' and 'that which is manufactured for sale' (1994: 414), I would urge us to trace out the specific contours of these processes in particular spaces and places. A more complex cultural politics might then emerge with which concerns about the ever-increasing penetration of the market might be more critically addressed.

Domesticating public space is an understandable reaction to the perceived inci-vility of urban life. But transforming the public space of the high street into the privatised commercial space of the shopping mall has clear social costs in terms of democratic access and public accountability. While we can exaggerate the democracy and freedom of the city's streets both now and in the past, the domestication of space through purification and privatisation necessarily involves increased social exclusion and heightened inequality. How much are we prepared to surrender the control of public space to market forces? How much freedom will we trade for increased security? And who will pay the costs for such perceived benefits in terms of increased surveillance or outright exclusion? Writing of the transformation of public space in Toronto's 'underground city', Hopkins (1996) argues such places inevitably involve morally charged, ideologically-laden questions of social justice. Drawing on evidence from recent research in north London, this chapter has argued that this is no less the case when we examine the contested spaces of the high street and the mall.

ACKNOWLEDGEMENTS

This paper draws on research that was funded by the ESRC under award number R000 234443. Thanks to my colleagues on that project (Bev Holbrook, Daniel Miller, Michael Rowlands and Nigel Thrift) and to Kay Anderson and Susan Smith who helped me to develop the argument about the domestication of space.

NOTES

1 By contrast, Marshall Berman usefully reminds us that 'public space has a dark and check-ered past', that 'Greek agoras, Italian piazzas, Parisian boulevards ... were turbulent places, and needed large police forces on hand to keep the seething forces from exploding' (Berman, 1986: 480–1).

2 The project involved survey work, focus groups and ethnographic research. Besides the author, the project team consisted of Daniel Miller and Michael Rowlands (Anthropology, UCL), Nigel Thrift (Geography, Bristol) and our Research Assistant, Bev Holbrook (now in Social and Administrative Studies, Cardiff).

3 Evocations of other places at WEM include idealised representations of New Orleans' Bourbon Street, a European boulevard, a New York shopping street ('Park Lane'), a Las Vegas casino (Caesar's Palace) and themed hotel rooms representing Egypt, Hollywood, Rome and Polynesia.

4 Sharon Zukin uses a similar metaphor in her account of the restoration of Bryant Park in New York which she cites as an example of 'domestication by cappuccino' (Zukin, 1995: xiv). She describes the same process, later on, as 'pacification by cappuccino' (28). While she does not discuss the implications of the metaphor in any detail, 'domestication' has a range of meanings all of which are appropriate, in varying degrees, to the processes described in this paper. These include the taming and civilisation of wild or savage environments, the privatisation or interiorisation of space, the enclosing and private owner-ship of land, and the feminisation of nature, making a 'home' or familial place from what was previously foreign or hostile territory. For a critical introduction to the domestication of nature, see Anderson (1997).

5 For an assessment of the continued demand for additional retail space, see Reynolds and Howard (1993).

6 These kinds of arguments are often associated with exaggerated accounts of consumer 'resistance', where shopping malls are represented as 'arenas of struggle' or even as 'the terrain of guerrilla warfare' (Fiske, 1989: 14). The basis of such claims is the 'oppositional cultural practices' of those who know the rules by which such centres work but use them

to subvert their intended meanings (browsing without buying, 'hanging out' without shopping, etc.).

7 Estimates of such costs vary but evidence from inner London suggests that 60 percent of trips to a Sainsbury superstore were made by car compared to only 8 percent of trips to a nearby high street store, while in Newcastle, 27 percent of town-centre shoppers travelled by car compared to almost 80 percent of those shopping at the out-of-town Metro Centre (Raven *et al.*, 1995: 10).

8 Discussions lasted for between one and a half and two hours, following a loose agenda drawn up by Bev Holbrook and myself. The discussions were tape-recorded and transcribed in full. Here, for the sake of simplicity, quotations have not been attributed to particular individuals but are identified by the name of the group in which they occurred. The method is described in more detail in Holbrook and Jackson (1996b).

9 Declining crime levels were reported at a Wood Green Open Forum addressed by the local police on 4 September 1995 and were subsequently publicised by the Town Centre Manager, Richard Thomas.

10 Jon Goss writes, for example, of the 'high-cultural disdain for conspicuous mass-consumption resulting from the legacy of a puritanical fear of the moral corruption inherent in commercialism and materialism ... sustained by a modern intellectual contempt for the consumer society' (Goss, 1993: 18). He goes on to identify discourses of dissociation, illusion and disguise, all of which serve to obscure the actual social relations of production that lie behind the commodity form and its associated cultures of consumption.

REFERENCES

Anderson, K. (1997) A walk on the wild side: a critical geography of domestication. *Progress in Human Geography* 21(4): 463–85.

Beck, A. and Willis, A. (1995) *Crime and insecurity: managing the risk to safe shopping.* Leicester, Perpetuity Press.

Berman, M. (1986) Take it to the streets: conflict and community in public spaces. *Dissent* 33(4): 470–94.

Berman, M. (1988) *All that is solid melts into air: the experience of modernity.* Harmondsworth, Penguin.

Bourdieu, P. (1984) *Distinction: a social critique of the judgement of taste.* London, Routledge & Kegan Paul.

Christopherson, S. (1994) The Fortress City: privatised spaces, consumer citizenship, in A. Amin (ed.) *Post-Fordism: a reader.* Oxford, Basil Blackwell, pp. 409–27.

Davis, M. (1990) *City of quartz: excavating the future in Los Angeles.* London, Vintage.

DoE (1988) *Major retail developments.* Planning Policy Guidance 6. London, Department of the Environment.

DoE (1993) *Town centres and retail developments.* Planning Policy Guidance 6 (revised). London, Department of the Environment.

Fiske, J. (1989) Shopping for pleasure. In J. Fiske *Reading the popular.* London, Unwin Hyman, pp. 13–42.

Goss, J. (1993) The 'magic of the mall': an analysis of form, function, and meaning in the contemporary retail built environment. *Annals, Association of American Geographers* 83: 18–47.

Harrington, C. and Crysler, G. (eds) (1995) *Street wars: space, power and the city.* Manchester, Manchester University Press.

Haug, W. F. (1986) *Critique of commodity aesthetics* (trans. R. Bock). Cambridge, Polity Press.

Holbrook, B. and Jackson, P. (1996a) The social milieux of two north London shopping centres. *Geoforum*, 27: 193–204.

Holbrook, B. and Jackson, P. (1996b) Shopping around: focus group research in north London. *Area* 28: 136–42.

Hopkins, J. S. P. (1990) West Edmonton Mall: landscape of myths and elsewhereness. *Canadian Geographer* 34: 2–17.

Hopkins, J. S. P. (1994) A consumption of resistance: challenging the socio-legal identities of malls. Paper presented to the Association of American Geographers annual meeting, San Francisco.

Hopkins, J. S. P. (1996) Excavating Toronto's underground streets, in J. Caulfield and L. Peake (eds) *Critical perspectives on Canadian urbanism*. Toronto, University of Toronto Press.

House of Commons Select Committee on the Environment (1994) *Shopping centres and their future* (Fourth Report, 1993–4 Session). London, HMSO.

Jackson, P. and Holbrook, B. (1995) Multiple meanings: shopping and the cultural politics of identity. *Environment and Planning A* 27: 1913–30.

Jackson, P. and Thrift, N. J. (1995) Geographies of consumption, in D. Miller (ed.) *Acknowledging consumption*. London, Routledge, pp. 204–37.

Nead, L. (1997) Mapping the self: gender, space and modernity in mid-Victorian London. *Environment and Planning A* 29: 659–72.

Raven, H., Lang, T. and Dumonteil, C. (1995) *Off our trolleys? Food retailing and the hypermarket economy*. London, Institute for Public Policy Research.

Reynolds, J. and Howard, E. (1993) *The UK regional shopping centre: the challenge for public policy*. OXIRM Research Paper A28. Oxford, Templeton College.

Sennett, R. (1970) *Families against the city: middle class homes of industrial Chicago, 1872–1890*. Cambridge, Mass., Harvard University Press.

Sennett, R. (1977) *The fall of public man*. New York, Alfred A. Knopf.

Sennett, R. (1993) *The conscience of the eye: the design and social life of cities*. London, Faber & Faber.

Shields, R. (1992) (ed.) *Lifestyle shopping: the subject of consumption*. London, Routledge.

Sibley, D. (1988) The purification of space. *Environment and Planning D: Society and Space* 6: 409–21.

Smith, S. J. (1985) News and the dissemination of fear, in J. Burgess and J. R. Gold (eds) *Geography, the media and popular culture*. London, Croom Helm, pp. 229–53.

Smith, S. J. (1987) Fear of crime: beyond a geography of deviance. *Progress in Human Geography* 11: 1–23

Stansell, C. (1986) *City of women: sex and class in New York, 1789–1860*. New York, Alfred A. Knopf.

URBED (1994) *Vital and viable town centres: meeting the challenge*. Report to the Department of the Environment by the Urban and Economic Development Group. London, HMSO.

Valentine, G. (1989) The geography of women's fear. *Area* 21(4): 385–90.

Walkowitz, J. (1992) *City of dreadful delight: narratives of sexual danger in late-Victorian London*. London, Virago.

Whatmore, S. (1995) From farming to agribusiness: the global agro-food system, in R. J. Johnston, P. J. Taylor and M. J. Watts (eds) *Geographies of global change*. Oxford, Basil Blackwell, pp. 36–49.

Zukin, S. (1995) *The cultures of cities*. Oxford, Basil Blackwell.

13

FOOD AND THE PRODUCTION OF THE CIVILISED STREET

Gill Valentine

●

From medieval times onwards street foods have been a common feature of the British urban landscape (Tinker, 1987). With few lower class households having a kitchen, sending meat out to be cooked in an oven or buying hot foods from cookshops and street vendors were common practices, predating contemporary fast foods by several centuries (Mennell *et al.*, 1992). The development of taverns, coffee houses, 'private hotels', workplace canteens and restaurants in the following centuries all popularised and refined the notion of 'eating out'. In the 1980s and 1990s there has been a particularly marked growth in the contemporary UK 'foodscape' (Yasmeen, 1992). Food is omnipresent on our streets (Finkelstein, 1989) – from takeaway burger bars and fish and chip shops to ice cream vans and chocolate vending machines, yet eating on the street has historically been taboo. Eating, like the other sin of the flesh, sex, has been constructed as a notoriously privatised activity. It belongs in the home, the institution (school or workplace) or the privatised 'public' space of the restaurant and occasionally in the appropriate culturally (and seasonally) sanctioned outdoor space of the park or seaside, but not on the street. This has been one of the unwritten rules that has shaped the nature of 'public culture' and social interaction on British streets over the last four centuries, but is this moral geography of polite food consumption in decline, and if so what does this mean for the street?

This chapter uses empirical material from a study of food, place and identity,[1] to explore the relationship between public acts of consumption and the production of the public space of 'the street'. It begins by outlining why, since the sixteenth century, the British street has been constructed as a 'civilised space' where public eating is out of place. It then goes on to consider contemporary attitudes to eating on the street and 'street culture' in a British northern city. It is therefore a very culturally specific chapter. The evidence of other studies is that 'foodscapes' vary widely between countries and different cultural contexts, as do the histories, traditions and practices of 'public eating' (see for example, Yasmeen 1992).

THE CIVILISED STREET

Food has historically played a key part in defining what it is acceptable to do in the street and when behaviour is out of place in the street. In medieval times it was commonplace for people to eat, belch, fart, spit, shit and so on in public. There were no social or moral codes about the presentation of the self. In 1530, Desiderius Erasmus produced a short treatise on manners titled *De Civilitate Morum Puerilium*. This set out a new concept of bodily propriety, which according to the German Sociologist Norbert Elias (1978/1982), triggered a gradual but none the less fundamental shift in public behaviour, culminating in the development of notions of self-restraint, embarrassment and shame. Initially it was the elite who adopted these new ways of behaving – Elias documents the role of court society in promoting 'civility' – but eventually, these social 'rules' trickled down the social strata and were practised by all citizens, becoming part of our taken-for-granted everyday behaviour. Appropriate ways to consume food formed a part of these moral codes. Margaret Visser argues:

> [P]eople increasingly obeyed these rules (there were fluctuations and differ-ences among groups, and the changes came about in the course of three centuries) not primarily because they were conscious of constraint, but because they were genuinely convinced that no other ways of eating would 'do'. *Civilised* people behaved like this: those guilty of infractions were merely showing how uncivilised they were. The new standards, which began to be introduced in the Renaissance, became gradually internalized, which means that, once learned in childhood, people took the rules for granted; they never thought about them – unless they were suddenly confronted with an action that was 'unmannerly'.
>
> (Visser, 1992: 59)

The custom of civilised eating served to regulate the boundary between 'public' and 'private'. One of the most unmannerly acts for a civilised person to perform was, and for many people still is, to eat in the street. Kass (1994), an opponent of eating on the street explains his revulsion thus:

> Eating on the street – even when undertaken, say, because one is between appointments and has no other time to eat – displays in fact precisely such lack of self control: It betokens enslavement to the belly. Hunger must now be sated it cannot wait. Though the walking street eater still moves in the direction of his [sic] vision, he shows himself as being led by his [sic] appetites. Lacking utensils for cutting and lifting to the mouth, he [sic] will often be seen using his teeth for tearing off chewable portions, just like any animal. Eating on the run does not even allow the human way of enjoying one's food, for it is more like simple fuelling; it is hard to savor or even to

> know what one is eating when the main point is to hurriedly fill the belly,
> now running on empty. This doglike feeding, if one must engage in it,
> ought to be kept from public view, where even if *we* feel no shame, others
> are compelled to witness our shameful behaviour.
>
> (Kass, 1994: 149)

The act of sitting down to eat at a table with cutlery, according to writers such as
Kass, transcends our animality. We are not slaves to our bodies constantly responding
to their greedy demands for food. Rather, it is through struggling against our bodies
by resisting their demands for instant gratification and postponing the moment of
consumption that we demonstrate our humanness. Kass argues that through customs
of 'civilised' eating we suppress our 'nature', we deny our bodies and elevate ourselves.
For example, Erasmus's code of manners was replete with references to controlling
the body, particularly the hands and the mouth, when eating. Indeed Mennell (1985)
claims that many of the taboos about eating and toileting were a product of concerns
about being 'civilised'. They were not motivated by concerns about hygiene. These
fears about food poisoning, body boundaries and 'pollution' of the inside by consumption of the 'outside' came later (Bell and Valentine, 1997a).

To eat at the table necessitates the ability to recognise different tastes which is
reflected in the structuring of the meal into different courses. This in turn requires
the use of different cutlery for different foods. These practices embody human rationality and disguise the animal need for food (Kass, 1994). Visser (1986: 12) states
that 'Civilisation entails shaping, regulating, constraining and dramatizing ourselves;
we echo the preferences and the principles of our culture in the way we treat our
food'. While Curtin points out that 'Food structures what counts as a person in our
culture' (1992: 4).

Taking a meal at the table with others defines and promotes commensality and
'community' at all scales from home to the nation (Bell and Valentine, 1997a). It
marks a commitment to social forms of behaviour – in particular, the meal at the
table has come to symbolise 'the family' (Charles and Kerr, 1988). The sociality of
the shared meal also distances us from the act of feeding.

> Paradoxically, eating with others enables us to draw attention away from our
> neediness and from the power that hunger exercises over us. Having company
> in meeting our necessity permits selective inattention to the brute fact of
> necessity. But when we see others staring at our food – or at our biting and
> chewing – we cannot hide from ourselves both that we are feeding ourselves
> and also what it means – including our enslavement to appetite, our submis-
> sion to the largely involuntary acts of biting, chewing and swallowing.
>
> (Kass, 1994: 147)

Thus to eat in the street is to give in to animal urges. It has connotations of
primitiveness, bestiality, the wild. To eat on the street is to graze, to use hands rather
than cutlery, to threaten social rules concerning the mouth – to lick, to chew nosily,
to drop food from the mouth and so on. It is to feed rather than to eat (Kass, 1994).

In addition to articulating a nature/culture dichotomy, taboos about eating on the street also articulate a particular understanding of 'public' and 'private'. The street may be a site of consumption but only a particular disembodied form of consumption is civilised – tomato sauce dripping down the chin is not an appropriate public spectacle. Erasmus's concept of bodily propriety necessitated that the body should be controlled and hidden. All signs of bodily functions should take place in 'private' space. The gaze – famously identified by Foucault (1977) as a mechanism of disciplinary power – plays an important role in producing appropriate public bodily performances. Outlining the development of penology in *Discipline and Punish*, Foucault argues that the gaze is potent because it gives 'power of mind over mind' so that individuals exercise self-discipline, producing subjugated, what he terms 'docile' bodies. He writes:

> [T]here is no need for arms, physical violence, material constraints. Just a gaze. An inspecting gaze, a gaze which each individual under its weight will end by interiorising to the point that he is his own overseer, each individual thus exercising surveillance over and against himself.
>
> (Foucault, 1977: 155)

This fear of the 'public gaze' – of being seen to eat in public – has served to put a moral brake on the pleasures of street food for many potential consumers. Women in particular can experience an extra dose of anxiety about being seen to be eating in the street because it has been constituted as an unladylike practice, women being constructed as genteel and delicate in terms of manners and as having smaller appetites and being more self-restrained than men (Lupton, 1996). Women's bodies have also been subject to a range of discourses which have defined geographically and historically specific norms about what the ideal feminine body shape should be (Garner *et al.*, 1980; Charles and Kerr, 1988; Bell and Valentine, 1997a). These discourses have particularly elided representations of slim female bodies with health and sexual attractiveness in what Lupton (1996: 137) has termed the 'food/health/beauty triplex'. Thus some women – particularly those who are overweight – avoid eating in public, especially on the street because of all its unfeminine connotations of animalistic greed.

While eating on the street can inspire us to feel shame and embarrassment at our selves, public eating has also been considered inappropriate because of its capacity to embarrass or revolt others forced to witness the spectacle of this intimate, embodied activity. Erasmus's treatise on manners required that 'You were controlling yourself, so as to prevent other people from being disgusted or "shocked"' (Visser, 1992: 60) by your behaviour. Such sensibilities around civilised eating were particularly apparent in Victorian Britain where eating in public was forbidden because 'to eat or be forced to see another eat promiscuously or immodestly constituted a kind of social obscenity' (Kasson, 1987: 139, quoted in Lupton, 1996: 22). Kass describes this, in his terms, offensive practice:

> [E]ating is *out of place* [my emphasis] in public, except on those public occasions explicitly convened to include it (like public festivals) or in those

public places set aside for eating (like restaurants or picnic areas). A man [sic] eating as he walks down the street eats in the face of passers-by, who must either avert their gaze or observe him objectively in the act. Worst of all from this point of view are those more uncivilised forms of eating, like licking an ice-cream cone.

(Kass, 1994: 148)

Polite behaviour on the street therefore is a ritual performed largely for the sake of others and for the sake of our relationships with them (Visser, 1992). Visser explains 'Politeness forces us to pause, to take extra time, to behave in pre-set, pre-structured ways, to "fall in" with society's expectations' (Visser, 1992: 40). Following the publication of Erasmus's treatise a person's public performance of the self, particularly their skills at body management and controlling emotions, became increasingly crucial to their social standing and success. Mennell (1985, 1991) argues that while in the Middle Ages insecure food supplies meant that the population was either feasting or fasting, so that feasting was a sign of wealth; by the eighteenth century improvement in food supplies contributed to the gradual civilising of appetite, so that 'Elegance and refinement, one's distance from the lower classes, came to be represented by the delicacy of food eaten and the moderation of the appetite' (Lupton, 1996: 21).

Similarly, Elias's (1982) work demonstrates that 'the purpose of manners and modes of etiquette was to distance oneself from the potentially menacing "other" in order to distinguish oneself from them' (Finkelstein, 1989: 166). The bourgeoisie, because of their desire to move up the social ranks, had to be particularly vigilant so as not to 'slip up' in their performance of the self and undermine their own social position, while simultaneously using 'good manners' to protect themselves from the greed, and lower instincts of others, making it difficult for the lower classes to cross social boundaries.

A more contemporary analysis of the social importance of an appropriate performance of the self is provided by the French sociologist, Pierre Bourdieu. He argues that 'the body is the most indisputable materialization of class taste' (Bourdieu, 1984: 190). He developed the concept of *habitus* to describe distinguishing aspects of behaviour, taste and consumption which coalesce to create flexible, rather than rigid categories of 'class', 'taking account of different concentrations of economic and cultural capital' (Shurmer-Smith and Hannam, 1994: 194). His argument is summarised by Featherstone thus:

Classes reproduce themselves by their members' internalization and display of certain tastes, which then mark only some for distinction. At the foundation of these tastes is the body. Taste is *embodied* being inscribed onto the body and made apparent in body size, volume, demeanour, ways of eating and drinking, walking, sitting, speaking, making gestures etc.

(Featherstone, 1987: 123)

Good taste does not include eating on the street. This is a 'common' and 'uncouth' practice. Walton (1992) for example, describing the quintessentially British fast food, 'fish and chips', argues that from their appearance towards the end of the nineteenth century to the present day, fish and chips have been seen as 'common' – as food for the working classes.

An important part of 'civilising' middle-class children throughout the past four centuries has therefore involved teaching them how to eat correctly because when they go out they are representing the family. Their behaviour has the power to bring embarrassment and shame on their parents and to undermine the parents' social identities. Visser (1992: 48) for example notes the Victorians had a proverb: 'every meal is a lesson learned'. Middle-class children were often not allowed to eat with or in the sight of adults until they had learned to eat properly. Elias claims that contemporary children now have 'in the space of a few years to attain the advanced level of shame and revulsion that has developed over many centuries' (Elias, 1982: 38).

Taboos about eating on the street are therefore bound up with social and moral understandings of humanness, 'civility', gender, and 'class' but they have also been captured to some extent in actual legislation, established by the Victorians. In the Victorian era middle-class anxieties about the working classes were expressed as a strong fear of spontaneity, disorder, uncleanliness, and a desire for rationality, control and order. At this time roving food vendors were a common part of the streetscape. To the Victorian middle classes they represented all the dangers of the working class. They were disorderly, polluting and associated with mass-consumption which to the Victorians stood for destructiveness and waste. As a result suburban fears about the menace of roving food vendors produced legislation on street trading which has contributed to contemporary popular conceptions of the moral geography of polite public food consumption (Burnett, 1983; Bell and Valentine, 1997a).

Food has therefore played an important part over the last four centuries in producing the street as a particular social environment. Society creates social expectations about how we should behave in different places. Some of these are written in law, others are unstated, taken-for-granted assumptions which become normative. These define what it is acceptable to do in a place and when behaviour is 'out of place' (Cresswell, 1996). We in turn read or decode the spaces we are in and respond to them. In this way 'Place is produced by practices that adhere to (ideological) beliefs about what is the appropriate thing to do. But place reproduces the beliefs that produce it in a way that makes them appear natural, self-evident. . . . Place constitutes our beliefs about what is appropriate as much as it is constituted by them' (Cresswell, 1996: 16).

Social understandings of human 'civility', class and gender in relation to food and eating have played an important part in producing the street as a civilised space. Social expectations about eating have determined that the street should be a space of self-restraint, culture and civility where the appetite is suppressed rather than a place where we are free instantly to gratify our so-called 'natural', 'animal' urges to eat; that it should be a 'public' place where intimate, 'private' bodily matters, such as the act of eating, are not on display; and that it should be an ordered place where the mess of consumption is out of sight and out of mind, rather than a space which is polluted by the disorderly behaviour of 'public' grazers. In turn the relative absence of people publicly eating in the street has in the past signified the inappropriateness of this activity in this space – making those who can't resist a bite on the move feel socially uncomfortable and 'out of place'. But the evidence of this research[1] is that attitudes to eating in the street – particularly by the young – may be changing and that this in turn may be leading to a re-definition of the nature of 'public culture' and hence the place of the street.

THE SELF-INDULGENT STREET

The last decades of the twentieth century have been marked by a rapid change in the pace of life. While the 'family' evening meal appears to remain an important social institution in many households (Whitehead, 1984; Charles and Kerr, 1988; Bell and Valentine, 1997b), people appear to have less time for the rituals of civilised dining during the day. There appears to be increasing 'disorganisation' in what constitutes a meal and the structure and timing of meal cycles (Herpin, 1988). The traditional British meal cycle of three main meals a day: breakfast, lunch and dinner (Goode *et al.*, 1984) has lost its midday anchor point. The culture of long business lunches has disappeared. Work pressures mean that employees in a range of walks of life find themselves forced to 'snatch' what they can in their break times (Valentine, 1997). For mothers in full-time paid employment the stress of trying to juggle 'home' and 'work' responsibilities means that the lunch hour is often used to complete domestic or personal chores, so that lunch becomes a sandwich consumed on the run rather than a 'meal' break. These two interviewees describe how the tempo of their work-places robbed them of the chance to sit down for a 'civilised' meal at lunch time.

> When you were working with clients, it just depends what sort of client it was, ehm, some of them we'd end up for a meal at a local restaurant or in the pub. That was, well when I started there that was quite normal, but by the time I'd finished there, the pressure and budgets had got to the stage where you couldn't usually, couldn't spare a lunch time to actually go out to eat, even to the pub. Ehm, plus the whole attitude towards drinking at lunch time changed quite a bit as well, because people got under more and more pressure they didn't want people out at the pub.
>
> (Woman, in her thirties, former accountant)

> I always seem to be busy in my lunch hours from work. I've always thought of my lunch hours from work as time to get things done rather than time to eat. So I've always tended to eat my lunch while I'm working or on the run and then use the lunch break to go off and get things done . . . you know the administration, you know everything, the running of things [the household], so yeah, I tend to do those things in my lunch hour.
>
> (Woman, in her forties computer operator)

A similar picture is also evident in children's lives. The premium placed on space and time, in many large comprehensive schools means that schools no longer have any choice but to switch from 'family service' style school dinners – where children replicate the 'family' meal by sitting at a table together, taking turns to wait on each other and abiding by social conventions and manners – to 'fast food' style cafeterias. As a result the social conventions of the 'family' meal have been substituted by more uncivilised – what Kass (1994) might even term – 'animalistic' practices of solo eating, grazing and canteen style fast foods. The social 'norm' for these children is not to consume food with cutlery but rather to use their fingers. Morrison describes her research on food in schools:

> If all pupils were to decide to make the canteen their regular lunch-time venue there would be insufficient places to accommodate them. As importantly, the process time required to consume canteen meals contradicts the clock time available. Fast eating is, therefore, an institutionalised necessity. Among the pupil adaptations to this is widespread eating without a knife or a fork, from paper bags and cans, and, as importantly, on the move. . . . Whilst there may be important reasons for young people to disregard what adults might consider appropriate 'table manners', there are very practical reasons, namely a scarcity of space and time, for pupils to eat quickly and without implements.
>
> (Morrison, 1995: 243)

The introduction of compulsory competitive tendering in schools, also means that now school meals providers must compete for children's custom with local chip shops and burger bars. Not surprisingly, many children prefer to 'graze' at their leisure on the High Street away from the surveillant gaze of teachers, rather than to eat in the hurly burly atmosphere of the canteen.

At lunch time the street is now a landscape populated by individual snackers, eating at different times, snatching different bits of food in different venues in between workplace meetings, school lessons and so on. Fine and Leopold (1993) for example claim that surveys suggest that in the UK, 75 per cent of adults and 91 per cent of children eat snacks at least once a day, often instead of a meal. They argue that 'Food is eaten more frequently but in smaller quantities. As a result there is a tendency to skip regular meals' (Fine and Leopold, 1993: 165). Goody (1982) argues that such shifts in eating habits are usually strongly related to changes in food production and distribution practices. Certainly the last decade has been a period characterised by economic restructuring and social change during which technology used in the production, storage, preparation and retailing of food, and patterns of food consumption, have significantly evolved. In particular, developments in the food-processing sector have expanded and promoted the range and availability of convenience and fast foods on the street.

The foodscape – the type and range of places from which foods are available outside the home – has also drastically expanded as the food industry has become more sophisticated at tapping into new markets. Finkelstein (1989) argues that food has been recast as fashion and entertainment – with internationally themed fast-food restaurants synthesising global and local cultures to provide foods from around the world to suit local tastes and habits. She notes that in 1986 there were over 140,000 different kinds of fast-food restaurant in the US alone, with McDonalds accounting for 17 per cent of the eating out market. Many of these fast foods, like McDonald's, are designed to be eaten without cutlery, to be consumed at speed, perhaps fleetingly at a plastic table or more likely on the street.

With these changes in the foodscape have gone some of the social stigma of responding to bodily demands for food with instant gratification. The culture of the 'civilised meal' has been usurped by the more informal social codes that belong to the contemporary world of fast food – eating in the street, using hands rather than cutlery, enjoying the messy pleasures of takeaway foods as this woman describes:

> 'I'm a mucky git of course I eat in the street [laughter]. A lot of people find that very rude innit while you're walking and taking and eating . . . I'm always forever doing that, I am you know, if I'm hungry I won't wait until I'm sat on the bus . . . No if I'm hungry I will get it out there and then . . . It's how you feel. And I'm not, you see I'm not a kind of shy person . . . It's very rare like I would get sommat and say well let me finish me shopping.
>
> (Woman, in her thirties, homemaker)

In the past the localised scale of neighbourhood shopping where proximity equalled intimacy meant 'public' behaviour was under the constant scrutiny of friends, neighbours and relatives. Close-knit school and workplace communities, often articulated in dress codes and uniforms, also contributed to making people conscious that their 'public' performances were under the surveillant gaze of others as this woman describes.

> Nah, I wouldn't feel right eating, Naa bar a chocolate, Naa I wouldn't eat a sandwich or anything walking down the street. [*Interviewer: why not?*] I don't know really. Probably how I feel about it. Unless it goes back to when I were doing my [nursing] training, and we weren't allowed to eat. Because we, er, we had an outdoor uniform that we wore. And if you were in uniform you weren't allowed to eat while you were in uniform.
>
> (Woman, in her forties, nurse)

These social pressures served (and in some cases still do) to ensure a civilised performance of the self in the street, but the anonymity offered by the scale and bustle of the contemporary city street has effectively removed this regulatory framework for many people. Today, in schools at least, teachers no longer carry out lunch-time duties and it is less common for schoolchildren (particularly in the upper years) to wear identifying uniforms. Most young people no longer feel any shame or embarrassment about publicly performing intimate bodily functions on the street, such as eating, nor are they concerned about whether such displays will offend other people's sensibilities. Rather than a sign of rudeness, eating on the city street is now understood as a pleasurable everyday activity (although the power of the surveillant gaze to produce a 'civilised street' may still have resonance in close-knit neighbourhoods and villages).

I'll happily walk down the High Street with one of the vegetable pasties from the Bakers Oven with bits of flaky pastry pouring over my coat as I go. I mean I'll sort of yeah, I will wipe it off as I go. I'm probably a bit more aware that you do look a bit of a state but which is still like a big contrast to like my Mum who was brought up that it was really awful to be seen eating in public and she still cringes if I walk down the street having a packet of crisps which seems to me a fairly moderate thing to eat in public. And I think she was brought up to believe that it sort of showed you were greedy somehow and you couldn't wait till you got home and you were stuffing your face at every opportunity possible, which I probably am, so she's probably right! But she saw that as like yeah something you just wouldn't let the neighbours see you doing.

(Man, in his twenties, office worker)

My grandmother is very funny about people eating in the street – she doesn't think it's the done thing at all, but I've just never been able to see what the problem is [*Interviewer: why doesn't she like it?*]. I think it's a very traditional view that it's impolite. I don't really understand why she doesn't like it, because to me if you're hungry and you happen to be in the street you eat there. I happen to quite like, perhaps if you're out shopping getting some food and sitting down on a bench and eating it, because to me, I like something to watch and sitting watching the world go by while you're eating to me seems quite nice.

(Woman, in her thirties, environmental worker)

Indeed, for teenagers the street is often the only private space they can carve out for themselves away from the regulatory gaze of family and teachers. Hanging round fish and chip shops and burger bars, eating on the street and carelessly tossing litter into people's gardens has become one way that young people can resist adult productions of space (Valentine, 1996a, 1996b).

Eating on the run offers equal possibilities for adults to breach social conventions because it transgresses all the carefully constructed class, gender distinctions and 'normative' constraints of 'manners and social graces' that are played out over the dinner table in the consumption of shared meals. Alfresco eating creates unpredictable spaces of freedom for the consumer – Zukin (1991: 235) for example talks about the 'illicit pleasures of mobility'. Many fast foods are also claimed to be more tasty when consumed on the street from paper or wrappers rather than when served on a plate at the table.

Fish and chips, that's the thing you can eat in the street and really enjoy it . . . It just tastes better when you eat it out of paper in the street. I don't know why, but it does. People have said this for years but it definitely does taste better . . . I think more so when you are away, you know if you are on holiday at the seaside for a day, or something like that. I mean, I know whenever we've gone to the seaside for the day you know . . . take me to the seaside or sort of somewhere and they're cooking fresh doughnuts. You know like in that little machine where they're cooking, the smell, I love doughnuts.

(Woman, in her forties, unemployed lone parent)

Sharon Zukin (1991: 39) argues that in contemporary Western cities a 'blurring of distinctions between many of the categories of space and time that we experience is taking place'. Home, work, school and the street were once very distinctive consumption landscapes – with eating being confined to the more 'privatised' spaces of home, canteen and restaurant. Now changing practices in each of these locations are increasing the role street foods play in our diets and are contributing to the revision of our understandings of 'civility'. In turn, the practice of eating, or 'grazing' on the street appears to be breaking down some of the social codes and that helped to maintain boundaries between distinctive spaces and times. The city street no longer appears to be a space where 'private' bodily propriety is quite so rigidly regulated by the 'public' gaze. Rather the city street appears to be a more liminal space somewhere betwixt and between home, canteen and restaurant (although this may be less true of the street in close-knit neighbourhoods and rural communities).

The wide range and hybrid quality of food available to eat on the British city street appears to have helped to create a variety of public cultures (high/low/ethnic etc.) and to have contributed to the relaxation of some of the social codes around performances of the self. Rather than being produced as a 'formal' space governed by codes of 'civility', 'privacy' and 'order' where we have to be watchful to maintain appropriate 'class', 'gender' and even 'human' performances, the street is being produced, through people's grazing and snacking habits, as a more informal, more democratic space – or perhaps some cynics might argue just a more consumption oriented space – where people no longer feel 'out of place' engaged in the self-indulgent practice of sating their appetites!

ACKNOWLEDGEMENTS

I am grateful to the Leverhulme Trust for funding the research upon which this chapter is based. I also wish to thank David Bell and Beth Longstaff for being such efficient, fun and stimulating research partners on this project.

NOTE

1   This involved case study research conducted in a northern British City with twelve households, two schools and a prison. Participants were interviewed five times each and were asked to keep food diaries, shopping receipts and to participate in constructing their own accounts of their eating through drawings, photographs and videos. The quotations taken from interview transcripts are verbatim. Ellipsis dots are used to indicate where an edit has been made. I am grateful to the Leverhulme Trust for funding this study.

REFERENCES

Bell, D. and Valentine, G. (1997a) *Consuming Geographies: we are where we eat.* London: Routledge.

Bell, D. and Valentine, G. (1997b) 'Eating identities: food, "family", home and the lifecourse'. Paper available from the authors.

Bourdieu, P. (1984) *Distinction: A social critique of the judgement of taste.* London: Routledge.

Burnett, J. (1983) *Plenty and Want: a social history of diet in England from 1815 to the present.* London: Methuen.

Charles, N. and Kerr, M. (1988) *Women, Food and the Family.* Manchester: Manchester University Press.

Cresswell, T. (1996) *In Place/Out of Place: geography, ideology and transgression.* Minneapolis: University of Minnesota Press.

Curtin, D. W. (1992) 'Food/Body/Person' in D. W. Curtin and L. M. Heldke (eds) *Cooking, Eating, Thinking: transformative philosophies of food.* Bloomington: Indiana University Press, pp. 3–22.

Elias, N. (1979/1982 [orig. 1939]) *The Civilising Process,* vol. I: *The History of Manners,* vol. II: *State Formation and Civilisation.* Oxford: Basil Blackwell.

Featherstone, M. (1987) 'Leisure, symbolic power and the lifecourse' in J. Horne (ed.) *Leisure, Sport and Social Relations.* Sociological Review Monograph 33.

Fine, B. and Leopold, E. (1993) *The World of Consumption.* London: Routledge.

Finkelstein, J. (1989) *Dining Out: a sociology of modern manners.* Cambridge: Polity Press.

Foucault, M. (1977) *Discipline and Punish.* London: Allen Lane.

Garner, D., Garfinkel, P., Schwartz, D. and Thompson, M. (1980) 'Cultural expectations of thinness', *Psychological Reports* 47: 483–91.

Goode, J. G., Curtis, K. and Theophano, J. (1984) 'Meal formats, meal cycles and menu negotiation in the maintenance of an Italian-American community', in M. Douglas (ed.) *Food in the Social order: studies of food and festivities in three American communities.* New York: Russell Sage.

Goody, J. (1982) *Cooking, Cuisine and Class. A study in comparative Sociology* Cambridge: Cambridge University Press.

Herpin, N. (1988) Le repas comme institution: compte rendu d'un enquete exploratoire', *Revue Francaise de Sociologie* 29: 503–21.

Kass, L. (1994) *The Hungry Soul: eating and the perfecting of our nature* New York: Free Press

Kasson, J. (1987) 'Rituals of dining: table manners in Victorian America, in K. Grover, (ed.) *Dining in America 1850–1900.* New York: University of Massachusetts Press, pp. 114–41.

Lupton, D. (1996) *Food, the Body, and the Self.* London: Sage.

Mennell, S. (1985) *All Manners of Food: eating and taste in England and France from the Middle Ages to the present.* Oxford: Basil Blackwell.

Mennell, S. (1991) 'On the civilising of appetite', in M. Featherstone, M. Hepworth, and B. Turner (eds) *The Body: social process and cultural theory.* London: Sage, pp. 126–56.

Mennell S., Murcott, A. and van Otterloo, A. H. (1992) *The Sociology of Food: eating, diet and culture.* London: Sage.

Morrison, M. (1995) 'Researching food consumers in school: recipes for concern', *Educational Studies* 21, 2: 239–63.

Shurmer-Smith, P. and Hannam K. (1996) *Worlds of Desire, Realms of Power: a cultural geography.* London: Edward Arnold.

Tinker, I. (1987) Street foods: testing assumptions about informal sector activity by men and women. *Current Sociology* 35, 3: 232–53.

Valentine, G. (1996a) 'Angels and devils: moral landscapes of childhood', *Environment and Planning D: Society and Space* 14: 581–99.

Valentine, G. (1996b) 'Children should be seen and not heard: the production and transgression of adults' public space', *Urban Geography* 17, 3: 205–20.

Valentine, G. (1997) 'The business lunch: geographies of the body in the workplace'. Paper available from the author.

Visser, M. (1986) *Much Depends upon Dinner*. London: Penguin.

Visser, M. (1992) *The Rituals of Dinner*. London: Viking.

Walton, J. K. (1992) *Fish and Chips and the British Working Class, 1870–1940*. Leicester: Leicester University Press.

Whitehead, T.L. (1984) '"Sociocultural" dynamics and food habits in a Southern community', in M. Douglas (ed.) *Food in the Social order: studies of food and festivities in three American communities*. New York: Russell Sage Foundation, pp. 55–81.

Yasmeen, G. (1992) 'Feminine foodscapes: women and food vending in Thailand'. Paper presented at the Southeast Asian Studies Summer Institute Graduate Student Conference, University of Washington, Seattle. Available from the author.

Young, I. M. (1990) *Justice and the Politics of Difference*. Oxford: Princeton University Press.

Zukin, S. (1991) *Landscapes of Power: from Detroit to Disneyworld*. Oxford: University of California Press.

Zukin S. (1995) *The Cultures of Cities*. Oxford: Basil Blackwell.

14

THE CULTURE OF THE INDIAN STREET

Tim Edensor

•

In this chapter, I examine the culture of Indian streets to provide a contrast to the Western streets considered elsewhere in this volume. It is important that explorations of the street should not blunder into the ethnocentric pitfalls of so many social and cultural theories, which examine distinct Western contexts and produce ideas that are taken as universally applicable. My principal aim is to highlight the increasingly regulated qualities of Western street life by examining the rich diversity of social activity in Indian streets. It is not my intention to idealise or romanticise the Indian street as a space of the 'Other' but I realise that my position as a Western scholar will leave me open to the charge of 'othering'. I recognise that Westerners seek out the different experience offered by the Indian street partly because they have consumed fantastic narratives and images of India. However, I go on to argue that these socially constructed preconceptions may be mediated or undermined by the sensual and social experience of space. This is part of a wider argument which insists that streets are not merely texts to be read. Those passing through, living and working in streets interpret their experience through social, sensual and symbolic processes. Thus, whilst the description I provide of the Indian street is necessarily general, it is not intended to convey any ideal, and although it may seem as if I am reinforcing a binary distinction between West and 'Other', I insist that the material and social distinctions between Western and Indian streets do exist, but they exist within an uneven global process whereby space is becoming more commodified and regulated.

After a discussion about the social practices, forms of movement, regulation, and sensual experience of Indian streets, there is a comparative section on the forms of social life and regulation of the Western street. I then explore the relationship between Western and Indian streets, arguing that the latter are 'othered' partly because similar material and social qualities have been expunged in the West by the intensification of consumer capitalism and the Appollonian urge to rationalise and regulate. Accordingly, this 'overdevelopment' has meant that 'other' spaces such as Indian streets retain a fascination for Westerners hungry for temporary disorder. I end the chapter by interrogating some of the most influential metaphorical concepts used in cultural geography, namely the *flâneur*, 'heterotopia' and pedestrian 'tactics' to show how their application can be widened and contextualised in different processes of social spatialisation.

SOCIAL PRACTICES

The Indian street is part of a 'spatial complex' which also comprises the bazaar and the fair and together they constitute an unenclosed realm which provides a 'meeting point of several communities' (Chakrabarty, 1991: 23). Thus, the street is located within a cellular structure that suggests a labyrinth, with numerous openings and passages. The flow of bodies and vehicles criss-cross the street in multidirectional patterns, veering into courtyards, alleys and cul-de-sacs. The busiest streets, the main arteries of this spatial network, are never merely 'machines for shopping' but the site for numerous activities.

This is reflected in the diverse spaces in and around the Indian street. Shops co-exist alongside work places, schools, eating places (see Figure 14.1), transport termini, bathing points, political headquarters, offices, administrative centres, places of worship and temporary and permanent dwellings. The multifunctional structure of the street provides an admixture of overlapping spaces that merge public and private, work and leisure, and holy and profane activities. This diversity contains a host of micro-spaces: corners and niches, awnings and offshoots.

In the bazaar a sense of familiarity is maintained through particular modes of address, types of economic exchange and the maintenance of formalised and convivial obligations. These strategies for dealing with the unfamiliar contribute to the formation of a gregarious environment which privileges speech and removes barriers between backstage and frontstage so that visual and verbal enquiry is facilitated. This provides a congenial environment for economic exchange, typified by barter, which, as Buie describes, is a sensual as well as economic activity; an 'art', a 'ritual' and a 'dance of exchange' (Buie, 1996: 227). Besides this particular form of economic activity, the proliferation of spaces provide contexts for a range of social practices that range from the commercial to the recreational, and from the industrial to the ritual. Such streets are 'centres of social life, of communication, of political and judicial acvtivity, of cultural and religious events and places for the exchange of news, information and gossip' (ibid.).

As a commercial realm, the street is occupied by diverse enterprises, organised according to a variety of time-space constraints. Whilst there are fixed shops, the street is also the work place of mobile providers of services such as dentists, fortune-tellers, shoe-shiners, barbers (see Figure 14.2), letter-writers, shoe repairers, bicycle fixers and tea-wallahs, as well as mobile stalls of all kinds. Moreover, the open fronts of most workshops mean that the activities of engineers, smiths, potters (see Figure

FIGURE 14.1 Cooking on the street, Agra, Uttar Pradesh. Source: author

14.3), bookbinders, metal workers and others spills out onto the side of the street, further blurring frontstage and backstage realms and activities.

As well as being a social space for transactions of news and gossip, particularly organised around particular micro-spaces such as rickshaw termini and tea stalls, the street is a site for announcement, and is host to adverts transmitted visually or by loudspeaker. For instance, vans publicise the current movie attractions with samples of the soundtrack, and when there are elections or local political disputes, loudspeaker vans broadcast political slogans. Demonstrations by political parties, and religious processions, theatrically transform the street into a channel of embodied transmission, and striking workers hold meetings and occupy spaces. The street thus becomes a temporary stage where political dramas and religious observances are played out.

As a site for entertainment, children make their own amusement, playing cricket and other games, whilst adults play cards, chess and *karam*. Moreover, travelling entertainers such as musicians, magicians and puppeteers set up stall and attract crowds. Besides these travellers, there are disparate hawkers and beggars as well as bands of religious adherents, *saddhus* and holy men, occasionally performing acts of abstinence and endurance. There is thus a constant stream of temporary pleasureable activities, entertainments and transactions. But there are also more mundane social activities such as loitering with friends, sitting and observing, and meeting people that also form distinct points of congregation.

Since many dwellings are located at the side of the street, it is also the site for domestic activities such as collecting water, collecting dung for fuel (see Figure 14.4), washing clothes, cooking and child-minding. For the pavement dwellers, the street is also a temporary home, necessitating the carrying out of bodily maintenance such as washing. Such temporary sites and activities dissolve preconceived notions of ownership, and question the distinction between private and public (Chandhoke, 1993: 69).

This proliferation of multi-use spaces can be dramatically contrasted with colonial attempts to demarcate single-purpose spaces, dividing cities into industrial, commercial and domestic areas, and more dramatically, constructing a physical separation between colonisers and colonised. Central to European concerns was the perceived erasure between public and private realms: colonisers were affronted by the ways in which open space was used for the domestic tasks and rituals of washing, changing, sleeping, urinating and cooking. The colonial enclaves built by the British testify to the urge to reconstruct urban and suburban aesthetics and order upon what was imagined as urban chaos. The erection of private bungalows, gardens, administrative buildings, and the laying out of parks and leisure facilites such as tennis courts, gymkhanas and golf courses, impose an alternative metropolitan spatial order wherein a network of manicured, broad avenues are marked against the imagined disorder of the 'native' quarter. Today, in many Indian cities, the colonial quarter has been reclaimed by bourgeois, commercial and administrative groups who attempt to re-imprint a power-in-spacing by appropriating these boundary-marking distinctions.

The range of social activities and demands in the bazaar tends to deny the pedestrian the option of seeking refuge in a distanced disposition; the social immersion that such an environment demands disrupts any lofty detachment.

MOVEMENT

It is difficult to move in a straight line on an Indian street. The pedestrian has to weave a path by negotiating obstacles underfoot or in front, avoiding hassle and teasing, and remaining alert about the hazards presented by vehicles and animals such as monkeys, buffaloes (see Figure 14.5), cows, pigs and dogs. Walking down the street cannot be a seamless, uninterrupted journey but is rather a sequence of interruptions and encounters that disrupt smooth passage.

The abundant simultaneous cross-cutting journeys means that purposive travel towards an objective must take account of others who will cross one's path. Rapid progress is usually frustrated. The variety of activities that are played out on the street are enacted at different speeds. Some linger or lounge, others gather in groups for long spells. Given the diversity of social activities played out in the street, there are a host of differently constituted time-space paths as people pursue diverse aims.

FIGURE 14.2 Street barbers and cycle rickshaws, Agra. Source: author

FIGURE 14.3 Roadside pottery, Agra. Source: author

FIGURE 14.4 Collecting dung for fuel, Agra. Source: author

The miscellaneous collection of vehicles that use the street: bullock-carts, cars, bicycles, motorbikes, auto- and cycle-rickshaws, buses and other diverse forms of transport, all move at different speeds as they manoeuvre for space, providing an ever-changing dance of traffic which contrasts with the controlled flow and pace of traffic movement on Western thoroughfares.

Thus passage is marked by disruption and distraction, not only by the exigencies of avoidance and the physical collision with others, but also by the distractions and diversions offered by these heterogeneous activities and sights. The choreographies of the street, with intersecting movements differing in direction and tempo, and constituted by humans, vehicles and animals, continually change, incorporating the necessarily contingent character of the pedestrian's dance.

REGULATION

The bazaar and street are subject to regulation but this is contingent, contextual and local. Rather than security guards, video surveillance and policing (see Chapters 15 and 17 of this volume), local power holders exercise policies of exclusion and control. Overall, however, surveillance is rather low-level. Whilst there are formal traffic rules, the various species of vehicles pay little heed to them as they jostle for position. The street performers, beggars and touts are rarely advised to 'move on' and the mentally

FIGURE 14.5   Driving buffaloes through a main thoroughfare. Source: author

and physically handicapped are not confined to institutions. The domestic, stray and wild animals that share the streets with people may suffer cruelly but there are few systematic attempts at controlling their movements or numbers. As I have mentioned, in most urban areas, small shops and makeshift dwellings spring up overnight on the borders of streets without seeking planning permission.

In a similar fashion, streets and bazaars are not subject to aesthetic control other than by force of convention. Streets are rarely planned to convey a particular overall impression or theme, and neither are street dwellings and other buildings policed to maintain an 'appropriate' appearance, with *ad hoc* signs, embellishments and crumbling masonry usually permitted.

This seeming disorder and lack of regulation disguises the forms of power that are played out in the street. For instance, a gendered distinction between private and public is evinced in that most of the shopkeepers and artisans in the public realm are male since it is generally considered unsafe and unrespectable for women to spend much time in certain public spaces. Similarly, in many villages and towns, the spatial divisions of caste are rigidly adhered to, although this is less marked in large urban areas. However, the demarcation of religious quarters can be rigidly maintained and the brutal communalist policing of religious others may flare up in times of political tension, as in the recent spate of 'fundamentalist' Hindu attacks on Muslim areas in mixed urban areas following the demolition of the Babri Masjid in Ayodhya. Power also works its way onto the street in less obvious ways. Bribes and favours are often needed to secure commercial sites and violence may be held in reserve to control lower castes and religious minorities from occupying particular domestic and work areas. Even in the most seemingly chaotic spaces of the shanty town, slum lords may wield control (Chandhoke, 1993: 70).

But even in the most regulated spaces, the 'unintended city' of the 'shanty town' insistently projects into and subverts 'planned urban spaces', challenging the spatial ordering of cities and hence, the social order. Chandhoke argues that the 'urban poor make and remake space . . . seize spaces and reshape in this way the entire urban form';

> They intrude into individual consciousness at traffic crossings . . . they inform us that cities are unequally constructed and maintained . . . (they) disrupt the coherence of the planned urban landscape, they retaliate and talk back to history and geography by making the homelessness of these people dramatically visible.
>
> (Chandhoke, 1993: 64)

Whilst norms of movement, activity and appearance exist and are mediated by power, the elastic attitudes to regulating them means that official intervention in one's trajectory through the street is less likely than that of the contingent decision of local power holders to exercise regulation over what might locally be regarded as inappropriate.

SENSUAL EXPERIENCE

I particularly want to bring out the rich sensual encounter that is promoted by the aforementioned processes of structuring, moving through, performing in and regulating the Indian street. The relationship between sensual experience, and spatial form and practice, has been barely touched upon and represents a rich field for further exploration (although see Porteous, 1990; Rodaway, 1994). Material spaces provoke particular forms of sense and feeling, and are themselves produced out of local social practices and meanings, including those which account for the senses. It is my contention that the pedestrian enjoys an infinitely more vivid sensual experience in the Indian street than in the Western street.

I have discussed the divergencies of movement in the Indian street, the crosscutting interplay of bodies and machines in motion. This panoply of living motion against a backdrop of randomly arranged buildings and objects produces an evershifting series of juxtapositions. Unforeseen assemblages of diverse static and moving elements provide surprising and unique scenes. Such haphazard features and events dis-order the gaze and spatial regularity. The flow of distracting sights negates scopic surveillance and easy visual consumption as the eye continuously swivels, alighting on changing episodes to the left and right, far ahead and close at hand. The norms of pleasurably jostling in the crowd, moreover, engender a haptic geography wherein there is continuous touching of others and weaving between and amongst bodies. The different textures brushed against and underfoot and the heating of one's skin

from nearby stoves render the body aware of diverse tactile sensations which interrupt concentrated gazing.

Visual imperialism is also denied by the powerful combination of other stimuli.
The 'smellscapes' of the Indian street are rich and varied. The jumbled mix of pungent
aromas – sweet, sour, acrid and savoury – produces intense 'olfactory geographies'.
Equally diverse is the soundscape which combines the noises generated by numerous
human activities, animals, forms of transport and performed and recorded music, to
produce a changing symphony of diverse pitches, volumes and tones.

By looking at the experience of, and negotiation with, modes of activity, movement, regulation and sensual experience, it seems that the body passing through the
Indian street is continually imposed upon and challenged by diverse activities, sensations and sights which render a state at variance to the restrained and distanced
distraction of the Western street. Here, the imaginative, improvisational predilections
of the pedestrian are stimulated into unexpected flights of fancy, and the passage
through the street is rhizomic rather than linear.

REGULATING THE WESTERN STREET

According to Chakrabarty, colonial, Western notions have become part of a globalising
discourse, of 'civic consciousness' and 'an order of aesthetics'. Indeed, in India, a process
is occurring wherein 'the thrills of the bazaar are traded in for the conveniences of the
sterile supermarket' (Chakrabarty, 1991: 29). Certainly, in most contemporary Western streets, notably high streets which were previously symbolic spaces for the
production and transmission of local identity, their reconstruction or disappearance has
resulted in the erasure of much social, sensual and rhythmic diversity in urban space
(see Chapter 12 of this volume). The imperatives of modernist planning and consumer
capitalism have tended to transform symbolic streets into functional spaces for maximising consumption and faciltitating transit (see Chapter 4 of this volume). A battery
of concepts and metaphors has been developed to account for these transformations in
public space. For instance, Sennett has referred to the growth in 'dead public spaces'
(Sennett, 1994), Auge has coined the phrase 'non-place' (Augé, 1995), Mitchell uses
the term 'pseudo-public space' (Mitchell, 1995) and Boddy refers to the systems of
bypasses, malls and subways that constitute the 'analogous city' (Boddy, 1992).
Whether there is a measure of hyperbole in these accounts or not, Indian streets seem
vibrant, multifarious and exciting compared to Western streets.

The delimiting processes whereby the range of social practices in Western streets
are reduced is captured by Sibley's notion of 'strongly classified' space (Sibley, 1988),
constructed out of an aesthetics and rationale which fears mixing of functions and the
disintegration of boundaries such as those between private and public, holy and profane
and backstage and frontstage. The rise of guarded residential communities, shopping
centres and private retirement homes reproduces the multiplication of pseudo-public
space which regulates entry, activity and rhythm. Augé argues that these urban spaces
are not 'relational, historical and concerned with identity'. Rather, they are realms of
'transit' as opposed to 'dwelling', sites of 'interchange' rather than a meeting place

or 'crossroads', where 'communication (with its codes, images and strategies)' is practised, rather than affective and convivial language (Augé, 1995: 107–8).

Street activity is monitored through surveillance and by what is considered 'appropriate'. Uncontrolled social interaction such as congregating, sleeping, 'hanging out', lounging on the pavement, and washing are all deterred. Moreover, heterogeneous commercial activities are discouraged by the dominance of large corporate retail outlets who control the management of space and refuse entry to smaller stalls, peddlers and mobile services. This also reduces the relations of barter and vocal enquiry in the process of consumption, mechanising and speeding up the relations of exchange. Whilst these commodified landscapes appear to promise a cornucopia of infinite variety, this is a manufactured and 'controlled diversity' rather than a realm of 'unconstrained social differences' (Mitchell, 1995: 119).

This triumph of 'non-space' depoliticises the street, forcing forms of resistance to adopt more covert strategies. Subaltern social movements depend upon the temporary seizure and transformation of public space in order to transmit alternative symbolic meanings, which the regulation of spaces for representation denies (Mitchell, 1995: 18). Whilst the presence of 'mall rats', beggars and shoplifters testify to certain forms of resistance, such opposition seems fleeting and gestural in the face of intensive surveillance.

In the 'expressway world', Western streets are increasingly organised as channels for unidirectional movement by reducing points of entry and exit, so that pedestrians and traffic may move rapidly and safely through, unhindered by idiosyncratic distractions. As Sennett exclaims, 'as urban space becomes a mere function of motion, it thus becomes less stimulating in itself; the driver wants to go through the space, not be aroused by it' (Sennett, 1994: 14). Likewise, the desensitised pedestrian, with little time to linger in the quest for commodities and experiences, marches ahead with no obstructions to prevent passing shops 'in review'. The similar tempos of the consuming pedestrian and the regulated flow of traffic facilitate the uninterrupted, anaesthetised passage through the contemporary urban landscape, reducing the diversity in the rhythms and choreographies of the street.

In her famous account of the New York neighbourhood where she lived, Jane Jacobs described the way in which the neighbourhood surveilled itself (Jacobs, 1995). In the contemporary carceral city regulation is systematic, rationalised and centralised rather than local, contingent and customary. Mike Davis has most vividly exposed how the 'new megastructures and supermalls have supplanted the traditional street and disciplined their spontaneity' (Davis, 1990: 356). Such spaces are increasingly subject to surveillance and are policed to exclude 'undesirable elements'. The gaze of police and security forces, and close-circuit TV systems accompany a reflexive control of the self, the body and the emotions.

The containment and commodification of the new streets encourages a 'controlled de-control of the emotions' (Featherstone, 1991: 105). The range of objects of desire and transgressive practices is reduced and unlike the carnivalesque spaces of Indian streets, these spaces are *designed* to stimulate desires: to escape, to meet 'others' and to transcend everyday existence. Yet this theming directs desire into the cul-de-sac of consumption.

These contemporary bureaucratic and capitalist imperatives of structuring and regulating streets powerfully influence the sensual experience of the pedestrian. Often, the very commodity promised by the marketers of these public spaces is a memorable, sensual experience. Thus the multisensual, complex and direct experience of the Indian street is replaced by a mediated and simulated sensual experience (Rodaway, 1994: 173). Sensually, the Western street is primarily a place for gazing rather than communicating. Theming has imposed a visual order; a predictable spectacle of few visual surprises, generated by the need for the large retail outlets to capture the attention of consumers. Accordingly, the pedestrian's gaze is directed to large window displays and slogans and away from the street (see Sorkin, 1992; Zukin, 1995).

Along with aesthetic policing and the imposition of design codes, the Western street is comprehensively deodorised, and sometimes re-odorised with commodified smells. The ordering of smell accompanies the process of removing perceived dirt and clutter which clears the street of 'surplus' stimuli and redirects attention towards products in the themed shopping environment. Likewise, the raucous cries of traders, political speakers and recorded or performed music are banned or kept at appropriate volume, reducing the soundscape to the similar rhythms emanating from boutiques. Also, the narrow scope for improvised or contingent movement minimises haptic experience and bodily contact and the smooth continuity of the flooring texture regulates the sense of touch, 'weakening the sense of tactile reality and pacifying the body' (Sennett, 1994: 17). Marked by deprivation in all sensory capacities, contemporary Western streets are marked by their non-sensuality.

TOURIST SPACES AND REGULATION

Having sketched the contrasts between Indian and Western streets, I now want to suggest that in some ways they are flip sides of the spaces of modernity, namely Appollonian cultural aspects which affirm 'structure, order and self discipline', and its opposite, Dionysian culture, representing 'sensuality, abandon and intoxication' (Rojek, 1995: 80). These modern tensions are highlighted by the proclivities of subjects on the one hand, to demand and impose epistemological, social and spatial order, and on the other, to long for disorder and transgression. Whilst the dominant urge is to seek refuge in reconstituted regimes of order in the face of continual change, the desire to transcend regulated minds, bodies and environments constantly bubbles below the disciplined surface of everyday life and finds various outlets (Cohen and Taylor, 1992).

In the same way that the colonisers of the European quarters of colonised cities used to look for excitement, often surreptitiously, in the 'native quarters', it seems that the dwellers and pedestrians in the over-regulated streets are drawn, fascinated by less ordered spaces. The tension between the two spaces is marked by the common-sense, rational appeal of instrumental bureaucratic and capitalist power to control and commodify spaces, and the desires of people to enter magical, Dionysian space. This is nowhere more apparent than in areas designated as tourist space where two forms of tourist space can be identified; organised and disorganised tourist space.

In the production of contemporary leisure, the tendency is for commercial interests to attempt to satiate the desire for otherness and sensuality. The dream-machines
of consumer capital provide imagined realms of 'otherness' (Rojek, 1995: 89). Such
spaces attempt to arouse the imagery and ambience of the carnival but can serve as
no more than 'sites of ordered disorder'.

Early modern forms such as carnivals and fairs have been largely superseded by
theme parks, shopping malls and festive marketplaces. A common theme for these
new spaces is this marketing of 'exotic otherness'. However, the provision of a safe
and familiar environment, and the regular codes of representation minimises the
disruption and excitement provoked by confrontation with difference. The tourist
gaze is structured by a repertoire of design codes which excludes supposedly surplus
elements, and provides a soupçon of exotica and a few key images. Whilst these
spaces represent *virtual* 'others', dominant heirarchical systems of spatial classification
also construct marginal spaces which are imagined to contain *actual* 'others'. But
whilst rhetorically typified as chaotic and dirty, they are also imagined to be spaces
of desire, permitting interconnection and hybridity, pregnant with possibility. Here
then are the two modalities of tourist space. Organised tourist space is the realm of
manufactured otherness, whilst disorganised tourist space is not dominated by
touristic commodifying imperatives at all but embraces a diverse range of relatively
unregulated activities, people and stimuli (see Edensor, 1998). It is both object of
desire and fear.

It is my contention that very few disorganised tourist spaces exist in the West.
If we consider what is innappropriately termed the 'developing world' to be at the
margins of a global tourist system, then 'otherness' exists in the peripheral regions
of global tourist space. Most Western tourists to the non-West are pulled into increasingly standardised tourist space of package tours, air-conditioned travel, consumption
and simulated local culture. However, the more intrepid are attracted by the less
regulated streets, bazaars and villages of disorganised tourist space. To be sure, the
narratives and images of colonial heritage continue to titillate the desires of *all*
Western tourists. India, for instance, is fantasised as more 'authentic' than the West,
a more spiritual, natural and unchanging realm, and tourist itineraries are shaped
according to these preconceptions. However, Indian streets exert a fascination for
Western tourists, not only because they contain perceived otherness, but because in
their social and cultural organisation, they contrast with the regulated streets of the
West and their replanted formations in organised tourist space.

While 'organised' tourist spaces are 'enclavic', shielding tourists from contact
with the local populace through surveillance and eliminating potentially offensive
sounds and smells, in the disorganised tourist space of the Indian street, services tout
for business on the street and cause a certain amount of 'hassle' for tourists. Through
haggling and repudiating advances, tourists and locals must mingle with each other.
Thus, opportunities for dialogue and exchange are capacitated so that the tourist
experience of 'otherness' is potentially of a different quality to the enclaved tourist.

By dwelling and moving through Indian streets, tourists' senses are excited by
a more variegated set of stimuli: sights, smells, noises, movement, which do not

accord with mass produced tourist imagery. Within disorganised tourist space, there is more interaction with the local population in shops and restaurants, with beggars and touts, and with people interested in sharing ideas and information with the tourist. The ability to remain shielded from local life in local space is not possible, but in any case, it is these streets that are sought by backpackers in search of 'authenticity' and the thrill of encountering the 'other'. In fact, the less-travelled path has a high degree of mystique and status-conferral.

Admittedly then, the experience of Indian streets sought by Western tourists is mediated by located expectations and presuppositions. Yet such 'escape attempts' (Cohen and Taylor, 1992) may be partially succesful because they challenge the physical and mental dispositions, deadened by passage through Western streets and organised tourist space, by confrontation with different orders of sensory experience, social interaction, regulation and movement. This experience feeds into the contemporary need to reinstate desire, disorder and unpredictablity into life.

THE *FLÂNEUR*, HETEROTOPIA AND TACTICS

In the light of the above, I now want to consider three frequently used figures of the modern city, the *flâneur*, Foucault's notion of 'heterotopia' and de Certeau's heroic 'tactics' of the pedestrian.

The flâneur

In its original conception, the figure of the *flâneur* is somewhat elitist, distanced from the crowd by his superior aesthetic sensibilities, a detached and self-contained poetic soul 'botanising on the asphalt'. However, the concept has been democratised and now often stands as a metaphor for the contemporary urban dweller, moving through the flux and transience of the city. Smart argues that *flânerie* is now a mode of being in the world rather than the provenance of a marginal character (Smart, 1994: 162) an assertion that seems to be instantiated in the organised and commodified spectacles of the Western city. The inference is that in the modern shopping centre or heritage park we are all *flâneurs*, *homo ludens* consumed by the dazzling spectacles and commodities on show. Yet whilst the democratisation of the concept avoids the elitism of the original dilettante of the street, the notion of these 'postmodern *flâneurs*' misses the central idea that to be a *flâneur* is typified by wallowing in flux, observing the fleeting and the transitory, witnessing unique juxtapositions and incidental meetings.

The random elements that so stimulated Baudelaire's hero are now rarely experienced in Western streets as I have argued, so rather than being at home 'in the ebb and flow, the bustle, the fleeting and the infinite' (Baudelaire, 1972: 399), the contemporary pedestrian is motivated by the purposive acquisition of commodities and commodified sights. In Western urban space, haphazard diversity has been 'liquidised into the lubricant of profit-making contraptions' (Bauman, 1994: 151). The shifting variety that so stimulated the early modern *flâneur* is replaced by the chanelling of the gaze to a reduced series of signs that delimits the object of his pleasure. Moreover, the freedom to loiter, the 'reprieve from time' (ibid.: 140) so critical to witnessing and interpreting the momentary passing scenes and incidents is denied by the policing of activity and the intensified speed of movement. Popular fear of downtown streets means passage through them is swift. The early modern vitality of the city that so stimulated the *flâneur* has become domesticated by imperatives to seek order, convenience, speed and the rapid turnover of commodities.

I suggest that the experience of the Indian street is in many ways akin to that of the early modern European metropolitan street. Accordingly, the diverse social activities, forms and styles of movement, types of regulation and above all, the sensuality of the Indian street, provide a rich environment for *flânerie*. The unpredictable juxtapositions, the fleeting occurrences, the disparate rhythms and multifarious sights, smells and noises facilitate the enjoyment of the urban realm. And the absence of any channelling of the gaze, overarching surveillance, disciplining of movement permit the *flâneur* to savour and contemplate the sensual urban experience.

Heterotopia

The term 'heterotopia' has been widely used in recent years but remains somewhat vague and carelessly applied to a range of discrete phenomena (Hetherington, 1996: 158). This is largely because Foucault's all too brief introduction of the term (Foucault, 1986) tends to suggest a range of exciting possibilities but is suitably unspecific. Rather than straining to identify Foucault's 'authentic' intention, I intend to utilise the concept to highlight the properties of heterotopia that challenge dominant modes of spatial ordering. In heterotopias, the random juxtapositions of disparate objects, activities and people not normally found together challenge hegemonic modes of regulating and representing space. The convergence of such miscellaneous and discordant sights erode epistemelogical and ontological security, disrupting the common-sense meanings of space. This transgressive potential infers that heterotopias continually speak back to dominant modes of power-in-spacing, interrogating the normativity of their disciplinary regimes and functional purpose.

Accordingly, the power-laden processes of classifying, spectacularising and commodifying difference in the disciplined streets of the West are revealed by the actual admixture and changing juxtapositions of difference found in the heterotopic Indian streets. The pedestrian is an essential part of this heteroglossia of 'otherness' rather than the distanced specator of manufactured spectacle. Enmeshed in its sensuality, he/she is denied an imperialist subjectivity.

As marginal spaces, heterotopia are placed on the borders of the normal systems of spatial representation. In one sense, India remains on the edge of the global tourist economy, yet has partially been incorporated into organised tourist space with the imposition of tourist enclaves. However much Indian streets are reclaimed within dominant systems of representation by tropes of exoticisation or moralising pity (see Hutnyk, 1996), actual passage through them reveals their disruptive heterotopic qualities wherein the flow of expectations learnt in the sterile spaces of the West are shattered. As heterotopias, Indian streets are situated within a global system of spatial ordering and it is the alternative set of street activities, forms of regulation, rhythms and sensual stimuli that mark their disjuncture from the over-regulated spaces. But in any case, besides this relational distinction, there remains a dis-ordering logic in their material and social organisation that affirm the anarchistic spirit of the heterotopia.

Symbolically then, heterotopic Indian spaces provide an escape route, or labyrinth, an alternative system of spatial (dis)ordering where transitional identities may be sought, sensual and imaginative experimentation indulged, and the Western hegemonic power/knowledge axis bewildered and challenged.

Pedestrian Tactics

Finally, I want to consider the highly influential ideas of de Certeau with regard to the foregoing discussion. De Certeau makes a distinction between the 'strategies', the normative, rationalised practices employed by the powerful in their (re)production of 'technocratically constructed, written and functionalised space' (de Certeau, 1984: xviii), and the contingent 'tactics' enacted by pedestrians to escape these carceral networks. These improvisational tactics, according to de Certeau, 'trace out the ruses of other interests and desires that are neither determined nor captured by the systems in which they develop' (ibid.). He privileges walking as a particularly inventive process through which pedestrians construct stories, thereby weaving places together in improvisational narratives. Thus despite the reduction of the street to a disciplined and sterile space, the pedestrian can reclaim, albeit contingently and fleetingly, a measure of control over material and symbolic space through an escape into memory and imagination.

Whilst de Certeau has been criticised for his optimistic construction of an heroic pedestrian (Rojek, 1995: 106), the suggestion that walking is an inventive activity is none the less intriguing. His evocation of the plethora of desires that are stimulated through the relationship between sensual bodily movement, fantasy and reverie convincingly refute deterministic notions of pedestrians being shaped mentally and physically by urban space and its control.

However, the problem with his account is that de Certeau seems to envisage his technocratic space as universal. Thus it is necessary to reinstate the distinct material and symbolic forms of urban space and regulation. The notion of a homogeneous carceral network suggests a kind of abstract space irrespective of time or place. What I am arguing is that the specific form of urban space influences the degree of tactical innovation and empowerment mobilised by pedestrians. Certain less regulated and commodified spaces facilitate these imaginings, epistemological dislocations and memories better than others. The more heterogeneous Indian street, less circumscribed and framed by the power of capital and bureaucracy, is a 'weakly classified space' (Sibley, 1988: 412) in which different people and activities mingle.

For instance, the Indian street offers an environment where smells, sounds and indefinable 'atmospheres' are apt to stimulate involuntary memories. Moreover, the movement towards the 'other' which de Certeau argues is part of the enactment of 'tactics' is also encouraged by the presence of so many actual 'others' in the street.

The disruption of movement, the jostling of bodies and the necessary sudden swerves similarly may awaken a heightened sensory awareness and disrupt patterns of thought. Likewise, unusual temporary juxtapositions of people, objects and animals may jolt the observer out of distraction and escape normative visual codes. The narratives and desires that de Certeau refers to remain located in the pedestrian but are enabled by the distinctive material, symbolic and social form of the Indian street.

CONCLUSION

The processes of logocentric spatialisation, regulation, commodification and the confining of difference are uneven processes. Simmel (1995) highlighted the juxta-position of multiple cultural forms and social practices that characterised the rhythm of the modern metropolis. Yet these dynamic elements are increasingly being extin-guished from the urban realm, or pushed out to marginal locations. Late capitalism has rendered street life predictable and marked by sensual deprivation by reducing difference to commodified sameness. The destruction of the functional and cultural diversity of the street has thwarted human contact, the desire for difference, and the need to wallow in the obscure and confusing. Appollonian ambition has predominated over Dionysian desire.

In this chapter, I have portrayed the diverse social activities, choreographies, regulatory regimes, and sensual experience of an idealised Indian street in order to reveal the richness of weakly defined, heterotopic spaces and their contribution to human pleasure and an understanding of difference and 'otherness'. This existing other space can and does serve as a refuge from the overdetermined, single-purpose streets of the Western metropolis, satisfying the lust to experience sensuality, the unclassifiable and the ever-changing. In these uncommodified spaces, at the margins of the global tourist economy and barely incorporated into the 'ideoscapes' and 'mediascapes' (Appadurai, 1990) of dominant representation, social and sensual confrontations can dissolve hegemonic preconceptions and disrupt notions of smooth passage, unhindered gazing, detached self-containment, convenience and antiseptic sterility so entrenched in Western regimes of urban spatialisation.

REFERENCES

Appadurai, A. (1990) 'Disjuncture and difference in the global cultural economy', in M. Feather-stone (ed.) *Global Culture: Nationalism, Globalisation and Modernity* (Sage, London), pp. 295–310.
Augé, M. (1995) *Non-places: Introduction to an Anthropology of Supermodernity* (Verso, London).
Baudelaire, C. (1972) *Selected Writings on Art and Artists*, trans. and ed. P. E. Charvet (Penguin, Harmondsworth).
Bauman, Z. (1994) 'Desert spectacular', in K. Tester (ed.) *The Flâneur* (Routledge, London), pp. 138–57.
Boddy, T. (1992) 'Underground and overhead: building the analogous city', in M. Sorkin (ed.) *Variations on a Theme Park* (Hill and Wang, New York), pp. 123–53.
Buie, S. (1996) 'Market as mandala: the erotic space of commerce', *Organisation*, 3(2): 225–32.
Chakrabarty, D. (1991) 'Open space/public space: garbage, modernity and India', in *South Asia*, 16: 63–73.

Chandhoke, N. (1993) 'On the social organisation of urban space: subversions and appropri-
    ations', *Social Scientist*, 21: 541–7.
Cohen, S. and Taylor, L. (1992) *Escape Attempts* (Routledge, London).
Davis, M. (1990) *City of Quartz* (Verso, London).
de Certeau, M. (1984) *The Practice of Everyday Life* (University of California, Berkeley).
Edensor, T. (1998) *Tourists and the Taj Mahal* (Routledge, London).
Featherstone, M. (1991) *Consumer Culture and Postmodernism* (Sage, London).
Foucault, M. (1986) 'Of other spaces', *Diacritics,* 16(1): 22–7.
Hutnyk, J. (1996) *The Rumour of Calcutta* (Zed, London).
Hetherington, K. (1996) 'The utopics of social ordering: Stonehenge as a museum without
    walls', in S. Macdonald and G. Fyfe (eds) *Theorising Museums* (Blackwell, Oxford), pp 153–76.
Jacobs, J. (1995) 'The uses of sidewalks', in P. Kasinitz (ed.) *Metropolis: Centre and Symbol of
    Our Times* (Macmillan, London), pp. 111–29.
Kasinitz, P. (ed.) (1995) *Metropolis: Centre and Symbol of Our Times* (Macmillan, London).
Mitchell, D. (1995) 'The end of public space?: People's Park, definitions of the public, and
    democracy', *Annals of the Association of American Geographers*, 85(1), 108–33.
Porteous, J. (1990) *Landscapes of the Mind: Worlds of Sense and Metaphor* (Toronto University
    Press, Toronto).
Rodaway, P. (1994) *Sensuous Geographies* (Routledge, London).
Rojek, C. (1995) *Decentring Leisure* (Sage, London).
Sennett, R. (1994) *Flesh and Stone* (Faber, London).
Sibley, D. (1988) 'Survey 13: purification of space', in *Environment and Planning D: Society and
    Space*, 6: 409–21.
Simmel, G. (1995) 'The metropolis and mental life', in P. Kasinitz (ed.) *Metropolis: Centre and
    Symbol of Our Times* (Macmillan, London), pp. 30–45.
Smart, B. (1994) 'Digesting the modern diet: gastro-porn, fast food and panic eating', in
    K. Tester (ed.) *The Flaneur* (Routledge, London), pp. 158–80.
Sorkin, M. (1992) 'See you in Disneyland', in M. Sorkin (ed.) *Variations on a Theme Park* (Hill
    and Wang, New York), pp. 205–32.
Tester, K. (ed.) (1994) *The Flaneur* (Routledge, London).
Zukin, S. (1995) *The Culture of Cities* (Blackwell, Oxford).

Part III

CONTROL AND RESISTANCE

15

POLICING CONTESTED SPACE

ON PATROL AT SMILEY AND HAUSER

Steve Herbert

•

Modern policing is both pervasive and authoritative. The contemporary police officer is a *patrol* officer, able and willing to insert state authority into the everyday of life on the street, to win compliance to order through displaying and reinforcing the power inherent in the uniform, the badge and the tools of coercion (Rubinstein, 1973; Walker, 1989). Police officers represent the most visible face of state authority, and work to achieve a seeming ubiquity across the space of the city. The patrol car symbolizes the free-floating nature of modern state power; police officers assert sovereign claim to the streets through the simple act of cruising.

This more visible and pro-active force is a consequence, in part, of the historical movement toward police professionalism. Begun in the early 1900s, the professional movement sought to create more rigidly controlled and hence less corrupt forces (Fogelson, 1977). Police departments in the United States were reconstructed along strict bureaucratic lines, and civil service protections were instituted to insulate key officials from outside political influence. And the aura of professionalism was meant to cleanse police forces of their dominant public image of incompetence.

This more palatable public face was enhanced by efforts to make police departments more technologically sophisticated. Police forces sought not only to be seen as organizationally efficient, but also as open to scientific developments. These emerging technologies had more than just public relations appeal; they also expanded the scope and intensity of police power. Abetted by a complex that links telephones, radios, computers and automobiles, modern urban patrol officers can map, encircle and infiltrate with increasing sophistication (Herbert, 1996a). Patrol officers thus can construct a mobile and expansive net of state power. And in any encounters with the citizenry, officers expect that their commands will be obeyed, that they will, according to Bittner (1990: 234), 'coerce a provisional solution upon emergent problems without having to brook or defer to opposition of any kind'. Police authority shall hold ultimate sway.

There is therefore a geopolitics to policing the street, a set of ongoing tactical efforts to ensure that police presence shall seem all-encompassing and ultimately all-powerful. Police officers regularly exercise territoriality (Sack, 1986); they seek to influence social action through controlling space. Like other modern systems of social control, police power is realised in and through space (Hannah, 1993; Lowman, 1986; Ogborn, 1993). Without the capacity to survey and interlink, to co-ordinate and capture, police officers' attempts to establish authority would be anaemic. Modern policing thus rests quite fundamentally upon a political geography (Fyfe, 1991), a need to create order through controlling space (Herbert, 1996b).

Police efforts to claim sovereignty over the street, however, are always subject to contestation. In dramatic cases, such as the urban rebellions that occurred in Britain in the 1980s (see Keith, 1993) or in Los Angeles in 1992 (Baldasare, 1994), this sovereignty is overtly and violently contested. But resistance to the police, like resistance more generally, is rarely so cataclysmic; it typically involves more subtle evasions and resistances (Scott, 1985). Whether spectacular or indirect, resistance is something that patrol officers neither suffer happily nor ignore easily.

At times, resistant populations are clustered in particular locales. In London, these are sometimes referred to as 'no-go areas' (Keith, 1991). In Los Angeles, they are referred to by the police as 'anti-police' areas, locations where residents are not to be trusted and where danger may easily befall the unsuspecting officer. Officer reactions to 'anti-police' areas make plain the often unstated police imperative for geopolitical control. When police dominion is regularly challenged, officers react with disdain and often with organized measures to reclaim a sense of tactical superiority. Take, for example, the Los Angeles street corner of Smiley and Hauser.

THE 'SMAUSER'

The Los Angeles Police Department (LAPD) divides the vast territory for which it is responsible into a series of eighteen patrol divisions. Each division is responsible for handling all the requests for police assistance within its territory and for investigating all crimes that occur there. The Wilshire Division covers a demographically diverse population situated in the centre of the city. It was the location of eight months of fieldwork spent observing a wide array of officers engaged in various aspects of police operations. The fieldwork consisted primarily of ride-alongs with two groups of officers. The first group, patrol sergeants, are responsible for supervising the two-person patrol car teams that handle radio calls. The second group, Senior Lead Officers, are responsible for community relations and for monitoring particular locales of concern to residents, such as street corners where drugs are being sold or encampments of homeless people. I observed several other types of officers on a more limited basis, including officers working on specialised units that examined narcotics, vice and gangs. The data and incidents upon which I draw below were all culled from the fieldnotes generated from the observations.

Smiley and Hauser is situated in a minority dominated neighbourhood in the southern end of the division. It lies just south of the Santa Monica freeway, which is typically considered the northern boundary of South Central Los Angeles, now notorious as the locus of much of the urban unrest in 1992. Indeed, Wilshire officers characterise the Smiley and Hauser area in these terms: as Wilshire's piece of South Central. This is not only because of the demographics of the area – mostly minority and lower income – but also because of a history of police–community antagonism.

Like many parts of South Central, the outward appearance of Smiley and Hauser belies its image as dangerous. Most of the structures are modest single-family homes, with an occasional four- or eight-unit apartment building. Some of the buildings are in some disrepair – stucco is chipped, a porch is collapsed on one side, a fence is sagging in the middle. But all the structures are occupied and many are in a condition that suggests ongoing care by their inhabitants. Lawns are generally well tended, and large palm and yucca trees line the street. The only vivid evidence that fear pervades the neighbourhood is the ubiquitous iron lattice that graces windows and doors, there to thwart unwanted entries.

Despite its mostly innocuous appearance, Smiley and Hauser is clearly the most vilified street corner in the Wilshire Division. Ask any officer in the division about areas of concern, and the 'Smauser', as it is commonly known, will top the list. The officers' reasoning for bestowing this honour on Smiley and Hauser is invariant: residents of the neighbourhood, sometimes referred to as 'terrorists' by Wilshire cops, have engaged officers in gunplay with some frequency in recent years. One incident involved a pair of patrol officers who pursued a car allegedly involved in a drive-by shooting. Just east of the intersection of Smiley and Hauser, the car suddenly stopped and two men jumped out. As they ran from the car, the men fired their pistols at the officers, seriously wounding one of them. Another incident involved an undercover narcotics officer who was engaged in a 'buy-bust' operation. Posing as an interested drug customer, the officer sought to close a deal with a street seller on the corner. The vendor became suspicious during the encounter and threatened violence. Ultimately, the two engaged gunfire, although neither was wounded seriously. In a third case, an officer on routine patrol encountered rifle fire that only succeeded in marring the physical appearance of his cruiser.

As a result of these events, officers feel quite justified in castigating the area, because it symbolises a threat to their personal safety. But the gunplay is understood as just the most significant aspect of a wider pattern of illegal activity and hostility to the police. Gunfire is an obvious contestation of police authority, but more subtle tactics of resistance include taking advantage of knowledge of local geography to successfully flee approaching officers or retreating into private homes to escape detection.

Of particular concern to the officers are the members of one particular street gang, the Geer Street Crips. (Geer Street is one block south of Smiley.) A relatively small group of African-Americans, the Geer Streeters are believed responsible for the gunfights with officers, and typically congregate in front of a handful of apartment buildings on Smiley. Of lesser but not insignificant concern is another street gang in the area, a Latino group known as 18th Street. Members of 18th Street do not engage the police with any frequency, and do not have a reputation for violence. Neither gang is especially aggressive towards each other or towards other gangs; neither displays any desire to control anything other than a few street corners for drug sales. The main point of antagonism in the area is between the LAPD and the Geer Streeters.

The gunplay at Smiley and Hauser frustrates officers because it endangers their safety and represents overt resistance to their authority. When gun battles occur, Wilshire officers move to assert a visible presence in an attempt to reclaim dominion, often through overtly geopolitical tactics.

It is less than a week since a Wilshire officer was seriously wounded in the first incident described above. The shooters involved are still at large. The sergeant expresses frustration at the 'smug looks' he sees on young men in the area, which he interprets as a boast that they got the best of the recent encounter. He is determined, however, not to sit idly by.

He parks his patrol car four blocks from the intersection and radios for all available cars to join him. He succeeds in gathering two other sergeants and three patrol car teams. They hunch over the back of a car and sketch their strategy on a piece of paper. A group of officers will approach a cluster of young men gathered in front of an apartment building on Smiley. Another group of officers will post themselves one block directly south, because they believe the young men will flee in that direction. Their goal is to question as many of the men as possible and to run checks on them for outstanding warrants. If such warrants exist, the men will be arrested. The ultimate goal is to place enough pressure on the group that someone will ulti-mately 'give up' the identity of the shooter(s).

I move with the group of officers posted to the south. Over the radio, I hear the other group of officers co-ordinating themselves. 'Okay,' orders one officer, 'lead the charge.' 'Okay, we're going in,' is the reply.

The manoeuvre is not especially successful. Many in the group do indeed flee to the south, but manage to elude capture by emerging well to the west of where the offices are posted. Of those who are detained, only one possesses an outstanding warrant.

Though these tactics proved only marginally effective, they clearly stem from the officers' desire to regain an advantage over their 'terrorist' opponents, to reclaim sovereign control of an 'anti-police' area. Though they net only one arrest, the officers at least make a symbolic attempt at regaining the upper hand, revealing to the neigh-bourhood, and to themselves, that violent resistance to police power will not be lightly tolerated.

Indeed, the inability to capture those responsible for the shooting continued to rankle Wilshire officers for months afterwards. They regularly complained about the ineffectual efforts of the specialised unit, located downtown at police headquarters, which was responsible for handling police shootings; the Wilshire officers wondered aloud why the investigators were neither visible nor successful. (A sergeant complained that they probably 'don't even know where Smiley and Hauser is'.) One watch commander – responsible for overseeing the patrol operation through one of the three daily shifts – decided after a few weeks to take matters into his own hands. He required each patrol team at the start of watch to head to Smiley and Hauser and do everything possible to obtain an arrest of anyone found on the streets there.

This more aggressive strategy was reminiscent of what some officers referred fondly to as the 'old days' of the LAPD, the period before the beating of Rodney King and the attendant controversy it created about excessive use of force (Independent Commission, 1991). It may perhaps be difficult to believe today, but the LAPD was once heralded as a model police department in the United States, the one seen broadly as the most successful in realising the professional ideal. Professionalism was the explicit goal of longtime LAPD chief William Parker, who served from 1949 to 1966. Parker restructured the internal organization of the department, and successfully promoted the LAPD as an aggressive, pro-active force. Such aggressiveness became quite well-established within the LAPD subculture (Cray, 1972; Escobar, 1993; Fogelson, 1971; Raine, 1967; Schiesl, 1990; Turner, 1968), helping create what one sergeant referred to as an 'ass-kicking' department.

In the 'old days', officers said, a relentless and aggressive patrol presence in Smiley and Hauser would have continued unabated until the shooters were captured. The inadequate efforts from downtown were seen as yet further evidence that the LAPD was becoming softer in the post-King era, and hence that the officers' ultimate authority in the streets was becoming more perilous. More than one officer complained regularly about the different attitude they found on display in the streets after the King beating, and longed wistfully for the institutional permission they formerly received for more aggressive tactics.

MORALITY IN POLICE TERRITORIALITY

The subcultural ethos of police officers is heavily moralistic (Herbert, 1996c; Reiss and Bordua, 1967). LAPD officers for example regularly castigate residents of Smiley and Hauser and other notorious areas as 'bad guys', as evil 'predators' against whom they struggle for control of Los Angeles. Implicit in this image is a self-construction as a morally driven defender of the good, the protector of order and peace in the streets. One officer for example referred to himself as a 'mercenary', a do-gooder for hire, paid by Angelenos to protect them from the demons who would otherwise roam unhampered. Policing is thus elevated to a plateau well above the mere mechanics of simple law enforcement, and instead understood as part of a more epic quest to restore tranquility and justice to Los Angeles, to cleanse the streets of the stain of criminalistic evil (see also Reiner, 1992).

This pervasive and pungent morality stems from the inherent insecurities of the police (Herbert 1996c). Officers are placed in the middle of a series of irresolvable dilemmas: they are asked to combat crime when they are unable to do so; they are asked to restore order in inchoate situations whose determinants are elusive and not easily rectified; and they often face the difficult choice of whether to use coercive, and potentially lethal, force. In the face of these dilemmas, officers retreat into a more simplistic world of good and evil, where police actions, no matter how ineffective or questionable, are cast more broadly as aspects of a grand and virtuous struggle. In this moralistic universe, police officers can understand and forgive themselves and one another because, at the end of the day, they are ultimately good and their actions ultimately defensible.

These moralistic self-constructions imply a particular reaction to places like Smiley and Hauser. When Wilshire officers aggressively patrol in response to a shooting of a fellow officer, it is not just an effort to reclaim sovereignty but, beyond that, a required response to re-establish peace and order to a neighbourhood plagued with terrorists so craven they will actually attack police officers. The self-imposed police mandate virtuously to protect the good from predators mandates an ardent response, an organized strategy to put the good guys back in charge. To allow the evil of resistance to roam wildly in the streets runs counter to the police's virtuous self-image.

Police territoriality is thus infused with more than just tactical considerations. Confronting young men on the corner of Smiley and Hauser is a required act, one impelled by the morally defined imperative to uphold peace and order. To capture the people responsible for shooting officers and otherwise polluting Smiley and Hauser is to cleanse the area of the stain of evil, and to re-install the good guys as the dominant authority.

But even cops compelled by a morally constructed need to establish sovereignty over places of resistance recognize that their efforts are not always successful, that they are engaged in an ongoing game of cat-and-mouse, an ever-shifting struggle where they do not always win. At times, it even seems as if the two sides are engaged in nothing much more than a sporting event, although one with occasional violent consequences.

PLAYING COPS AND ROBBERS

On another night, a group of officers are again attempting to detain, question and potentially arrest a group of young men hanging out in front of the notorious apartment buildings on Smiley. This time they succeed in detaining a group of five young men. Another group of men had successfully fled when the officers approached.

As the officers are searching and questioning the young men, the two sides begin mildly taunting each other. They engage in an ongoing debate about whether a rookie officer, fresh out of the police academy and ready for adventure, could have caught those who fled. The officers, of course, argue for the rookie, and discuss the eagerness and rambunctiousness of young officers. The young men naturally respond by sticking up for their speediest member, arguing that he can elude any of the police's best. Both the officers and the young men rise to the verbal challenge of countering each argument, and both clearly enjoy themselves; there is much joking and laughing.

Like two groups of fans of rival sports teams, both the officers and the young men argue for the comparative merits of their fastest and most aggressive runners. Whatever else might be at stake at Smiley and Hauser, in this revealing moment both sides openly accept that there is a purely physical element in the contest, that they play a well-armed game of hide and seek. Indeed, both sides seem to embrace the physicality of the struggle and to enjoy the moves and countermoves that are part of it. The young men at Smiley and Hauser quite loudly brag of their ability to elude capture and taunt the officers who are too slow to keep up with their swiftness. The officers respond by touting the merits of their eager young recruits in an attempt to remind the young men that they can only escape the police for so long. The game of cops and robbers will continue.

In fact, officers see evidence of gang members' resistance and derision throughout the neighbourhood. While driving down Hauser one afternoon, a sergeant noticed that the insignia of the 18th Street Gang – XVIII – was written in chalk in large numerals on the street. He hypothesised that the numerals had been placed there to be seen by the LAPD helicopter officers who fly over the area regularly. In other words, the act of inscribing the insignia so grandly was a gesture of territorial marking, an act of nose-thumbing at LAPD efforts to control that space through its high-tech gadgetry. On another night, an officer is cruising through the area and notices that a few of the street signs have been taken down. He believes that the gang members are responsible, that they want to make it harder for the police to find their way around the area and thus harder for them to co-ordinate their efforts to encircle and capture.

Implicit in the cops' talk and tactics is a grudging respect for the gang members and their tactics. While they vilify their opponents as predatory trouble-makers, many officers will also admit that they are smart and thus hard to capture One officer compared them to young entrepreneurs who, in their drug sales activity, know how to secure a market and protect their interests. (This theme of gang members as entrepreneurs is supported by such ethnographic studies as Jankowski, 1991 and Padilla, 1992.) Indeed, the cleverness of the young men explains why many officers will use all encounters with them as potential opportunities to gain invaluable intelligence. Officers will ask those they encounter about particular individuals and their current activities and locations. They will try to learn something about gang members' escape routes and hiding places.

After stopping an automobile near Smiley and Hauser that had a burnt-out headlight, the officers involved begin questioning those in the car, which include three men and two women, all of whom appear to be in their late teens or early twenties. One officer focuses particular attention on one of the young men. The officer believes he recognizes the young man and wonders where he spends his time. The young man is evasive at first, but then mentions something about the 'drain', an area near a set of run-off ditches that flow beneath the freeway. The officer smiles in recognition and says that is right, that in their previous encounter the young man stayed behind but his buddy escaped. The officer then begins asking questions about which direction his buddy ran, trying unsuccessfully to get intelligence about escape routes.

The conversation soon turns to a young man named Detwan, whom the officers are trying to locate. The young men warm to this conversation and inform the officers that Detwan has recently been beaten, and discuss how he has managed to get himself in trouble with various people in the neighbourhood. Both sides find much to laugh about in the stories of Detwan's troubles, and the conversation takes on a relaxed air. The officers manage to learn much about Detwan's present activities and favoured locations.

Of course, many officers who take a more macho attitude towards police work (see Fielding, 1994; Herbert, 1996b) affect a tough posture toward the young men who hang out near Smiley and Hauser. This more aggressive stance is met with an identical response from the residents, and little information is passed along to the officers. Indeed, one of the main criticisms of the more aggressive, pro-active patrolling practices of departments like the LAPD is precisely the evaporation of community co-operation that typically results (Lea and Young, 1985; Skolnick and Bayley, 1988). In Los Angeles and elsewhere, the orientation of the police to the community is the object of considerable efforts at reform.

GEOPOLITICAL CONTROL IN THE ERA OF COMMUNITY POLICING

While the professional model dominated reform movements through much of the twentieth century, it has been supplanted by the move toward 'community policing'. Now ubiquitous as an organizational mandate amongst police forces, community policing implies a set of mechanisms to improve relations between cops and citizens (see Greene and Mastrofski, 1988; Moore, 1992). These mechanisms take various forms, but typically focus on increasing community access to, and oversight of, the police. More frequent and fruitful interactions between community and police will, the logic goes, resolve the tensions created by the more detached, professional and pro-active officers of old, and increase the amount of information flowing to the cops. This information can then be used by officers to capture criminals and create order.

The LAPD has joined the rush towards community policing. Indeed, a reform-minded police chief, Willie Williams, came into the LAPD from Philadelphia in 1992 as an avowed proponent of community policing, and explicitly attempted to reform the department along those lines. He met with considerable resistance from a rank and file accustomed to being experts and to expecting deference, not advice, from the community (Herbert, 1996b). Still, in many divisions, upper-level personnel are recognizing that the political winds blow in a new direction, and are steering their ships accordingly. This, too, has had implications for policing Smiley and Hauser.

Smiley and Hauser lies just to the south of that portion of the Santa Monica Freeway that collapsed during the 1994 Northridge earthquake. Because the freeway is a federal highway, the Wilshire Division received extensive funding from the federal government to ensure the security of the detour constructed to route the diverted freeway traffic along city streets. The captain of the division chose to use much of that money to fund a 'task force' to concentrate on the Smiley and Hauser area. The task force consisted of about a dozen officers and a patrol sergeant whose permanent assignment was Smiley and Hauser.

There were two main foci of the task force. One was to concentrate on those suspected of criminal activities. This involved co-operating with Federal Bureau of Inspection officers interested in major drug dealers, gang unit officers who regularly track the two gangs in the area, and probation officers interested in those recently released offenders who might be violating conditions of their probation. Task force officers gathered information about potential suspects from all these quarters, and used that to locate and arrest as many people as possible.

The other main focus was the broader community. Task force officers went door-to-door announcing their arrival in the neighbourhood. They were attempting to help the residents 'take back' the neighbourhood by getting pro-actively involved in policing the community themselves. The officers organised and hosted several community meetings designed to solicit concerns from residents and to develop strategies for addressing them. The sustained presence of the task force, the officers hoped, would help overcome fears of retaliation that might have suppressed residents from previously working with the police to help uncover the activities of the gang members. The task force, in other words, more overtly embraced the philosophy of community policing, it recognized the police's need for community co-operation and sought to create venues to develop it.

The fieldwork ended before the task force dissolved, so the results of its efforts are unclear. Research on community policing, however, consistently gives little optimism that such efforts as this task force succeed. At times, such efforts succeed in reducing fear of crime (Skogan, 1994), but rarely actual levels of crime (Rosenbaum and Lurigio, 1994). Citizens are often reluctant to participate due to either mistrust of the police or fear of retaliation from gang members or others targeted by the officers (Grinc, 1994).

But the effort to enlist community support, no matter how effectual, clearly underscores the ongoing imperative felt by police departments to reclaim control of resistant spaces. The very language of such efforts – the desire to help communities 'take back' their neighbourhood – underscores the drive toward sovereignty that dominates everyday police activity. No matter how one understands the purpose and promise of community policing, it clearly represents the latest attempt to make police authority as pervasive and trenchant as possible.

CONCLUSION

To a police officer, the street is the most visible and important place where their authority is established (Van Mannen, 1974). To take a drive is to assert sovereignty, to mark the street as a space of police supremacy. As Rubinstein (1973: 166) boldly puts it, 'For the patrolman the street is everything; if he loses that, he has surrendered his reason for being what he is.'

This mandate for territorial control of the street is made evident when police authority is contested. Smiley and Hauser is demonised by Wilshire officers because the resistance there shatters the illusion of police pre-eminence. The resistance, however, must be answered; the police must continue to insist on having the last word. Driven by an abiding sense of moralistic fervour and a love of the sheer physicality of the chase, officers deploy in manoeuvres to try to ensure that the 'good guys' win. The shift toward community policing may change the tactics the officers employ, but the ultimate task is the same: to ensure that police authority on the street is pervasive and basal.

But police hegemony, like hegemony more broadly, is ever unstable. In most major cities, and especially in Los Angeles, police legitimacy is hardly robust. Resistance comes from many quarters, not just from those who are indeed engaged in criminal activity. Thus, police officers charge themselves with a task that they can never ultimately accomplish; they will achieve sovereignty over only parts of Los Angeles, and only for limited periods of time. Police officers can control none of the social forces – namely dynamics of race and class that affect employment, education and housing – that most profoundly shape life at Smiley and Hauser. Indeed, officers are regularly seen from the street as symbols of the wider systemic forces that structure and constrain opportunities. In seeking to be more pervasive symbols of authority, officers merely remind many on the street of broader forces of oppression. Given this, resistance to the police is inevitable. And thus the game of hide and seek will continue indefinitely.

REFERENCES

Baldasare, M. (ed.) (1994) *The Los Angeles Riots: Lessons for the Urban Future*, Boulder, Colo.: Westview Press.

Bittner, E. (1990) *Aspects of Police*, Boston: Northeastern University Press.

Cray, E. (1972) *The Enemy in the Streets*, Garden City, NY: Anchor Books.

Escobar, E. (1993) 'The dialectics of repression: the Los Angeles Police Department and the Chicano movement', *The Journal of American History* 79: 1483–514.

Fogelson, R. (1971) 'White on black: A critique of the McCone Commission's Report', in T. Platt (ed.) *The Politics of Riot Commissions, 1917–1970*, New York: MacMillan, pp. 307–34.

Fogelson, R. (1977) *Big-City Police*, Cambridge, Mass.: Harvard University Press.

Fielding, N. (1994) 'Cop canteen culture', in T. Newburn and E. Stanko, (eds) *Just Boys Doing Business?*, London and New York: Routledge, pp. 46–63.

Fyfe, N. (1991) 'The police, space and society: the political geography of policing', *Progress in Human Geography* 15: 249–67.

Gazell, J. (1976) 'William H. Parker, police professionalization and the public: an assessment', *Journal of Police Science and Administration* 4: 28–37.

Greene, J. and Mastrofski, S. (eds) (1988) *Community Policing: Rhetoric or Reality?*, New York: Praeger.

Grinc, R. (1994) 'Angels in marble: problems in stimulating community involvement in community policing', *Crime and Delinquency* 40: 437–68.

Hannah, M. (1993) 'Space and social control in the administration of the Oglala Sioux', *Journal of Historical Geography* 19: 412–32.

Herbert, S. (1996a) 'The geopolitics of the police: Foucault, disciplinary power and the tactics of the Los Angeles Police Department', *Political Geography* 15: 47–59.

Herbert, S. (1996b) *Policing Space: Territoriality and the Los Angeles Police Department*, Minneapolis: University of Minnesota Press.

Herbert, S. (1996c) 'Morality in law enforcement: chasing "bad guys" with the LAPD', *Law and Society Review*, 30(4): 799–818.

Independent Commission on the Los Angeles Police Department (1991) *Report,* Los Angeles: City of Los Angeles.

Jankowski, M. (1991) *Islands in the Street: Gangs and American Urban Society*, Berkeley and Los Angeles: University of California Press.

Keith, M. (1991) 'Policing a perplexed society?: No-go areas and the mystification of police–black conflict', in E. Cashmore and E. McLaughlin (eds) *Out of Order?: Policing Black People*, London and New York: Routledge, pp. 189–214.

Keith, M. (1993) *Race, Riots and Policing: Lore and Disorder in a Multi-racist Society*, London: UCL Press.

Lea, J. and Young, J. (1985) *What Is To Be Done About Law and Order?*, Harmondsworth, Middlesex: Penguin.

Lowman, J. (1986) 'Conceptual issues in the geography of crime: toward a geography of social control', *Annals, Association of American Geographers* 76: 81–94.

Moore, M. (1992) 'Problem-solving and community policing', in N. Morris and M. Tonry (eds) *Modern Policing*, Chicago: University of Chicago Press, pp. 99–158.

Ogborn, M. (1993) 'Ordering the city: surveillance, public space and the reform of urban policing in England 1835–56', *Political Geography* 12, 505–21.

Padilla, F. (1992) *The Gang as an American Enterprise*, New Brunswick, NJ: Rutgers University Press.

Raine, W. (1967) *Los Angeles Riot Study: The Perception of Police Brutality in South Central Los Angeles*, Los Angeles: Institute of Government and Public Affairs.

Reiner, R. (1992) *The Politics of the Police*, Toronto: University of Toronto Press.

Reiss, A. and Bordua, D. (1967) 'Environment and organization: a perspective on the police,' in D. Bordua, (ed.) *The Police: Six Sociological Essays*, Wiley: New York, pp. 25–55.

Rosenbaum, D. and Lurigio, A. (1994) 'An inside look at community policing reform: definitions, organizational changes, and evaluation findings', *Crime and Delinquency* 40: 299–314.

Rubinstein, J. (1973) *City Police*. New York: Farrar, Straus and Giroux.

Sack, R. (1986) *Human Territoriality: Its Theory and History*, Cambridge: Cambridge University Press.

Schiesl, M. (1990) 'Behind the badge: the police and social discontent in Los Angeles since 1950', in M. Schiesl and N. Klein (eds) *20th Century Los Angeles: Power, Promotion and Social Conflict*, Claremont, Calif.: Regina Books, pp. 153–94.

Scott, J. (1985) *Weapons of the Weak: Everyday Forms of Peasant Resistance*, New Haven, Conn.: Yale University Press.

Skogan, W. (1994) 'The impact of community policing on neighborhood residents: a cross-site analysis', in D. Rosenbaum (ed.) *The Challenge of Community Policing: Testing the Promises*, Thousand Oaks, Calif.: Sage, pp. 167–81.

Skolnick, J. and Bayley, D. (1988) *The New Blue Line: Police Innovation in Six American Cities*, New York: Free Press.

Turner, W. (1968) *The Police Establishment*, New York: G.P. Putnam's Sons.

Van Maanen, J. (1974) 'Working the street: a developmental view of police behavior', in H. Jacob (ed.) *The Potential for Reform of Criminal Justice*, Beverly Hills, Calif.: Sage, pp. 83–130.

Walker, S. (1977) *A Critical History of Police Reform*, Lexington, Ken.: Lexington Books.

Walker, S. (1989) 'Broken windows and fractured history: the use and misuse of history in recent police patrol analysis', in R. Dunham and G. Alpert (eds) *Critical Issues in Policing*, Prospect Heights, Ill.: Waveland Press, pp. 382–94.

Woods, J. (1973) 'The progressives and the police', unpublished PhD dissertation, Los Angeles: University of California.

16

URBAN RENAISSANCE AND THE STREET

SPACES OF CONTROL AND CONTESTATION

Loretta Lees

•

In more cynical (but not any the less real) accounts of the postindustrial/postmodern city, the city (typically a US city) is represented as a combat zone. Urban researchers appear as war correspondents reporting from the front. Mike Davis's (1992: 154) Los Angeles 'bristles with malice. The carefully manicured lawns of the Westside sprout ominous little signs threatening "ARMED RESPONSE!" Wealthier neighbourhoods in the canyons and hillsides cower behind walls guarded by gun-toting private police and state-of-the-art electronic surveillance systems.' In Neil Smith's (1992: 62) New York City, '[t]here were cavalry charges down East Village streets, a chopper circling overhead, people out for a Sunday paper running in terror down First Avenue'. Urban violence is real enough, but the rhetoric surrounding it is even more militant. Whether described by the police or by critical geographers, the street is presented as a battlefield. In the case of Neil Smith's Lower East Side, this is more than simply metaphor, as demonstrations against gentrification escalated into a mêlée with renegade cops, but descriptions of war in the streets are much more pervasive than actual street fighting. Ironically, advocates of more police on the beat and their critics both see urban public space as under siege, even if, in the contemporary city of antagonisms, they cannot agree whether the threat comes principally from crime, disorder, and moral decline or from unemployment and uneven development.

Yet despite the lip service so often paid to the embattled public space of city streets, it is not always clear what is so special or so public about them. As Keith (1995: 297) notes, 'the street has occupied a cherished place in the lexicon of urbanism'. It is alternatively '[r]omanticised as the site of authentic political action, celebrated and reviled as the font of low culture or feared as a signifier of dangerous territorialisation'. Each of these different images of the street suggests a different aim and object for the struggle to control the street.

Many commentators have focused on the street as a site and symbol of democratic protest and politics. Though often somewhat short on definitions, they suggest that public space is 'open-minded' space (see Walzer, 1986, on open-minded spaces and single-minded spaces) in which members of the public can gather freely to discuss and debate their political beliefs (see Sorkin, 1992). To this view, the street represents the consummate democratic public space. It provides an urban space/place where 'subversive forces, forces of rupture, ludic forces act and meet' (Barthes, 1986: 96; see also Sennett, 1990) and where all of a city's 'inner contradictions can express and unfold themselves' (Berman, 1986: 484; see also Berman, 1982). Indeed, Don Mitchell (1995: 124) claims that 'revolutions entail a taking to the streets and a taking of public space'. Only on the streets can marginalised groups make themselves publicly (and thus politically) visible enough 'to be counted as legitimate members of the polity' (115).

There are a number of problems with this account of the politics of public space, not least of which is its under-theorisation of who and what come to count as being truly 'public' and/or 'political' as well as how and where they can come to count. Geographers have been in the forefront of efforts to reconsider the multiplicity of publics and politics involved with public space(s) (see Howell, 1993; Watson and Gibson, 1995; Urban Geography, 1996). Critics have discussed the exclusions of women (Deutsche, 1990a; Fraser, 1990; Young, 1990; Wilson, 1991; Ruddick, 1996a) and children/youth (Ruddick, 1996b; Valentine, 1996) built into so-called universal public spaces. They have also emphasised the ambiguity and complexity of public space and moved us towards an appreciation of the ways in which both the 'public' and its activities as well as the city and its spaces are mutually constitutive.

Despite these important revisions to conventional notions of urban public space, there is at least some continuing value to traditional romanticism about the universalist and emancipatory potential of taking to the streets. In particular, it focuses critical attention on the marginalisation as well as the surveillance and police control of the street (Jackson, 1992; Campbell, 1993; Allen and Pryke, 1994; Fyfe, 1995; Graham and Marvin, 1996: 220–7; Herbert, Chapter 15 of this volume; Fyfe and Bannister, Chapter 17 of this volume). Is it true, as Mike Davis (1992: 155) claims, that the 'universal consequence of the crusade to secure the city is the destruction of any truly democratic urban space'?

Davis (1990, 1992) is not alone in condemning urban renaissance/gentrification efforts, which he likens to an escalating 'cold war on the streets' (1990: 234), for shattering the once open public space of the city and erecting in its place an exclusive and 'forbidding corporate citadel, separated from the surrounding poor neighborhoods by battlements and moats' (1992: 154). Neil Smith (1992, 1996) also contends that gentrification is designed to reduce diversity on the streets, domesticating urban space and rendering it safe for reinvestment and resettlement by the wealthy and well-heeled. For these critics, the control and pacification of unruly streets that comes with urban revitalisation plans suffocates what Davis (1990: 231) calls the vital 'democratic admixture' of different segments of the public interacting on city streets and in parks.

Heavy police presence on the street may well inhibit some forms of sociability and legitimate political activity, but the corollary – that its removal will liberate the public and the street – does not necessarily follow. While it is important to recognise the political significance of the street, it is also important not to romanticise it excessively. Taking to the streets, whether by the police or by members of the public, is usually about asserting control and dominance. Beatrix Campbell (1993) demonstrates this superbly in her exposition of lawless masculinity in Britain's dangerous places. Young British men, unemployed and marginalised, take to the streets, claiming the space of *their* council estates by joyriding, rioting, and terrorising their neighbours. In short, a taking to, and of, the street is not an inherently democratic or emancipatory act. In the wrong circumstances it can destroy the 'publicness' of public space just as surely as excessive police control and surveillance.

Gentrification, although often condemned for homogenising the street and stifling its vitality, is underwritten by the same utopian images of urban public space and the streets upheld by its critics. In the gentrification literature, the street is celebrated as a kind of permanent festival (Kasinitz, 1984; Boyer, 1994). The diversity and variety of lively downtown streets are favourably contrasted with the sterile homogeneity of the suburbs. But as middle-class security anxieties illustrate, celebration of difference and diversity only goes so far. While gentrification celebrates diverse city streets, it also pacifies and represses them, in order to make them feel 'safe' for a middle class public. Gentrification is a contradictory process. It promotes and enlivens the public space(s) of the street at the same time as it encourages and legitimates withdrawal from and control over them.

In this essay, I want to work through the complexities and contradictions of public space through a reconsideration of the gentrification process. Gentrification exemplifies many of the difficulties that trouble me about the debate now raging over the supposed 'end of public space'. While pessimistic authors like Sorkin (1992) bemoan the death of the street and the end of truly public space, others read the contemporary scene much more ambiguously. Goss (1996: 223) for example sees in new urban public spaces, such as festival marketplaces, the articulation of both 'genuine urban life' and 'rational strategies of social control'. Much of the confusion, I would suggest, stems from a singular understanding of the street, and of public space more generally, as either free and democratic or repressed and controlled. Public space is both at the same time. It is simultaneously a space of political struggle and expression and of repression and control. The challenge is to appreciate these complex modalities and to make the most of the positive while resisting the worst.

I will convene my discussion of how urban renaissance/gentrification affects urban public space through two case studies from downtown Vancouver: the first, Library Square – the site of Vancouver's new public library; the second, the southern end of Granville Street, located near to Library Square. By situating my empirical accounts in Vancouver, Canada, I hope to illustrate that 'these trends are actually more varied and complex than the usual reliance on American literature suggests' (Graham and Marvin, 1996: 186; Lees, 1997).

Although I agree that there may have been an increase in the control and surveillance of public space, it is important to emphasise that public space has always been controlled. It has never been truly 'free' and 'open'. I want to dislodge the end of public space thesis by illustrating two features: first, public space is being opened up in new and complex ways; second, the control of urban public space can almost always be countered, subverted, and resisted.

RETREATING FROM THE STREET: VANCOUVER'S NEW PUBLIC LIBRARY

Gehry's baroquely fortified Frances Howard Goldwyn Regional Branch Library in Hollywood (1984) positively taunts potential trespassers to 'make my day'. This is undoubtedly the most menacing library ever built, a bizarre hybrid (on the outside) of dry-docked dreadnought and Gunga Din fort. With its fifteen-foot security walls of stucco-covered concrete block, its anti-graffiti barricades covered in ceramic tile, its sunken entrance protected by ten-foot steel stacks, and its stylized sentry boxes perched precariously on each side, the Goldwyn Library . . . projects the same kind of macho exaggeration as Dirty Harry's 44 Magnum.

(Davis, 1990: 239)

The construction of Vancouver's new public library illustrates a number of the complexities and contradictions of public space at the end of the century. I want to pay particular attention to four aspects of the library: its role fostering gentrification and economic redevelopment in the Downtown; the displacement of the public space of the street by the privatised mall and surrogate street of the library's arcade (see too Rendell, Chapter 6 this volume); strategies of security and surveillance; and the potential of the imaginary, immaterial, and electronic spaces opened up by the new library.

Located in Library Square in downtown Vancouver (Figure 16.1), the Vancouver Public Library building opened in June 1995 as the city's newest civic landmark and megastructure. The building mimics the design of the Colosseum in Rome, but its appropriation of a classical European form is playful and seems better suited to Disney World than a serious civic structure such as a public library. The Colosseum design is merely an oval-shaped facade, a colonnaded wall that surrounds and protects the glass and concrete box that is the library proper.

The thick walls of the library facade are not inconsequential to the building's role in municipal plans for gentrifying the eastern section of downtown. City planners hope the library can serve both as 'a major public statement [and as] a potential catalyst for a new surge of urban development on the eve of the millennium' (McKenzie, 1993: 42). As a beach-head for the gentrification of the eastern section of downtown, the library will stabilise other, already gentrifying areas nearby such as Gastown, Chinatown, and Yaletown. Expressive of this purpose, the library coils inward, away from the street, demonstrating the kind of fortress-like, siege mentality that Davis (1992: 168–9) discusses with reference to the Goldwyn Branch Library

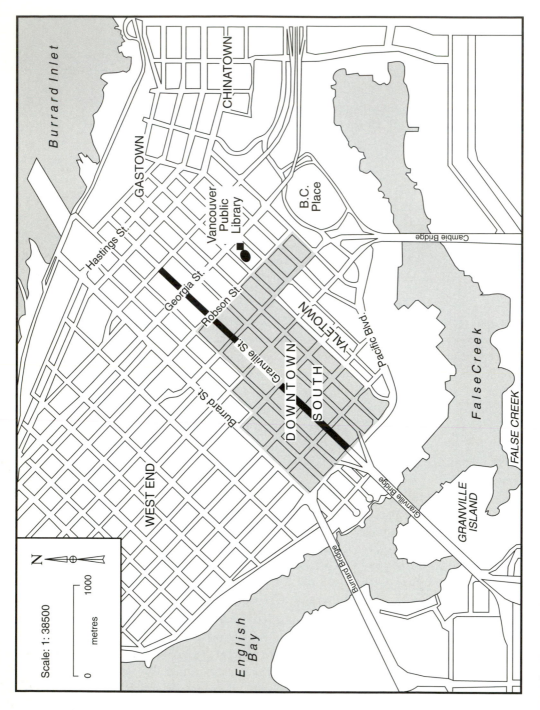

FIGURE 16.1 Downtown Vancouver

in Hollywood, California, which is also being used as a beach-head for gentrification. The oval shape of Vancouver's new library is introspective, and from the outside, it is not easy to determine where the entrance door is located. Its thick facade wall is set back from and dominates the street, segregating the library from the city outside (Figure 16.2).

Theme park type urban designs such as Vancouver's Colosseum style library have attracted much critical attention (see Zukin, 1991; Sorkin, 1992; Boyer, 1993; Knox, 1993; Christopherson, 1994). Davis (1990) argues that these new megastructures are taking over from and marginalising the street (see also Crilly, 1993). He is concerned that these valourised spaces displace the public space of the city and subject the public to control and surveillance by private police forces (Davis, 1990: 226).

Such criticism can certainly be directed at the new Vancouver Public Library. Instead of connecting with the city streets beyond, the library focuses in on the arcade sheltered inside its facade walls (Figure 16.3). Lined with retailing units such as McDonalds, Blenz Coffee and Yogenfruz, this surrogate street seems to play directly into Davis's (1990) thesis about the increasing marginalisation of the authentic public space of the street. Unlike the free-wheeling streets outside the library, where people interact as neighbours and citizens, the library arcade feels like a private space where visitors must behave as paying customers. This atmosphere is directly related to the

FIGURE 16.2 The fortress-like library. Source: author

FIGURE 16.3 The library arcade. Source: author

library's increasingly commercial functions. Within the library arcade, the seating is located to one side, adjacent to the retailing units. It looks and feels proprietary, and people are only encouraged to sit or occupy the space if they are consuming something from the shops. The result is the picture of civility – a bourgeois public eating and drinking in comfortable and cosmopolitan surroundings – but it excludes those who cannot afford the price of entry to this pleasant but restricted environment. The street kids and homeless to be found on Granville Street not far from the library are not much in evidence in the library arcade.

The new library is an excellent example of what Bianchini (1989: 5) calls a privately owned and managed public space offered for public use. The chief of security summarises this well when he states: 'this is a private space owned by the City . . . who allow the public to use it' (interview, July 1995, quoted in Lees, 1997: 338). A private security force, hired by the City, patrols the building controlling what members of the public may do within the library. Political activity, religious activity, and leafleting are forbidden within the library walls, and restricted within the court-yard around Library Square. These activities are perfectly legal in the public space of the streets surrounding the library, but the library, for purposes of political expression, is treated as private, not public property.

Thus far, the picture largely supports Davis's (1992: 155) claim that attempts to secure the city for gentrification and middle-class consumption lead to the destruction of any truly democratic urban space. But the public space of the library is more complex than epitaphs for the end of public space allow.

Though covered and enclosed, the library arcade is no less open or closed than the downtown streets. Anyone is at liberty to enter the public library and its surrogate street, as they are the streets outside. Whether they wish to, or would feel comfortable doing so, is a different story. Whilst the universally accessible public space of the street provides a strong anticipatory and counterfactual position from which to criticise the privatisation of city space, it may also blind critics to the possibilities of new environments like the library arcade. The library arcade is transgressive, liberally mixing traditionally separate public and private spaces with civic and commercial functions (for further analysis see Lees, 1997).

The effects of the library's security apparatus are also more ambiguous than they may seem at first glance. The concern with security is certainly about rational strategies of control, but this is not an entirely bad thing. The safety of women and children was a major consideration in the design of the new library. In the old library building, library employees and members of the general public alike complained about dark corners and dead ends they felt uncomfortable entering. These fears were both fanned and confirmed by a number of disturbing incidents in the old library: problems with theft and vandalism, instances of verbal and physical abuse, notes soliciting sex for money inside children's books, a man who streaked naked through the language and literature section of the library (Kines, 1995: F5).

The new library has been designed with these public safety concerns in mind. The number of guards has increased and the electronic security system protecting library collections from theft and vandalism has also been upgraded. The dark and unsafe spaces that troubled the old Vancouver Public Library building were also designed out of the new one. Its colonnaded facade and high arches shower the interior with natural light. Furthermore, unlike more traditional public libraries, in which the book stacks are hidden on the edges of the building, the new Vancouver public library is turned inside out: its shelving is located in the centre of the structure and the reading areas on the outside, next to the wide glass windows. This makes for a brighter and 'safer' reading environment.

The security of the washrooms, which in the old public library had been the site of many undesirable sex and drugs incidents, was also a major design consideration. In the new library, the forty-eight washrooms are located on the same side of each floor. They are well lit with an open floor plan and large gaps above and below the cubicle doors. Moreover, security guards patrol them regularly.

To safeguard youth and children in the library, the youth section has been placed on the ground floor of the library next to the bulk of the security apparatus, while the children's section has its own specially designed and protected/secured space below the ground floor.

These security measures make it possible for women, children, and the elderly, segments of the public who sometimes found the old public library unsafe, to enjoy the public space of Vancouver's new public library. But at the same time they also subject library patrons to new and insidious forms of surveillance and control. Surveillance cameras allow a guard sitting at a security console to monitor activity within the library twenty-four hours a day. As Foucault (1979: 172) noted of the tight discipline within prisons, these measures are designed to 'permit an internal, articulated and detailed control – to render visible those who are inside it'. Liberal and radical critics alike have attacked these schemes to control, regulate, and police the public spaces of the city, but their criticisms often discount the degree to which some security measures are necessary to make the public spaces of the library, and of the city more generally, safe and accessible to all members of the public. It may well be that in the particular case of the Vancouver Public Library some of the steps taken in the name of public safety are excessive, yet concern with public safety is also a necessary part of any programme for fostering truly democratic public space.

Library officials have also revised the code of conduct, and the effect of these measures is potentially quite worrying. As I mentioned before, leafleting and political speech-making, mainstays of democratic expression, are explicitly forbidden in the pseudo-street of the library arcade. Violators can be expelled from the library or even banned for anywhere from twenty-four hours to a few months. Library officials, though, are loath to invoke this sanction for any but the most stubborn offenders (Kines, 1995: F5). The rules are, for the most part, discretionary, and library employees are encouraged to use persuasion before calling security to deal with unruly library patrons. For more serious offences the security staff may issue a citizen's arrest and call in the local police force. By policing individual conduct, library officials reduce the potential of this public space to serve as a site for democratic protest.

There are, however, other spaces for contestation in this new library. The library offers access to new symbolic, intellectual and electronic spaces, spaces that have largely been ignored by those who announce the end of democratic public space. The volumes of books in the library provide knowledge and information to the reading public, as do the 250 computers from which people can use CD-ROMs or access the internet. As W. Mitchell (1995: 8) notes, these electronic spaces subvert, displace and radically redefine 'our received conceptions of gathering place, community and urban life'. Free from the watchful eye of the library's security apparatus, they have great potential to serve as the democratic public spaces of the future.

Unlike many public libraries, the Vancouver Public Library provides free access to cyberspace to anyone who walks in off the street or, as is becoming increasingly common, accesses its site remotely (see Graham and Marvin, 1996: 20, on pay per view or bit by bit in public libraries). Reading and computer literacy classes are also provided free of charge. The internet services available in Vancouver's new Public Library stand aside from arguments such as Davis's (1990) on the 'unwired' and thus disempowered ghetto. Constructing 'electronic public spaces', in this case a 'freenet', eliminates social isolation and combats social polarisation (Graham and Marvin, 1996: 359–60). Attempts to foster electronic public spaces in and through the public library are increasing. Microsoft, for example, is working with public libraries across the continent to increase public access to the internet in economically disadvantaged communities. Through its 'Libraries On Line' programme it is donating money as well as computer hardware, software and expertise to public libraries throughout North America (Lees, 1997).

In summary, the new Vancouver Public Library is an ambivalent space, a space in which there is a desire both to accommodate a pluralistic public and to control it through rational strategies of surveillance and discipline. The library and its arcade act as a beach-head for gentrification, while inside the library itself, public activity is subjected to new forms of surveillance and control. But the library also offers free public access to knowledge and to the uncontrolled space of the internet. These immaterial, but no less real, spaces provide new sites from which to achieve the utopian dreams of genuinely democratic public space. Although telecommunications have not yet rendered the street irrelevant to public life, as Dear (1995: 31) suggests, they have opened up exciting new avenues for civic life and democratic exchange.

ADVANCING ON THE STREET: GRANVILLE STREET, DOWNTOWN VANCOUVER

This conscious 'hardening' of the city surface against the poor is especially brazen in the Manichaean treatment of Downtown microcosms. . . . The persistence . . . of street people . . . sours the image of designer Downtown living and betrays the laboriously constructed illusion of a Downtown 'renaissance'. City Hall then retaliates with its own variant of low-intensity warfare.

(Davis, 1990: 232)

Once the proud centre of Vancouver's theatre district, Granville Street has, in recent years, become known as Vancouver's 'mean street'. Home to street kids, transients, drug dealers, and single room occupants in rooming hotels (usually single men with disabilities on fixed incomes), the southern end of Granville Street (see Figure 16.1) is notorious in Vancouver for decay and dilapidation:

> Granville Street. Where the cops stalk bars and bust peep shows and Burger King staff act as bouncers. Down the street at McDonalds the hash browns sell for 60 cents and a gram of hash changes hands under the table for $10. It's the blemish on beautiful Vancouver, the street that people love to hate.
>
> (Cox, 1987: B1)

Regular headlines, such as 'Street People Sink Granville in Shame' (*The Province*, 1973: 1), have decried this unseemly state of affairs and rallied city officials to periodic action. But nearly thirty years of revitalisation efforts have failed to dislodge the tawdry mixture of iron-gated pawn brokers, sex shops and cheque-cashing outlets from this portion of the downtown (Figure 16.4). Undaunted, city planners are at it again, and today the gritty southern end of Granville Street is at the centre of another redevelopment plan. This time, city officials and real-estate industry boosters hope to redevelop/gentrify the surrounding area into a high-density, middle-income, residential neighbourhood. This latest plan to 'clean up' Granville Street, like its

FIGURE 16.4 A junky downtown street. Source: author

predecessors, illustrates the complexities and contradictions of efforts to revitalise the public space of the street.

Over the years, local shopkeepers have tried a number of different measures to (re)gain control of the street and to rid Granville of its undesirables. Many businesses hired private security guards to discourage street kids and the homeless from gathering in and around their establishments. One security guard at the hotel Chateau Granville was so fearsome that the local police named him 'Sgt Slaughter' (Cox, 1987: B1). At the McDonalds on Granville, the manager went so far as to install an electrical control system in the washrooms, allowing employees to lock customers out of the washrooms at the touch of a button. In the event of trouble, the security system could also be used to lock the offending parties in until the police arrived (Dodd and Baglo, 1974: 6). McDonalds also put up signs proclaiming 'a $1 per person minimum charge' and 'we reserve the right to refuse service' (*The Vancouver Sun*, 1982: A3). By denying access to public facilities and refusing regular service to the poor, McDonalds and the other businesses along Granville Street tried to control, and thus effectively to privatise, the public space of the street. These efforts at control illustrate a disturbing sign for the future of public space, but the urgency of the business revitalisation campaign also speaks to a hidden resistance by those, especially the poor, the homeless, and the street kids, most marginalised by these plans.

The City of Vancouver has also tried to spruce up Granville Street. The first major effort began in 1973 when the City approved a $3.25 million plan to revitalise the commercial area at the north end of Granville Street. The plan closed six blocks of Granville to cars (but not taxis or buses), turning the street into a pedestrianised zone which planners dubbed the Granville Mall. Consciously rejecting the example of American urban redevelopment, which converted 'once vital pedestrian streets into traffic sewers' (Davis, 1990: 226), planners and reform politicians in Vancouver hoped the pedestrianised Granville Mall, serviced by public transit, would enliven downtown streets, bringing people back into the inner city, and thereby recreating the rich street life and urban culture of a bygone era (Ley, 1980). Vancouver's then-mayor Art Phillips described Granville Mall as 'the cornerstone to downtown redevelopment' (*The Vancouver Sun*, 1973: 11), but a pedestrian mall where citizens could mix and meet was not practical in a city that otherwise remained heavily dependent on the automobile. Without a local residential population, shops along Granville Mall were dependent on suburban visitors who could no longer drive downtown conveniently. Of those who did, many preferred the enclosed and protected atmosphere of the underground Pacific Centre shopping mall, which had just opened, syphoning customers away from the streetside shops on Granville Mall.

Despite (or, in fact because of) these efforts, the southern end of Granville Street fell into severe decline. It became a 'slum gap', an urban ghost town (Kalapinski, 1975: 19). Only the drug trade thrived. According to 1981 police statistics, 40 per cent of drug charges laid in Vancouver occurred in the 800 block of Granville Street, just south of Robson (Deacon, 1982: B1). Tensions were high, prompting one police officer to say: 'What you need is a D9L (Caterpillar bulldozer) down each side of the street. The only way you can clean it up is to level the damn thing' (quoted in Cox, 1987: B3). Merchants in the area were angry and set up the Theatre Row Business Association to lobby the police and the City to clean the street up.

In response, the police requested that the City remove benches from the street and install brighter lighting to discourage street kids and drug dealers from inhabiting the area. They also beefed up their patrols, focusing beat officers around the 800 block of Granville and other trouble spots. Unlike Smith's (1996) revanchist city there were no police sweeps to remove homeless people from the public space of Granville Street. Instead, the police focused on street-level drug traffic. One Granville drug dealer complained: 'it's like a war down here'. Vancouver officials could not have been more pleased. Police Superintendent Hank Starek boasted: 'We have gained some measure of control on the mall' (quoted in Deacon, 1982: B1). The drug trade retreated to other parts of the city, but no development followed in its wake to clean up the image of southern Granville Street.

By the late 1980s city officials agreed that the revitalisation of Granville required an economic change. Rather than trying to drive the street kids and homeless off Granville Street (and to some other part of downtown as it did with the drug trade), the City is now trying to attract some other kinds of activity to the street, thereby diluting the presence of street kids along Granville. It sponsored a Canada-wide competition among architects and planners to re-design the area around Granville.

The winner, architect Roger du Toit, argued that there should be a shift from service and commercial uses to residential, and in 1991 City Council put the finishing touches on this plan. Instead of merely giving Granville Street another facelift, the planners hope to 'organically transform' it, and to do so they have developed a more holistic approach. They have re-zoned the area – now dubbed Downtown South (nestled between the West End and BC Place, see Figure 16.1) – to allow residential development (City of Vancouver Planning Department, 1991). The City hopes to strike a balance between entertainment and residential uses. The two or three blocks south of Robson Street on Granville will be themed as an entertainment district with theatres, cafés and night clubs. The residential part of the plan includes a middle-class neighbourhood to accommodate local office workers. The City expects about 8,500 units to be built, with approximately 11,000 new residents moving into town-houses and apartments over the next ten years, half of which should be completed by the year 2000 (Ward, 1996: D3).

To accommodate the homeless and indigent population of southern Granville Street, the City is using a mixture of regulations and financial incentives to preserve as many of the rooming hotels as possible. It spent $1.9 million to buy a run-down Granville Street rooming house – The Gresham Hotel. It also has plans for facilities to help stabilise the long-term hotel residents who live in Granville's 1,000 low-rent rooms. The goal of these measures is to resolve low-tension class warfare by ensuring that low-rent rooms are protected. Whether it will succeed is another matter, but the picture in Vancouver is much less grim than suggested by gentrification critics like Kasinitz (1984), who argue that gentrification displaces the poor from the public space of the street and replaces them with the middle classes. Although the City of Vancouver plans to gentrify in Downtown South and to 'upgrade' Granville Street, it hopes to do so without destroying and/or displacing the community of old timers and poor who live there.

Business groups in the area are also starting to realise that they have a stake in preventing homelessness. Indeed, an alliance has formed between developers, local community groups, landlords, and tenants called the Downtown Planning Group. They have presented proposals to City Council aimed at providing low-cost housing for displaced residents and community services to deal with prostitution and street crime (Sarti, 1991: B2).

The Downtown South plan is for chic and grit to co-exist (Figure 16.5). It is a 'hybrid' that links social needs and redevelopment (see Ruddick, 1996b: 192–8). Even the presence of street kids has been mobilised to serve the cause of local boos-terism. As a store owner on Granville Street explained, 'So what if Granville is a little undesirable. If it wasn't like this, it wouldn't be cool. It would be Robson Street (Vancouver's Rodeo Drive)' (quoted in Usinger, 1996: 5). The City's vision depends on a liberal belief that with the right financial and institutional support different classes can live together. As one planner said, 'Granville serves so many needed roles that it's going to remain a complicated street' (Ward, 1996: D5). Whether Granville Street can actually remain affordable and a place where the poor will not feel like a blight on a middle-class landscape remains to be seen.

FIGURE 16.5 Granville Street with redevelopment looming in the background. Source: author

Much of the attention both in this essay and in more general discussions of the public space of Granville Street has focused on adults – their decisions, their conflicts, their stories. This is not surprising, for as Valentine (1996: 214) argues, unchaperoned teenagers are often considered a 'polluting presence on the street' and subject to adult controls and regulation. But Valentine, like the apocalyptic end of public space theorists who illustrate middle-class hegemony over the homeless and the poor, overemphasises control and underemphasises contestation. Although sympathetic to youth, she constructs a theory of power over them, not power for them (see Ruddick, 1996b: 194). Manichean dichotomies such as oppressor/oppressed are not questioned.

Street kids resist the spatial hegemony of the adult world in a variety of ways (see Winchester and Costello, 1995, on street kids' culture of resistance). One way in which the Granville street kids distinguish themselves from the adult world is by separating themselves from the adult homeless. They gather on Granville Street, which is perceived to be marginally safer than the 'skid row' area around Hastings Street in the Downtown Eastside – where the majority of adult homeless people are located. Their struggle has resulted in a concentration of services for them around Granville Street. 'The Gathering Place', for example, just off Granville, offers street kids a place where they can get cheap food and a shower, hang out in the TV lounge, take classes, and use the library.

Granville street kids define themselves not as homeless but as street kids. The name 'street kid' indicates possession of the space of the street. By asserting their identity in this way, they challenge their stigmatisation as homeless (Ruddick, 1996b: 92). This struggle for self-definition is also expressed through tactical inhabitation of space (see Ruddick, 1996b: 92). By sleeping in spaces such as the Capitol 6 (movie theatre) underground parkade (Canadian Broadcasting Corporation, 1996), they challenge the pregiven meaning of the public space of Granville Street. Knowing where to find these places of shelter, street kids interrupt the grand narrative of Downtown South. For this reason, Chang (1994: 114–5) likens them to an accelerated version of de Certeau's 'walkers' who rewrite the spatial order of the city. Even their selection of southern Granville can be read as a statement. There, at the gateway to the central city, Granville street kids make themselves visible to Vancouver's public, city officials and tourists alike, thereby dramatising the right of the poor not to be isolated and excluded or dispossessed of their home (see Deutsche, 1990b, on the Homeless Vehicle Project in New York City, and strategies of visibility).

For the street truly is their 'home'. This was powerfully exemplified in *Home Street Home*, an award-winning documentary produced by two pregnant street kids from Granville – Spice and Shorty – for the local Canadian Broadcasting Corporation's evening news (CBC, 1996; for another first-person account of life on Vancouver's streets, see Lau, 1989). Refusing to be leered at for the six o'clock news, Spice and Shorty took control of the camera and thus of what the public (at least the TV public) would see of life on the street.

Though often looked upon with contempt or pity by passers-by, Spice and Shorty see their fellow street kids as 'a family' on Granville Street. Like a family they 'help each other', perhaps much more effectively than the abusive families so many street kids left behind. For these street kids, the public space of Granville has a use value, in contrast to the exchange value it holds for businesspeople, planners, and developers. Granville Street is where they make their lives, and *Home Street Home* showed those of 'us' who are not street kids some of the very different ways they try to inhabit this space.

Granville Street then is both a space of control by adults – City planners, politicians, shopkeepers and the police, and a space of contestation by street kids and transients. Like all streets it is a complicated space with a variety of partial and negotiated stories to tell.

CONCLUSION: MAKING SPACE FOR THE STREET

What do these stories about Vancouver's new public library and Granville Street tell us about the experience and control of urban public space at the end of the century? The war rhetoric used by those who bemoan the death of public space does capture something. Library designers and Granville Street planners have gone to some length to patrol and secure these urban spaces in the name of public safety. But the overriding concern with 'control' tells us nothing about how space(s) can be appropriated and contested. For the public space of the street is not pregiven, in either its form or its meaning. It is produced through contestation and social negotiation. In the case of both the new public library and the renaissance of southern Granville Street, the resulting public space has had to accommodate the demands of a variety of different publics in a way that the end of public space thesis does not allow.

This leads to a second conclusion about the heterogeneity and local specificity of public space. The examples most often cited of the so-called end of public space are drawn from the United States and then universalised as a problem common to the entire continent and indeed the Western world more generally. By focusing on a relatively liberal city, with its own tradition of progressive urban planning and on two different kinds of public space within Vancouver, I have tried to show that public space is not a homogeneous entity. Public spaces differ depending on their social, cultural, economic, and symbolic functions, and, perhaps most importantly, depending on the meanings, contested and negotiated though they are, that different publics bring to them.

The public spaces in the two Vancouver case studies are contradictory. Partly this is because they serve different publics and thus accommodate different kinds of control and security concerns. But partly it is because of the different ways in which they both transgress the public/private dichotomy, making me uncomfortable about using the term 'public' to describe either one. This ambiguity is related to the chaotic nature of the gentrification process itself (Lees, 1996). Simultaneously embracing and withdrawing from the public spaces of city streets, gentrification is deeply ambivalent in its stance to urban life. Its attempts to foster genuine public culture on the street subverts that very goal, as efforts to secure urban space stifle its celebrated diversity and vitality. This ambivalence is expressed in the contradictions and ambiguities of Vancouver's new public library and its Granville Street redevelopment plans, which are themselves part and parcel of the paradoxical gentrification process itself. In conclusion, then, it might be said that the street is there for the making (Goss, 1996). It is always a site of control and contestation. Make of it what you will.

ACKNOWLEDGEMENT

I would like to thank The Leverhulme Trust who provided the funding for this research.

REFERENCES

Allen, J. and Pryke, M. (1994) The Production of Service Space, *Environment and Planning D: Society and Space*, 12: 453–75.

Barthes, R. (1986) Semiology and the Urban, in M. Gottdiener and A. Lagopoulos (eds) *The City and the Sign: an introduction to urban semiotics*, Columbia University Press: New York, pp. 87–98.

Berman, M. (1986) Take it to the Streets: conflict and community in public space, *Dissent* (summer) 33(4): 476–85.

Berman, M. (1982) *All That is Solid Melts into Air*, Simon and Shuster: New York.

Bianchini, F. (1989) The Crisis of Urban Public Social Life in Britain: origins of the problem and possible responses, *Planning Practice and Research*, 5(3): 4–8.

Boyer, C. (1993) The City of Illusion: New York's public spaces, in P. Knox (ed.) *The Restless Urban Landscape*, Prentice Hall: Englewood Cliffs, pp. 111–26.

Boyer, C. (1994) *The City of Collective Memory: its historical imagery and architectural entertainments*, MIT Press: Cambridge, Mass. and London, England.

Campbell, B. (1993) *Goliath: Britain's Dangerous Places*, Methuen: London.

Canadian Broadcasting Corporation (1996) *Home Street Home*, 13 March.

Chang, E. (1994) Where the 'Street Kid' Meets the City: feminism, postmodernism and runaway subjectivity, in P. Delany (ed.) *Vancouver: representing the postmodern city*, Arsenal Pulp Press: Vancouver, Canada, pp. 97–120.

Christopherson, S. (1994) The Fortress City: privatized spaces, consumer citizenship, in A. Amin (ed.) *Post Fordism: A Reader*, Blackwell: Oxford, pp. 409–27.

City of Vancouver Planning Department (1991) *Downtown South Community Plan*.

Cox, S. (1987) Heart of Darkness: groups seek solution for maligned mall, *The Vancouver Sun*, 23 April: B1.

Crilly, D. (1993) Megastructures and Urban Change: aesthetics, ideology and design, in P. Knox (ed.) *The Restless Urban Landscape*, Prentice Hall: Englewood Cliffs, New Jersey 07632, pp. 127–64.

Davis, M. (1990) *City of Quartz: excavating the future in Los Angeles*, Vintage Books: New York.

Davis, M. (1992) Fortress Los Angeles: the militarization of urban space, in M. Sorkin (ed.) *Variations on a Theme Park: the new American city and the end of public space*, Hill and Wang: New York, pp. 154–80.

Deacon, J. (1982) 'Police Push Pushers in "Granville Maul"', *The Vancouver Sun*, 18 February: B1.

Dear, M. (1995) Prolegomena to a Post Modern Urbanism, in P. Healy, S. Cameron, S. Davoudi, S. Graham and A. Madani Pour (eds) *Managing Cities: the new urban context*, Wiley: London, pp. 27–44.

Deutsche, R. (1990a) Men in Space, *Strategies: A Journal of Theory, Culture and Politics*, 3: 130–8.

Deutsche, R. (1990b) Uneven Development: public art in New York City, in R. Ferguson, M. Gever, T. Minh-ha Trinh and C. West (eds) *Out There: marginalisation and contemporary cultures*, MIT Press: Cambridge, Mass. and London, England.

Dodd, K. and Baglo, G. (1974) Ugly Realities of Life on the Mall, *The Vancouver Sun*, 29 June: 6.

Foucault, M. (1979) *Discipline and Punish: the birth of the prison*, Vintage Books: New York.

Fraser, N. (1990) Rethinking the Public Sphere: a contribution to the critique of actually existing democracy, *Social Text*, 25/26: 56–80.

Fyfe, N. (1995) Controlling the Local Spaces of Democracy and Liberty?: 1994 British Criminal Justice Legislation, *Urban Geography*, 16, 3: 192–7.

Goss, J. (1996) Disquiet on the Waterfront: reflections on nostalgia and Utopia in the urban archetypes of festival marketplaces, *Urban Geography*, 17, 3: 221–47.

Graham, S. and Marvin, S. (1996) *Telecommunications and the City: electronic spaces, urban places*, Routledge: London and New York.

Howell, P. (1993) Public Space and the Public Sphere: political theory and the historical geography of modernity, *Environment and Planning D: Society and Space*, 11: 303–22.

Jackson, P. (1992) The Politics of the Streets: a geography of Caribana, *Political Geography*, 11, 2: 130–51.

Kalapinski, R. E. (1975) *A Survey of the Granville Mall Retail Sales, Phase II*, prepared for the City of Vancouver, Social Planning Department.

Kasinitz, P. (1984) Gentrification and Homelessness: the single room occupant and the inner city revival, *Urban Social Change Review*, 17: 9–14.

Keith, M. (1995) Shouts of the Street: identity and the spaces of authenticity, *Social Identities*, 1, 2: 297–315.

Kines, L. (1995) Life Inside Isn't All Words and Wisdom, *The Vancouver Sun*, 24 May: F5.

Knox, P. (ed.) (1993) *The Restless Urban Landscape*, Prentice Hall: Englewood Cliffs.

Lau, E. (1989) *Runaway: Diary of a Street Kid*, HarperCollins: Toronto.

Lees, L. (1997) Ageographia, Heterotopia, and Vancouver's New Public Library, *Environment and Planning D: Society and Space*, 15: 321–47.

Lees, L. (1996) In the Pursuit of Difference: representations of gentrification, *Environment and Planning A*, 28: 453–70.

Ley, D. (1980) Liberal Ideology and the Postindustrial City, *Annals of the Association of American Geographers*, 70, 2: 238–58.

McKenzie, S. (1993) Mistake or Masterpiece? *Vancouver* (January/February): 42–50.

Mitchell, D. (1995) The End of Public Space?: people's park, definitions of the public, and democracy, *Annals of the Association of American Geographers*, 85, 1: 108–33.

Mitchell, W. (1995) *City of Bits: space, place and the infobahn*, MIT Press: Cambridge, Mass.

Ruddick, S. (1996a) Constructing Difference in Public Spaces: race, class, and gender as interlocking systems, *Urban Geography*, 17, 2: 132–51.

Ruddick, S. (1996b) *Young and Homeless in Hollywood: mapping social identities*, Routledge: London and New York.

Sarti, R. (1991) Developers, Tenants Unite on Downtown South Woes, *The Vancouver Sun*, 15 May: B1–B2.

Sennett, R. (1990) *The Conscience of the Eye: the design and social life of cities*, Knopf: New York.

Smith, N. (1996) *The New Urban Frontier: gentrification and the revanchist city*, Routledge: London and New York.

Smith, N. (1992) New City, New Frontier: the Lower East Side as Wild, Wild West, in M. Sorkin (ed.) *Variations on a Theme Park: the new American city and the end of public space*, Hill and Wang: New York, pp. 61–93.

Sorkin, M. (ed.) (1992) *Variations on a Theme Park: the new American City and the end of public space*, Hill and Wang: New York.

The Province (1973), 16 April: 1.

The Vancouver Sun (1973) Mall to Open with Parade Activities, says Phillips . . . , 5 November: 11.

The Vancouver Sun (1982) Mall Problems 'Outside' McDonalds, 12 August: A3.

Urban Geography (1996) Special Issue 'Public Space and the City', 17, 2–3: 127–247.

Usinger, M. (1996) Cool Customers Reclaim Strip: Granville Street's visionary entrepreneurs undo decades of decay, *The Vancouver Courier*, 21 April: 1, 4–5.

Valentine, G. (1996) Children Should be Seen and Not Heard: the production and transgression of adults' public space, *Urban Geography*, 17, 3: 205–20.

Walzer, M. (1986) Pleasures and Costs of Urbanity, *Dissent* (summer) 33, 4: 470–5.

Ward, D. (1996) Vancouver's Last Frontiers, *The Vancouver Sun*, 2 March: D3–D5.

Watson, S. and Gibson, K. (eds) (1995) *Postmodern Cities and Spaces*, Blackwell: Oxford, England and Cambridge, Mass.

Wilson, E. (1991) *The Sphinx in the City: urban life, the control of disorder, and women*, Virago Press: London.

Winchester, H. and Costello, L. (1995) Living on the Street: social organisation and gender relations of Australian street kids, *Environment and Planning D: Society and Space*, 13: 329–48.

Young, I. (1990) *Justice and the Politics of Difference*, Princeton University Press: Princeton, NJ.

Zukin, S. (1991) *Landscapes of Power: from Detroit to Disneyworld*, University of California Press: Berkeley, Calif.

17

'THE EYES UPON THE STREET'

CLOSED-CIRCUIT TELEVISION SURVEILLANCE AND THE CITY

Nicholas R. Fyfe and Jon Bannister

•

INTRODUCTION: THE FORTRESS IMPULSE

The streets of central Glasgow have witnessed a spectacular renaissance. An ambitious plan to 'reaestheticise' key areas of the city centre begun in the mid-1980s (see Boyle and Hughes, 1995) has resulted in a dazzling infrastructure of new cultural institutions, high-class shopping malls and gentrified enclaves. But these 'playful spaces' of the 'New Glasgow' (Spring, 1990) are also 'fortified cells'. From the private security guards patrolling the malls to the remotely operated gates controlling access to courtyard residences, central Glasgow contains a panoply of human, physical, and technological methods to monitor and regulate the behaviour of its citizenry. And overseeing the city centre, quite literally, are thirty-two closed-circuit television (CCTV) surveillance cameras, located on six-metre poles and the sides of buildings, that swivel, tilt and pan across the public spaces of the streets below (see Figure 17.1). Welcome to the city transformed 'from industrial wasteland to post-industrial cultural centre' (Short *et al.*, 1993), welcome to the 'hard reality of administered space' in the 'Fortress City' (Christopherson, 1994: 409).

Glasgow is, of course, not unique in the way redevelopment and regeneration have been shadowed by enhanced social control over public and private space. The ability to maintain property values in the gentrified enclaves of inner cities and profits in the malls, restaurants and cultural centres of downtown, are increasingly bound up with questions of security (Christopherson, 1994). To use Ellin's (1996: 145) stark phrase, 'form follows fear' in the contemporary city. In North America, examples range from the 'strategic armouring' of city streets in Los Angeles (including the use of 'bumproof benches' and the bulldozing of public toilets; see Davis, 1992: 160–4) to the underground and overhead walkways that cocoon pedestrians from the

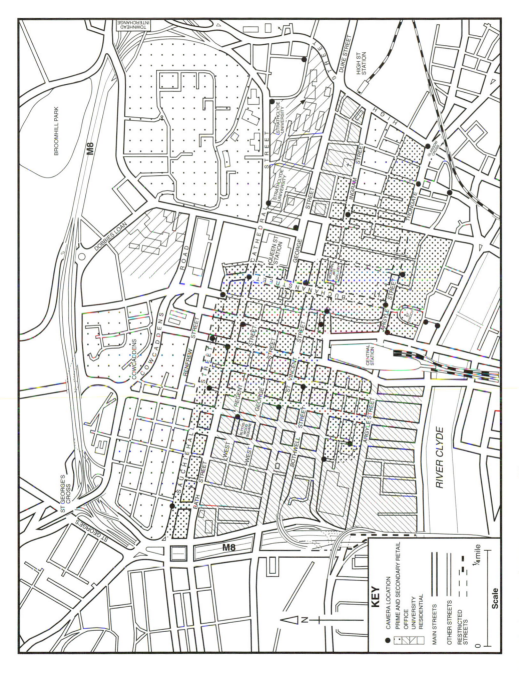

FIGURE 17.1 The location of CCTV cameras in Glasgow city centre

perceived dangers of the streets of Minneapolis, Calgary and Montreal (Boddy, 1992). In British towns and cities, too, the fortress impulse in urban design and management is becoming more prominent. Gated residential communities, the private policing of office and shopping spaces, local curfews to reduce the risk of public disorder at night in city-centre streets, and, within the last five years, the proliferation of public space CCTV surveillance systems are all increasingly common strategic responses to anxieties about crime and concern at declining consumer and business confidence in urban centres. Of course, the 'fortress impulse' in urban design is not new (see Bannister, 1991; Tilley, 1995) but the widespread introduction of CCTV surveillance cameras and other surveillance technologies has significantly increased what Rule (1973) calls 'surveillance capacity' in contemporary cities. To an unprecedented degree, people are now under surveillance in the routines of everyday life and thus more visible to invisible watchers than ever before. At work, numerous devices can monitor the duration of our working day and the length of our rest periods; when consuming goods, surveillance techniques evaluate our credit worthiness, record our purchases and analyse our preferences; and as we move through space, CCTV surveillance cameras (of which it is estimated there are currently more than 500,000 in public and private spaces across Britain; see Norris *et al.*, 1996) watch and record us on transportation systems, in shops and offices, and on the street. In short, someone 'going about his or her daily routine may be under watch for virtually the entire time spent outside the house' (Squires, 1994: 396).

Responses to the fortress impulse in urban design, and the broader 'surveillance society' (Lyon, 1994) of which it is a part, range from optimism at the discovery of potential technological fixes to chronic urban problems, to despair at the creation of an Orwellian dystopia. Lying between these extremes, however, is a middle ground characterised by a profound ambivalence about the impact of increased surveillance. Ellin (1996: 153) argues, for example, that while the gated residential communities, private policing and the surveillance systems do contribute to giving some people a greater sense of security, such developments can 'also contribute to accentuating fear by increasing paranoia and distrust among people'. Lyon (1994: 219) on the other hand urges us to recognise the Janus-like character of the 'electronic eye', arguing that surveillance 'spells control *and* care, proscription *and* protection'. Research on these themes is in its infancy but the rapid diffusion of CCTV surveillance means that these electronic 'eyes upon the street' (to draw on Jane Jacobs' (1961: 45) phrase) are quickly becoming just part of the urban infrastructure of British towns and cities, as familiar as telephone boxes or traffic lights. Of course, CCTV surveillance cameras are anything but 'just' infrastructure. As we highlight in this chapter, there are complex economic and political forces behind the expansion of CCTV surveillance in the public spaces of towns and cities and this is having significant intended and unintended consequences for social experience in urban areas.

'DOWNTOWN AS MALL'?: CCTV AND THE ECONOMIC AND POLITICAL RESTRUCTURING OF URBAN PUBLIC SPACE

Britain has more public space CCTV systems than any other advanced capitalist nation (Graham *et al.*, 1996: 2). The first system was launched in the south coast seaside resort of Bournemouth in August 1985. By August 1996 all Britain's major cities (with the exception of Leeds) boasted city-centre CCTV schemes and there are in excess of 200 police and local authority schemes operating in high streets and smaller towns (Norris *et al.*, 1996: 2). It should be of little surprise, then, that the use of such surveillance is now seen as 'inevitable' by most local authorities throughout Britain (see Reeve, 1996: 75). But what is surprising is that the extensive adoption and the continuing enthusiasm for CCTV has occurred against a background where there is little agreement among researchers about the effect of CCTV on recorded crime (see pp. 261–2 below). Claims by Government ministers and those responsible for CCTV systems have certainly created a perception that CCTV is a proven crime-reducing technology, effective regardless of the particularities of place, but as yet there is little consistent research evidence to sustain these claims. Indeed, reviewing existing evaluations of CCTV schemes, Short and Ditton (1995: 10) conclude that they are 'wholly unreliable' while Pawson and Tilley (1994: 294) describe them as 'post hoc shoestring efforts by the untrained and self-interested practitioner'. Thus the perception of CCTV as 'the ultimate technological fix' (Naughton, 1994) and the continuing widespread investment in CCTV surveillance in towns and cities is less the outcome of a careful evaluation of different crime reduction strategies and more (we argue) a recognition that CCTV 'fits' with a wider economic and political agenda to do with the contemporary restructuring of urban public space.

CCTV and the regeneration of urban economies

The 1980s witnessed increasing concern about the erosion of the economic role of town and city centres. De-industrialization and the rapid growth of out-of-town retail and business parks lead to the dereliction of many central urban areas and growing fears about the economic and social consequences of this spatial reorganisation of economic activity. In addition the growing flexibility and mobility of capital has meant that cities are increasingly competing against one another to secure footloose capital investment. Against this background of intra- and inter-urban competition, many local authorities are expanding their activities in the urban arena beyond tradi-tional managerial responsibilities of providing welfare to the local community, and playing a more active entrepreneurial role in trying to attract investment by private capital to stimulate local economic regeneration (Boyle and Hughes, 1995). This commitment to a more entrepreneurial position is evident in a variety of strategies being introduced to reverse the economic decline of town and city centres. Such strategies encompass city marketing campaigns, improvements in environmental urban design, the appointment of town centre managers, and, increasingly, improve-ments in security. Lying behind the concern for security is recognition that one of the perceived advantages of malls and out-of-town retail and business parks is their level of safety which traditional shopping streets have generally been unable to match. That sense of safety in the publicly accessible but often privately owned spaces of the mall and the retail and business park partly reflects the intensity of surveillance provided by both private security personnel and CCTV cameras. Keen to emulate these conditions and create a 'downtown as mall' (Mallett, quoted in Christopherson, 1994: 418), town centre managers view the installation of CCTV as of key impor- tance in improving the economic attractiveness of town centres (Reeve, 1996: 72).

This economic agenda underpinning the installation of CCTV is clearly illus- trated in Glasgow. A city where employment in manufacturing almost halved between 1981 and 1991, its economic prosperity is now heavily dependent on the service sector, particularly retailing and office employment concentrated in the city centre. As Spring (1990: 54) observes, 'The New Glasgow is a city of rampant consumerism – like Disneyland, a magic world of eternal consumption.' When, however, a routine survey of economic activity in the city revealed that the economic costs of crime (such as damage to and theft of property, and the deterrent effect of the fear of crime on visits by potential consumers) was a cause of 'business drift' from the city, the Glasgow Development Agency (GDA, a government quango for promoting economic development in the city) investigated ways of tackling this problem. Inspired by the apparent success of Birmingham's city centre CCTV scheme, GDA persuaded the city and regional councils to accept the introduction of cameras and then forged a partnership between the public and private sectors to finance the instal- lation and operation of the CCTV system, known as Citywatch. Indeed, in its campaign to persuade businesses to meet some of the capital and running costs, the slogan used was CCTV 'doesn't just make sense – it makes business sense'. It was claimed Citywatch would encourage 225,000 more visits to the city a year, creating 1,500 jobs and an additional £40 million of income to city-centre businesses.

Similar economic concerns with the commodification of public space have encour- aged the introduction of CCTV surveillance in other towns and cities. In King's Lynn, the benefits to retailers of a perception that 'the town would get a reputation as a secure shopping centre and there would be real prospects of increased custom' (King's Lynn and West Norfolk Borough Council, 1990: 4) was of key importance to the introduction of cameras in this small market town. Similarly in Liverpool where twenty cameras began operating in the city centre in July 1994, the 'bolstering of business and consumer confidence' was a significant factor in the introduction of CCTV given a perception that 'The core trading area can be seen by some as a dangerous place' (Coleman and Sim, 1996: 11–14).

'Law and order' politics and public space

The commodification of public space which has fuelled the expansion of CCTV sur- veillance is closely allied to the contemporary political context of public space at national and local levels. At a national level, the New Right doctrine of 'free economy- strong state' (Gamble, 1988: 28) which has underpinned the Conservative

Government's political programme since 1979, demands that the state play a strategic role in securing conditions under which commerce can flourish. Against a background of rising crime (recorded crime almost doubled between 1981 and 1991) and outbreaks of violent disorder without precedent in post-war Britain, the Government has introduced a series of neo-conservative measures to enhance the powers of the state to regulate activities in public space that might disrupt business and commerce and conflict with consumer citizenship (see Fyfe, 1995a, 1995b). Examples include the 1986 Public Order Act which introduced new conditions on marches and assemblies explicitly 'to reduce the serious disruption sometimes suffered by pedestrians, business and commerce' (Home Office, 1986: para. 4.22); and the 1994 Criminal Justice and Public Order Act which through the new offence of aggravated trespass would even criminalize consumers demonstrating outside shops (Liberty, 1995: 23). As the Home Office Annual Report makes clear, the development of CCTV in public space complements this legislation and has a central role in the Government's law and order strategy (Coleman and Sim, 1996: 27). Indeed, the Government has allocated £15 million to a 'CCTV Challenge Competition' (Home Office, 1995) for a further 10,000 cameras, published *CCTV: Looking Out for You* (Home Office, 1994) a manual on how to set up public space CCTV systems, and amended planning legislation to remove the need for planning permission for CCTV installations. The enthusiastic response of local authorities to these developments illustrates how elements of the national 'law and order' political agenda are mirrored at a local political level. Eager to 'manage out' behaviour which might undermine the economic potential of their area, many local authorities have not only installed CCTV cameras but introduced bylaws to outlaw 'anti-social behaviour'. Reeve's (1996: 75) survey of town centre managers found that a third of respondents had 'new powers to ensure that inappropriate activities and uses do not occur'. Many of these new powers focus on restricting or prohibiting drinking in public places. In Glasgow, for example, a blanket ban on drinking alcohol in public places was introduced in August 1996 because, as the Chair of the city's licensing board explained, 'The behaviour of some people drinking in the streets and parks is a nuisance to residents and shopkeepers' (*The Glasgow Herald,* 1996b: 9). Similarly, in Oxford there are proposals to make it an offence to spit in any street or to consume, or even give the impression of consuming, any intoxicating liquor in designated public spaces (Reeve, 1996: 75). The effective enforcement of such local legislation and the deterrence of proscribed behaviour is clearly enhanced by CCTV surveillance.

The absence at both national and local levels of any significant political resistance to the increasing bureaucratisation of space signalled by the widespread use of CCTV surveillance has no doubt reinforced central and local government commitment to this technology. In contrast to other European countries where civil liberties issues have impeded the development of such surveillance, Prime Minister John Major's concerns that 'we will no doubt hear some protest about a threat to civil liberties' (quoted in Groombridge and Murji, 1994: 283) have not been realised. Britain's civil liberties watchdogs, Liberty (in England and Wales) and the Scottish Council for Civil Liberties, have argued that while they are 'in principle opposed to people being spied on in public places' they recognise that 'it can help prevent and detect crime in certain clearly defined circumstances' (Scottish Council for Civil Liberty, 1995, p.5; see also Liberty, 1989). Their energies have, therefore, been directed towards devising rules governing the use and operation of CCTV, arguing for a balance between the right to privacy and the right to security, rather than actively opposing CCTV. At a local level, public opinion surveys conducted in areas planning CCTV systems have generally revealed high levels of public support: 95 per cent of those asked in Glasgow said they were in favour; in Sutton in south-east London 85 per cent welcomed the introduction of CCTV to the town centre; and in Airdrie (a market town near Glasgow), 89 per cent believed CCTV would reduce their fear of crime. But as Ditton (1996) has cogently argued, such surveys may be as unreliable and methodologically flawed as those claiming to reveal dramatic reductions in crime caused by CCTV (see below).

Where resistance is developing to CCTV it appears economically rather than politically motivated. The annual operating costs of town-centre CCTV systems (which one survey estimates at an average of £72,000 but in some places are as high as £250,000; Norris *et al.*, 1996) are borne in many areas by public–private partnerships. Such partnerships appear increasingly fragile. Less than eighteen months after introducing the cameras in Glasgow, headlines proclaimed 'Crimewatch spy cameras in cash crisis' (*The Glaswegian*, 1996: 1) and 'Spy camera cuts "could risk lives"' (*The Glasgow Herald,* 1996a). The shortfall in funding is blamed on non-contributing local businesses. However, defending their decision not to continue funding the system, many businesses argue that as part of the city's public infrastructure street cameras should be funded from business rates, while some retailers, with branches in other towns and cities, point to the ever-increasing costs of contributing to several CCTV schemes. In Liverpool similar tensions exist but are exacerbated by claims that cameras have displaced crime from city centre to suburban shopping centres, increasing costs to traders in these areas (Coleman and Sim, 1996: 20–1). More seriously, as Brown (1995: 1) notes, many retailers now appear unconvinced about the effectiveness of public area CCTV systems for increasing turnover or profits.

THE 'FEEL GOOD FACTOR'? CCTV AND THE URBAN EXPERIENCE

In the market town of King's Lynn, one of the first places in Britain to introduce public space CCTV, 'The surveillance system has grown', according to a local council officer, 'because of the 'feel-good factor' it has created among the public' (quoted in Davies, 1995: 60). Similarly, within a few weeks of Glasgow's CCTV system becoming operational, the assistant chief planning officer proudly proclaimed that 'CCTV is already creating a "feel-good factor" in the city centre' (*Planning Week*, 1994: 4). The veracity of these claims is less important here than what the statements indicate about the discourse being woven around CCTV by its proponents, namely that public space CCTV will create a 'feel-good factor' by 'reassuring town centre users that they are safer' (Home Office, 1994: 14). Of course, that 'feel-good factor' depends partly on reductions in crime and the fear of crime but it is also more than this. It includes the exclusion from public space of those unable or unwilling to subscribe to the norms of consumer citizenship. As the Home Office guidance on introducing CCTV makes clear, CCTV has the potential to deal with the 'problems' of 'Groups loitering', 'Drunkenness', 'Disorderly behaviour' or 'those whose behaviour is suspicious' (Home Office, 1994: 12–14). Indeed in Wolverhampton (part of the Birmingham conurbation), it was hoped that introducing CCTV would deter those 'large groups, usually of young single people, . . . [whose] mere presence is a nuisance to people who want to use the streets and shopping centres in more conventional ways' (Liberty, 1989: 1). This section examines some of these facets of the 'feel-good factor'.

CCTV and crime

It's nearly seven thirty on a winter's evening in Glasgow. In the city centre police station housing the control room of the CCTV surveillance system, two operators sit in front of a wall of colour television screens watching activity on the streets. One operator uses a set of controls to pan a camera along a main street crowded with late-night shoppers making their way home. His attention is caught by two youths standing in a shop doorway, only a few yards from the bustle of the street. He zooms the camera in and sees the younger of the youths taking a wallet out of his pocket and handing it to the other one. From the way the older youth appears to be shielding the boy from the street and the look of fear clearly visible on the younger boy's face, the operator becomes suspicious. With the camera fixed on the doorway, he contacts the adjacent police control room and relays the pictures to them. Now the younger boy is seen removing his ski jacket and the decision is taken to dispatch officers to the scene. Within two minutes a police car containing two officers arrives, one apprehends the bewildered assailant the other goes to comfort the victim.

(Authors' fieldnotes, November 1995)

Such an incident epitomises many of the benefits of city centre CCTV surveillance. Offenders can, quite literally, be caught in the act. And even if the operator fails to see an incident as it happens, once it is reported it may be possible to identify the offenders by simply looking through the recorded video footage of the relevant location. One consequence of CCTV therefore is improved clear-up rates for crime. In Airdrie town centre, where cameras have operated since 1992 and where one of the few pieces of independent research on the impact of CCTV has been conducted, the police 'cleared up' (that is to say, the offender(s) was(were) apprehended, cited, warned or traced) for 16 per cent more crimes and offences in the CCTV area in the two years following the installation of cameras compared to the two years before installation (Short and Ditton, 1996: 8). While improved detection is one important benefit, it is the deterrence effect of CCTV that is claimed to be 'its strongest feature' (Graham *et al.*, 1996:6). Although several 'before' and 'after' evaluations of the impact of CCTV on crime have been conducted, many are methodologically flawed and therefore unreliable (Short and Ditton, 1995: 10). Of the few evaluations which are methodologically robust and conducted by independent researchers, the study of Airdrie did yield encouraging evidence. It revealed a 21 per cent fall in the total number of recorded crimes and offences in the CCTV area in the two years after the system was established compared to the two years before its introduction, the greatest reduction occurring in the category of crimes of dishonesty (such as burglary, theft of and from motor vehicles and shoplifting) which fell by almost half in the study period. Similar reductions in recorded crime were found in Sutton town centre where total crime decreased by 20 per cent in the two months following the introduction of CCTV (Sarno, 1995, p.32). Such reductions do, of course, beg the question of whether crime has simply been displaced to other areas. In Airdrie, it appeared that 'the crimes prevented in the CCTV area did not, as far as can be ascertained, re-emerge in adjacent or nearby areas' (Short and Ditton, 1996: 1); by contrast, in Sutton the number of crimes out of camera range increased (Sarno, 1995: 32).

This equivocal evidence regarding displacement also characterises the results of research on the impact of CCTV on the fear of crime. In Birmingham, for example, Home Office research revealed that the percentage of those feeling safe or very safe in the street did increase, from 27 per cent before CCTV to 29 per cent afterwards, but so did the proportion of those feeling very unsafe, from 43 per cent to 45 per cent (Brown, 1995: 43). Indeed, when people in Birmingham were asked whether CCTV would make any difference to how safe they felt at night, over half of the respondents asked during the day said it would make no difference, rising to two-thirds of respondents asked at night (ibid.: 44). Unfortunately, this research does not reveal the impact of gender on perceptions of safety, but another study concluded that CCTV does little to allay women's anxieties about male behaviour in public space (Brown, 1996). This finding is important not simply because it calls into question claims by proponents of CCTV that it will create safer environments for women but also because it highlights the socially and spatially specific nature of any 'feel-good factor'. As Coleman and Sim (1996) cogently argue, 'the feel-good factor' focuses on a particular conception of danger and order within city-centre public space, marginalising other areas of crime and anxiety, particularly domestic environments and the experiences of women.

From privatisation to purification

The impact of CCTV extends wider than simply its influence on crime and the fear of crime. In Newcastle city centre, over a quarter of the incidents for which the CCTV system was used between 1993 and 1995 involved dealing with begging and vagrants or 'suspicious youths' (Centre for Research on Crime, Policing and the Community, 1993, quoted in Graham *et al.*, 1996: 20); and in Birmingham 17 per cent of incidents where CCTV was used to help the police concerned 'nuisance, drunks and begging' (Brown, 1995: 40). What this evidence suggests is that those perceived not to belong in commercial public spaces now risk being 'monitored and harassed, losing rights as citizens just because they aren't seen to be lucrative enough as consumers' (Graham *et al.*, 1996: 19). The result is a subtle privatisation of public space (ibid.) as commercial imperatives define acceptable behaviour, excluding those who detract from the consumption experience. As has been noted in relation to the private spaces of the mall, this can lead to 'the virtual disenfranchisement from city life of young people of low spending power and other – generally low income – residents whose appearance and conduct do not conform to the moral codes of well-ordered consumption enforced by shopping centre managers' (Bianchini, 1990: 5). Now that many of the design and management practices associated with privately owned shopping centres, including use of CCTV, have been uncritically adopted in the management of public space, 'the effective implementation of the values of private interest in town centres' (Reeve, 1996: 78) is well-advanced.

Such privatisation of public space raises important political and cultural issues which challenge CCTVs 'feel-good factor'. The use of town-centre CCTV to manage out 'inappropriate' behaviour provides stark confirmation of Davis's (1992: 156) contention that 'The universal consequence of the crusade to secure the city is the destruction of any truly democratic urban space.' Indeed, such use of CCTV completely destroys the vision of Berman (1986) and others (see Jacobs, 1961; Young, 1990) of the street as an 'open-minded public space'. Far from the streets being spaces that encourage 'encounters between people of different classes, races, ages, religions, ideologies, cultures, and stances towards life' (Berman, 1986: 484), the potential impact of CCTV is the imposition of 'a middle-class tyranny on the last significant urban realm of refuge for other modes of life' (Boddy, 1992: 150). Given the connections made between public spaces and the public sphere as a crucible of participatory democracy (Ruddick, 1996: 133), the political consequences of this purification of space of those 'troublesome others' are profound. As Iris Young argues

> The critical activity of raising issues and deciding how institutional and social relations should be organised, *crucially depends on the existence of spaces and forums to which everyone has access*. In such public spaces, people encounter other people, meanings, expressions and issues which they may not understand or with which they do not identify
>
> (Young, 1990: 240, emphasis added).

This need for heterogeneity has an important cultural as well as political dimension. The privatisation of the public realm through CCTV surveillance risks impoverishing the urban experience in ways which are potentially damaging to the collective emotional culture of urban communities (Robins, 1995). Indeed, more than twenty years ago Richard Sennett in *The Uses of Disorder* vigorously defended the need for heterogeneity in urban communities in the face of contemporary trends of attempts to purify the urban experience (see also Sibley, 1995). 'The great promise of city life', he contended, 'is a new kind of confusion possible within its borders, an anarchy that will not destroy men [sic], but make them richer and more mature' (Sennett, 1996: 108). Sennett was therefore dismayed at what was happening to the modern metropolis where urban renewal programmes were destroying traditional sites of social gathering and rigid land-use zoning robbing neighbourhoods of their social and functional differentiation. Moreover, for Sennett the purification of disorder and difference from space had important psychological and behavioural consequences. '[D]isorderly, painful events' in the city are worth encountering, he argued, because they force us to engage with 'otherness', to go beyond one's own defined boundaries of self, and are thus central to civilised and civilising social life (Sennett, 1996: 131–2). Without disorder and difference people do not learn to deal with conflict so that if conflicts do erupt they tend to be more violent. In a passage which seems to anticipate the contemporary phenomenon of 'road-rage', Sennett notes that because individuals now have 'so little tolerance of disorder in their own lives and having shut themselves off so that they have little experience of disorder as well, the eruption of social tension becomes a situation in which the ultimate methods of aggression, violent force and reprisal, seem to become not only justified, but life preserving' (Sennett, 1996: 44–5).

Whether public space CCTV will influence social interaction on the street in ways which accord with Sennett's thesis on the purification of space is unclear. At the very least, however, his arguments on 'the uses of disorder' highlight a critical absence from current urban regeneration agendas (Robins, 1995). Like Sennett (and before him Mumford, 1961), Robins argues that too little attention has been given to the collective emotional dimensions of urban culture. And like Sennett, Robins believes painful events are worth encountering because 'Fear and anxiety are the other side of the stimulation and challenge associated with cosmopolitanism' (Robbins, 1995: 48). Against this background, Robins is concerned about the potential impact of the fortification of space and the use of surveillance cameras on the collective emotional life of the city. The desire to purify space of any behaviour likely to provoke anxiety and to insulate ourselves from the 'complexities of the city' may in fact deny 'the emotional stimulus and provocation necessary for us if we are to avoid, both individually and socially, stagnation and stasis' (ibid.: 60). Indeed, for Robins one of the challenges of urban regeneration is to make a readiness to be out of control, combined with a maturity to handle the consequences, the basis of an urban public and political culture. The potential of public space CCTV surveillance systems to create a series of 'stagnant fortresses' (Smith, 1979) clearly flies in the face of this challenge.

CONCLUDING COMMENTS: CCTV AND 'THE PRODUCTION OF SPACE'

Research on CCTV surveillance has, to date, focused principally on evaluating the 'before' and 'after' impact of cameras on the level and distribution of crime. Although important, this research agenda has marginalised the broader economic, political and socio-cultural issues surrounding the development of CCTV surveillance in public space. These broader issues, highlighted in this chapter, are not simply of substantive significance but also of wider theoretical relevance. The economic and political forces fuelling the diffusion of CCTV across urban Britain are vivid examples of the more general processes of commodification (encompassing property relations and capital circulation) and bureaucratisation (associated with enhanced surveillance and regulation by the state) which, according to Lefebvre (1991) are of key importance to 'the production of space' in contemporary capitalism. Lefebvre's thesis identifies these twin processes as pivotal to the way 'abstract space' (dominated by the activities of the economy and the state and defined in terms of exchange value) is imposed on the concrete space of everyday life. But Lefebvre's ideas have a wider purchase on understanding the development and significance of CCTV surveillance. The colonisation of concrete space by abstract space is accomplished through the dominance of particular representations of space and spatial practices. The vision of a 'downtown as mall' examined at the beginning of the chapter has become just such a dominant representation of urban public space with its associated spatial practices of consumption and capital circulation. Other representations of the space of urban centres risk being marginalised. The vision of Jane Jacobs (1961), for example, with its emphasis upon an eclectic mix of land-uses in city centres generating street activity twenty-four hours a day and thus providing 'eyes upon the street' belonging to the 'natural proprietors of the street' (Jacobs, 1961: 45) to ensure the safety of residents and strangers, is one which is only receiving limited economic and political support at present. Moreover, in Britain (unlike other countries in Europe) there are few grass-roots 'revolts against the gaze' of CCTV cameras which might form the basis of what Lefebvre refers to as spaces of representation, spatial representations which challenge dominant spatial practices. Against this background, the diffusion of CCTV surveillance cameras is set to continue, extending via the panopticon principle a network of socio-spatial control and discipline across the streets of Britain. Although these electronic 'eyes upon the street' might reduce certain types of crime and increase business and consumer confidence in town and city centres, the price may be a high one. Under the constant gaze of CCTV surveillance cameras, Boddy's (1992: 123) claim that streets 'symbolise public life, with all its human contact, conflict and tolerance' will be difficult to sustain.

REFERENCES

Bannister, J. (1991) *The Impact of Environmental Design upon the Incidence and Type of Crime: A Literature Review*, Edinburgh: Scottish Office Central Research Unit.

Berman, M. (1986) 'Taking it to the streets: conflict and community in public space', *Dissent*, 333(4), 476–85.

Bianchinin, F. (1990) 'The crisis of urban public social life in Britain: origins of the problem and possible responses', *Planning, Policy and Research.* 5(3), 4–8.

Boddy, T. (1992) 'Underground and overhead: building the analogous city', in M. Sorkin (ed.), *Variations on a Theme Park: The New American City and the End of Public Space*, New York: Hill and Wang, pp. 123–54.

Boyle, M. and Hughes, G. (1995) 'The politics of urban entrpreneurialism in Glasgow', *Geoforum*, 25(4), 453–70.

Brown, B. (1995) *CCTV in Town Centres: Three Case Studies*, London: Home Office Police Department.

Brown, S. (1996) 'What's the problem girls? CCTV and the gendering of public safety', paper presented at the CCTV: Surveillance and Social Control conference, University of Hull.

Christopherson, S. (1994) 'The fortress city: privatized spaces, consumer citizenship', in A. Amin (ed.), *Post-Fordism: A Reader*, Oxford: Blackwell, pp. 409–27.

Coleman, R. and Sim, J. (1996) 'From the dockyards to the Disney store: surveillance, risk and security in Liverpool City Centre', paper presented to the Law and Society Association Conference, University of Strathclyde.

Davis, M. (1992) 'Fortress Los Angeles: the militarization of urban space', in M. Sorkin (ed.), *Variations on a Theme Park: The New American City and the End of Public Space*, New York: Hill and Wang, pp. 154–80.

Davies, S. (1995) 'Welcome home Big Brother', *Wired*, May: 58–62.

Davies, S. (1996) 'Big Brother, 1984 and all that', paper presented at the CCTV: Surveillance and Social Control conference, University of Hull.

Ditton, J. (1996) 'The public aceptability of CCTV', paper presented at the CCTV: Surveillance and Social Control conference, University of Hull.

Ellin, N. (1996) *Postmodern Urbanism*, Oxford: Blackwell.

Fyfe, N. R. (1995a) 'Law and order policy and the spaces of citizenship in contemporary Britain', *Political Geography*, 14(2): 177–89.

Fyfe, N. R. (1995b) 'Controlling the local spaces of democracy and liberty? 1994 criminal justice legislation', *Urban Geography*, 16(3): 192–7.

Fyfe, N. R. and Bannister, J. (1996) 'City Watching: closed circuit television surveillance in public spaces', *Area*, 28(1): 37–46.

Gamble, A. (1988) *The Free Economy and the Strong State: The Politics of Thatcherism*, London: Macmillan.

The Glasgow Herald (1996a) 'Spy camera cuts 'could risk lives'', 29 May: 8.

The Glasgow Herald (1996b) 'Drinking outside the law', 16 August: 9.

The Glaswegian (1996) 'Crimewatch spy cameras in cash crisis', 22 February: 1.

Graham, S., Brooks, J. and Heery, D. (1996) 'Towns on the Television: closed circuit TV in British towns and cities', forthcoming in *Local Government Studies*, 22(3): 3–27.

Gregory, D. (1993) *The Geographical Imagination*, Oxford: Blackwell.

Groombridge, N. and Murji, K. (1994) 'As easy as AB and CCTV', *Policing*, 10(4): 283–90.

Home Office (1986) *Review of Public Order Law*, Cmnd 9510, London: HMSO.

Home Office (1994) *CCTV: Looking Out for You*, London: Home Office.

Home Office (1995) *CCTV Challenge Competition 1996/7: Bidding Guidance*, London: Home Office.

Jacobs, J. (1961) *The Death and Life of Great American Cities: The Future of Town Planning*, Harmondsworth: Penguin.

King's Lynn and West Norfolk Borough Council (1990) *Report on Proposed CCTV System*, King's Lynn: King's Lynn and West Norfolk Borough Council.

Lefebvre, H. (1991) *The Production of Space*, Oxford: Blackwell.

Liberty (1989) *Who's Watching You? Video Surveillance in Public Places*, London: Liberty Briefing Paper No. 16.

Liberty (1995) *Defend diversity Defend dissent: What's Wrong with the Criminal Justice and Public Order Act 1994*, London: National Council for Civil Liberties.

Lyon, D. (1994) *The Electronic Eye: The Rise of Surveillance Society*, Cambridge: Polity Press.

Mumford, L. (1961) *The City in History*, London: Secker and Warburg.

Naughton, J. (1994) 'Smile you're on TV', *Life – The Observer Magazine*, 13 November.

Norris, C., Moran, J. and Armstrong, G. (1996) 'Algorithmic surveillance: the future of automated visual surveillance', paper presented at the CCTV: Surveillance and Social Control conference, University of Hull.

Pawson, R. and Tilley, N. (1994) 'What works in evaluation research?', *British Journal of Criminology*, 34(3): 291–306.

Planning Week (1994) 'PAN aims to boost city centre safety', 10 November: 4.

Reeve, A. (1996) 'The private realm of the managed town centre', *Urban Design International*, 1(1): 61–80.

Robins, K. (1995) 'Collective emotion and urban culture', in P. Healey *et al.* (eds) *Managing Cities: The New Urban Context*, Chichester: John Wiley, pp. 45–62.

Ruddick, S. (1996) 'Constructing difference in public spaces: race, class and gender as interlocking systems', *Urban Geography*, 17(2): 132–51.

Rule, J. (1973) *Private Lives, Public Surveillance*, London: Allen-Lane.

Sarno, C. (1995) 'Imapct of CCTV on crime', in M. Bulos (ed.) *Towards a Safer Sutton: Impact of Closed Circuit Television on Sutton Town Centre*, London: London Bororugh of Sutton, pp. 4–32.

Scottish Council for Civil Liberties (1995) *Rules for the Operation and Use of Closed Circuit Television (CCTV) in Public Places*, Glasgow: Scottish Council for Civil Liberties.

Sennett, R. (1996) *The Uses of Disorder: Personal Identity and City Life*, London: Faber and Faber.

Short, E. and Ditton, J. (1995) 'Does CCTV affect crime?' *CCTV Today*, 2(2): 10–12.

Short, E. and Ditton, J. (1996) *Does Closed Circuit Television Prevent Crime?*, Edinburgh: Scottish Office Central Research Unit.

Short, J. R., Benton, L.M., Luce, W.B. and Walton, J. (1993) 'Reconstructing the image of an industrial city', *Annals of the Association of American Geographers* 8(2): 207–24.

Sibley, D. (1995) *Geographies of Exclusion: Society and Difference in the West*, London: Routledge.

Smith, M. P. (1979) *The City and Social Theory*, Oxford: Blackwell.

Spring, I. (1990) *Phantom Village: The Myth of the New Glasgow*, Edinburgh: Polygon.

Squires, J. (1994) 'Private lives, secluded spaces: privacy as political possibility', *Environment and Planning D: Society and Space*, 12: 387–401.

Tilley, N. (1995) 'Seeing off the danger: threat, surveillance and modes of protection', *European Journal of Criminal Policy and Research* 3(3): 27–40.

Young, I.(1990) *Justice and the Politics of Difference*, Princeton: Princeton University Press.

18

NIGHT DISCOURSE

PRODUCING/CONSUMING MEANING ON THE STREET

Tim Cresswell

•

The built environment is built because it's been allowed to be built. It's been allowed to be built because it stands for and reflects an institution or a dominant culture. The budget for architecture is a hundred times the budget for public art because a building provides jobs and products and services that augment the finances of the city. Public art comes in through the back door like a second class citizen. Instead of bemoaning this, public art can use the marginal position to its advantage: public art can present itself as the voice of the marginal cultures, as the minority report, as the opposition party. Public art exists to thicken the plot.

(Acconci, 1990: 176)

The street is a realm in which a number of different kinds of writing[1], of inscription, take place. The space of the street is often a space in which we encounter words and pictures in voluminous quantities . . . advertisements, instructions, political messages, newspapers, illegal posters, monumental murals and messages, graffiti. My aim here is to provide an interpretation of 'illegal', 'illegitimate' and unsanctioned forms of text in the context of some thoughts about the street and its role in modern life. I take my title from an essay by Karrie Jacobs (Jacobs, 1992), in which she reflects on her observation that all kinds of political and risky words and images appear in New York's streets overnight. Graffiti, political messages, subversive art and recon-figurations of existing messages all appear in the morning, after the secretive curtain of night has been raised. These provocative and illegal posters, graffiti and pictures, she suggests, provide a richer and more controversial alternative to the words that appear during the day, those sanctioned forms of inscription dominated by contemporary advertising. She suggests that Night Discourse approximates the older ideal of a public realm – an arena in which members of the public meet to accommodate competing values and expectations and hence in which all goals are open to discussion and modification (Arendt, 1958; Sennett, 1977)

> As daytime discourse gets thinner and more banal, year by year, as fewer genuine sentiments or opinions are voiced by politicians, and bold political statements vanish from mainstream media, street posters are becoming the one medium in which controversial opinions can find a general audience. . . . Maybe Night Discourse is the last form of political debate.
>
> (Jacobs, 1992: 14)

Below I examine some of these forms of Night Discourse in relation to the street's nocturnal imaginings. First, however, a few warnings are in order. While it is certainly true that the time/space of the street at night has historically allowed any number of subversions, including those of 'Night Discourse', it is also true that this time/space is being progressively eroded by the ever-widening gaze of closed-circuit television with its disciplinary function. Increasingly the spaces of the street are far from secretive at night (Davis, 1992; Fyfe and Bannister, 1996). As we shall see, the origins of street lighting are embedded in the perceived need for order at night, a logic that is extended by CCTV. The second warning here is that the street at night is a particularly gendered space which has been constructed as a space for men which is unsafe for women (Valentine, 1989). Despite this, though, many of the producers of the best Night Discourse have been women, particularly those involved in the Billboard Banditary discussed below.

THE STREET AS A SITE OF DISCIPLINE AND DEVIANCE.

The street has played a central role in the history of modernism, but a role which is far from straightforward. Here I focus on the tension between the street as a site and sign of domination and order and as a site and sign of unrest, rebellion and disorder. Marshall Berman's book *All That Is Solid Melts into Air* (Berman, 1988) provides a good starting point. In it he has written of the Parisian boulevard and the Nevsky Prospect in St Petersburg as sites of contestation between a disciplinary and ordering 'Modernism from Above' and an everyday and chaotic 'Modernism from Below'. Just as Haussmann's boulevards cut open the working-class neighbourhoods of the city in order to provide access to the police and army, to subject the people to an authoritarian gaze, and to prevent the possibility of future barricades, the boulevard also presented new opportunities for chance meetings between the workers and bourgeoisie as evidenced by the poetry of Baudalaire and others. Similarly the Nevsky Prospect was the result of modern planning on a massive scale. As one observer put it, when thinking about St Petersburg:

> geometry has appeared
> land surveying encompasses everything
> Nothing on earth lies beyond measurement
> (Quoted in Berman, 1988: 177)

Here St Petersburg, and its grandest street, the Nevsky Prospect, is the site of modern rationality, imposed, in every way, from above. The rational space of the street 'was a distinctively modern environment in many ways. First the street's straightness, breadth, length and good paving made it an ideal medium for moving people and things, a perfect artery for the emerging modes of fast and heavy traffic' (Berman, 1988: 194). But the Nevsky Prospect was also the site of a remarkable series of fictional and real resistances. It was the only significant public space within which the government could not dictate actions and interactions. It emerged as a 'kind of free zone in which social and psychic forces could spontaneously unfold' (ibid.: 194). It was in this 'free zone' on 1 September 1861 that a mysterious horseman appeared flinging leaflets around for the pedestrians to read. The leaflet read:

> We do not need either a Tsar, an Emperor, the myth of some lord, or the purple which cloaks hereditary incompetence. We want as our head a simple human being, a man of the land who understands the life of the people and is chosen by the people. We do not need a consecrated Emperor, but an elected leader receiving a salary for his services.
>
> (quoted in Berman, 1988: 214)

This was followed three weeks later by a political demonstration the likes of which had never been seen before in St Petersburg. The appearance of uncontrolled writing on the street, in the form of the horseman's leaflets, led to the origin of the street as a modern political space. The street in the form of Haussmann's boulevard, the Nevsky Prospect and, later, the New York freeways imposed by Robert Moses, became simultaneously the site of power, of ritual, of parades and processions and the site of disrespectful carnival, of protest marches, of encounters with the newly visible, and often angry, other.

In addition to the street's role as site of power and site of resistance many commentators, including Berman, have noted how the street has become a dead space, a space within which all that is deviant and threatening is located by an imagination which finds the mixing of unlike categories distasteful and threatening. Jane Jacobs famously noted how the street, in modern New York, has become the place in which deviance is located, a guaranteed signifier of deviance and disrepute linking the prostitute (the streetwalker), the homeless child (the street arab), and mischievous youth (the street gang). In particular she notes the sinister tone which 'the street' invokes when it is used to signify the failure of planners. In relation to children she observes how 'keeping them off the streets' has become almost a moral imperative:

> These pale and rickety children, in their sinister moral environment, are telling each other canards about sex, sniggering evilly and learning new forms of corruption as efficiently as if they were in reform school. The situation is called 'the moral and physical toll taken of our youth by the streets'. Sometimes it is called simply 'the gutter'.
>
> (Jacobs, 1961: 74)

Zygmunt Bauman has also reflected on the city street of the 1990s, noting the stark difference it exhibits from the Boulevards and the arcades of Baudilarian Paris.

> To the innocent, who had to leave for a moment the wheeled security of their cars . . . the street is more a jungle than the theatre. One goes there because one must. A site fraught with risks, not chances; not meant for the gentlemen of leisure, and certainly not the faint-hearted among them. The street is the 'out there' from which one hides, at home or inside the car, behind security locks and burglar alarms.
>
> (Bauman, 1994: 148)

Tied up with the tension between order and control, between normality and deviance, in the modern street is the issue of temporality. The street has its rhythms and flows. The delicate sidewalk ballet that Jane Jacobs has written of varies through the day and through the year – the street is a site of rhythmic geographies. The starkest of these temporal variations and repetitions is that between day and night. It is the street at night especially which plays a central role in modern nightmares of urbanity.

> Nightfall brings forces very different from those that rule the day. In the symbols and myths of most cultures, night is chaos, the realm of dreams, teeming with ghosts and demons as the oceans teem with fish and sea monsters.
>
> (Schivelbusch, 1988: 81)

In seventeenth-century Paris signs of all kinds were removed from the street to produce smooth uniform planes of discipline. At roughly the same time (1667) street lanterns were introduced by decree. It was believed that who lights the streets rules them. Naturally the lighting of the street was administered by the police (Schivelbusch, 1988). The street, in these pre-Haussmann days, was already a dangerous place and social engineers were already changing the street to serve the purposes of the absolutist state. The street at night though was doubly terrifying for those with an interest in established order. From a very early time (1467 in England), anyone on the street at night was required to hold a torch so that they might be visible to the forces of order. Later street lighting was introduced partly for the same reasons.

It is not surprising that a popular form of revelry and rebellion was lantern smashing. In Paris in particular popular uprisings both big and small were often accompanied by the smashing of street lights and lanterns so that the life of the street was once again invisible and threatening (to the bourgeoisie in the comfortable, well-lit, drawing rooms). This smashing of lanterns provided a perfect addition to the mounting of barricades. Literal walls existed alongside walls of darkness.

In the rest of this chapter I focus on writing which appears at night on the street. The street, this site where deviance, political uprising and revolution is so often located, is also the site for a very public kind of discourse. The connection made between landscape and text (Duncan, 1990) is made very real by the sheer weight of words and images to be found on city streets. Perhaps the mysterious horseman of the Nevsky Prospect with his insurgent leaflets provided a foretaste of the multiplication of texts on the street. It is through such texts (and images) that it is possible to see the creative tension that occurs between the spaces of propriety defined by modernism from above and the mobile and ephemeral sniping of the outrageous that arise from 'below'.

ART ON THE STREET – NEGOTIATING DUALISMS

What happens when our expectations of the city are confronted with something which does not conform to common sense? I would suggest that such denials of expectations provide a potent challenge to invisible structures of power by making the urban dweller, however temporarily, aware of his or her surroundings. Public, activist, art often attempts to subvert the everyday spaces of the city to add hidden voices to political discourse. Public art transgresses many key socio-spatial divisions and thus presents a geographer with an intriguing set of issues. Art in public space, particularly when political or activist in nature, transgresses some long-held and almost invisible boundaries of what constitutes appropriateness. Art has, in the modern Western world, been constructed as the product of individual inspiration and genius that is understood and appreciated by generally well-educated people in rarefied spaces that stand at the top of a hierarchy of spaces imbued with what Pierre Bourdieu (Bourdieu, 1984) has called cultural capital – the galleries. Alternatively art is purchased for large amounts of money and displayed in private space as a sign of achievement, status and taste. Public, activist art cannot fit easily into this milieu. It is often anonymous, it is there for all to see, it exists in the open, on the street, in the spaces of the everyday. Public art is often transitory and cannot be bought.

> Indeed, to juxtapose the terms public and art is a paradox. Art is often said to be the individual inquiry of the sculptor or painter, the epitome of self-expression and vision that may challenge conventional wisdom and values. The term public encompasses a reference to the community, the social order, self-negation: hence the paradox of linking the private and public in a single context.
>
> (Hoffman, 1992: 113)

If the observations of Bauman and others (above) are at all accurate, the streets are the lowest spaces in the same hierarchy of which galleries form the pinnacle. The location of art on the street, then, is a contradiction in terms, in more ways than one. Art is supposed to be pure, the object of a detached and pure gaze. Indeed, the very value of art is supposed to be located in its purity – what Walter Benjamin has called an aura.

In so far as public art is meant to play an active role in political issues it is also transgressive. Politics, like art, has been constructed as the domain of the select few and it is supposed to occur in specialised spaces designated as political. Politics, since ancient Greece, has been public only in the older sense of a masculine domain of reason and argument. For art to be political then is to transgress the boundary between reason and emotion – to suggest that political thought is everyday thought which has an aesthetic value. Public art, then, challenges many of the spatialised assumptions that go under the heading of 'common sense' in the contemporary world:

> If anything unifies the whole range of contemporary sculpture from minimalism to the political agitation of Krzysztof Wodiczko, it is the idea enunciated best by Krauss, that contemporary sculpture takes as its subject 'the public, conventional nature of what might be called cultural space.'
>
> (North, 1992: 10)

PRODUCTION/CONSUMPTION

In *The Work of Art in the Age of Mechanical Reproduction* Walter Benjamin (1968) noted how the ability to reproduce original works of art might provide a catalyst for the internal self-destruction of capitalism. He described how a culture based on authority, autonomy and authenticity could be destroyed as art was able to appear in many contexts other than the solitary space of the original. By appearing in many contexts, he suggested, the authority of an 'auratic culture' would be undermined to such an extent that a new 'democratic culture' might appear in which the consumer, rather than the producer, would determine the meaning(s) of artistic endeavour. Benjamin might have been prophesying the arrival of the photocopying machine.

Indeed the dualism of production and consumption is another one which street art negotiates and subverts. Along with the idea of strategy and tactic, in de Certeau's lexicon, lies the language of production and consumption.

> To a rationalized, expansionist and the same time centralized, clamorous, and spectacular production corresponds another production, called 'consumption.' The latter is devious, it is dispersed, but it insinuates itself everywhere, silently and almost invisibly, because it does not manifest itself through its own products, but rather its ways of using the products imposed by a dominant economic order.
>
> (de Certeau, 1984: xii–xiii)

Public art can be thought of in terms of this simultaneous production and consumption. Night discourse is obviously a form of production – there is a material product in the form of the image itself and the meanings that it encourages. It is simultaneously a tactical consumption of the urban landscape. Most of us, as we make our way through the city, consume it in the sense that we interpret its meaning and act accordingly. But our consumptions, our meanderings, leave no trace. This is not the case with night discourse. Night discourse happens when people read (consume) the landscape and then leave evidence for their consumption. Night discourse is a production and a consumption of street space.

CULTURE JAMMING

Public art is potentially an endless catagory of public expression ranging from advertisements to graffiti, from public monuments of heroes on horses to guerrilla theatre.

In this essay I am mainly concerned with various subversive forms of public expression more or less conforming to a list provided by Arlene Raven (1993: 1): 'street art, guerrilla theatre, video, page art, billboards, protest actions and demonstrations, oral histories, dances, environments, posters, murals, paintings and sculpture . . . ' These kinds of actions can all be described by the term 'culture jamming'. Below I focus on billboard defacement and the subversive projections of Krzysztof Wodiczko.

Billboard Banditry

In his pamphlet titled *Culture Jamming*, Mark Dery (1993: 6) asks 'what shape does an engaged politics assume in an empire of signs?' His answer is a form of semiological guerrilla warfare (as envisioned by Umberto Eco (1967)) he calls Culture Jamming. Culture Jamming as a term originates from the words of a member of the band Negativeland who was describing billboard alterations: 'the skilfully reworked billboard . . . directs the public viewer to a consideration of the original corporate strategy. The studio for the cultural jammer is the world at large' (quoted in Dery, 1993: 6). In Culture Jamming the signs and significations of the mass media are hijacked and diverted to both draw attention to the original message and create new messages with radically different intent. Jammers, says Dery, are 'attempting to reclaim the public space ceded to the chimeras of Hollywood and Madison Avenue, to restore a sense of equilibrium to a society sickened by the vertiginous whirl of TV culture' (ibid.: 13).

Examples of Culture Jamming include the alteration of billboard space, wittily changing one message to another. San Francisco and Oakland have been popular locations for this billboard banditry where 'Tropical blend. The Savage Tan' becomes 'Typical Blend. Sex in Ads'. Artfux of New Jersey change a Coca-Cola Board to say 'Drink Coca-Cola – It Makes You Fart'. In London feminist bandits alter an ad for tights showing a pair of legs in high heels emerging from an egg with the words 'Legs as soft and smooth as the day you were born' with the large black words 'Born Kicking'. An ad for a Fiat car featuring a woman lying on top of a car saying, 'It's so practical darling' is altered with the addition of the words 'when I'm not lying on cars I'm a brain surgeon'. A Greenpeace ad features a woman dragging a fur coat with the words 'It takes up to 40 dumb animals to make a fur coat. But only one to wear it.' Overnight the words 'Men kill animals, Men make the profits, and Men make sexist ads' appears. This billboard banditry is but one form of Guerrilla semiotics that attacks under the cover of night to reinscribe the spaces of the city with messages that do not originate from points of money, power and privilege.

Culture Jamming, Guerrilla Semiotics, Night Discourse, whatever the label, potentially achieves two simultaneous political objectives. First it draws attention to the way the urban environment has been semanticised – given a language and meaning of power, authority and commerce which creates and reproduces expected ways of thinking and acting. Second it creates new and alternative meanings – it asserts the ability to read and write differently. As one billboard graffitist has put it:

The streets are public places. Graffiti is an expression of the experiences and ideas of people who live on those streets but don't own them or the houses or the businesses. Graffiti creates solidarity between all those people. It isn't academic, it's immediate and doesn't require money.

(Posener, 1986: 3)

Billboards effectively give language to the urban environment. Everywhere (but particularly in poor neighbourhoods) we see symbols of varying creativity and depth suggesting that a particular product is the one we want. Billboard banditry provides a weapon with which people with no money or access to the media (which amount to the same thing) can re-semanticise the advertising executive's facade cheaply and effectively. By refacing billboards the powerless turn the one-way communication of advertising into an interchange of ideas and images. As the Melbourne graffitist quoted by Posener states, 'I like watching reactions to my graffiti. Is it painted over, or added to? Sometimes the reactions make me think again, maybe add something else. It's like asking questions'(Posener, 1986: 5).

Most of the writing on the street comes from above and takes the form of directives and commands – 'buy this', 'cross the street now', 'open', 'closed'; a one-way flow of information and misinformation – the antithesis of a public realm in which discussion happens and agreements are made. Billboard banditry engages in a subversive dialogue by saying new things and pointing towards what the pre-existent advertisements are already saying. The signs and messages can no longer be taken for granted and the viewer, having seen the graffiti, has to take a thoughtful (rather than unconscious) stance. The original message is not erased but underlined.

The meaning of graffiti of all types is, in many ways, related to its temporary location (Cresswell, 1996). Billboard graffiti relies on a complicated interplay between the pre-existent space of the billboard ad and the message which is added. In de Certeau's terms the billboard is an established and legitimate(d) site of the proper, neatly slipping into the nexus of capital, consumption and patriarchy. The billboard bandit uses and subverts this established space and site of significance, opportunistically altering its meaning in a way which is dependent on the ad's original message which remains present even as it is subverted. Even more ephemeral are the urban projections of Krzysztof Wodiczko.

Wodiczko's Projections

Krzysztof Wodiczko is an artist of Polish origins and dual Polish-Canadian citizenship who spent many years working in Poland within the constructivist tradition designing propaganda for the Polish State. One influential constructivist dictum was that the aim of art is to organise the 'rhythms of life', a dictum that Wodiczko later rejected and reversed by suggesting that art should 'interrupt, interfere, and intervene in the already highly organised "rhythms of life"' (Wodiczko, 1986: 37). At some point in his life he began to construct a series of works which sought to comment on the discipline and surveillance of the Polish state. He did this work at night, after hours, when he had finished working for the state and was later to note that by pursuing such art at night he (and other artists) were acting as 'collaborators with the system not in the morning but in the evening' (in Wodiczko, 1992: 11, note 2) by underlining and reconfirming the division of day and night, proper and marginal that the Polish State had manufactured in the first place. Later he became interested in going outside the art gallery and questioning the assumptions implicit in the urban environment (see Deutsche, 1988; Hebdige, 1992; Phillips, 1992; Smith, 1993).

Wodiczko is best known for his 'projections'. These involve projecting images onto the public facades of the city. Examples include the projection of a swastika onto the South African embassy in London (1985), the image of businessmen's hands in a firm embrace onto the Massachusetts Institute of Technology (1981) and missiles onto war memorial columns (1983). His images are sometimes accompanied by words. During the Gulf war for instance he projected images of skeletal arms carrying guns and petrol pumps onto the Arco de la Victoria in Madrid (1991) accompanied by the word 'Cuantos?' asking both how much? (the price of petrol) and how many? (the numbers of dead).

Wodiczko's aim is to ask questions of public space by confronting people on the street with these images. Nobody pays to see them and they are not 'expected'. Under the cover of night he challenges the authority of these spaces. 'The attack,' he says, 'must be unexpected, frontal, and must come with the night when the building, undisturbed by its daily functions, is asleep' (Wodiczko, 1992: 84). Like numerous rebels before him Wodiczko uses the public space/time of the night to make his mischievous critique. The effect of the unexpected is to cause people to look again at the spaces he has temporarily defaced. Wodiczko knows that the spaces of the city give shape to ideology:

> Superficially we resent the authority of its massive monumental structure
> . . . yet in our heart of hearts . . . we will allow ourselves to become intoxi-
> cated by its structural ability to embody, and to artistically grasp our intimate,
> unspoken drive for the disciplined collaboration with its power.
>
> (Wodiczko, 1992: 2)

So he attacks these buildings with symbols, jarring our consciousness, making the familiar (and thus unnoticed) strange and worthy of attention. He also knows the limits of transgression. He knows that his images shock us into a new, more conscious relationship with urban space. He also knows that these symbol attacks can only be temporary for extended action would only result in the return of familiarity and neglect. Wodiczko's projections ape other forms of street art. Advertisements appear and then disappear to make way for new images. Bill posters for bands and political actions appear overnight to be covered up or ripped down almost instantly. Electronic signs flash momentary messages across squares and plazas. The leaflets of the Nevsky rider were picked up, thrown away, made into slush by rain and faded by sun. The proper, the artistic, the serious are marked as imagined eternities while the words on the street are fleeting and transient. Again the ideas of Michel de Certeau point towards the tensions between the 'proper' meanings of the established spaces of the street and the momentary incursions of confrontational words and images. To de Certeau the creation of the 'proper' in urban space is achieved through the victory of space over time. Temporality is expelled though the creation and maintenance of boundaries and territories. Temporality and fluidity are the enemies of the proper.

It is no surprise then that the tactics of the weak involve the use of time: 'Tactics are procedures that gain validity in relation to the pertinence they lend to time – to the circumstances which the precise instant of an intervention transforms into a favourable situation' (de Certeau, 1984: 38). The space on which such incursions are fleetingly inscribed is the space of the edifice. 'The space of the tactic is the space of the other. Thus it must play on and with a terrain imposed upon it and organized by the law of a foreign power' (ibid.: 37). It is in this manner that Wodiczko poaches on the public spaces of the city at night, having, as it were, a conversation with the spaces and the people who inhabit them.

> Wodiczko appears in the dark of night at noted sites to reinscribe selected surfaces of the city with penetrating, ethereal images of light and colour. His work does not, cannot, exist in the light of day. He and his art are part of the nocturnal culture of the city; the ideas are about the night side of things. The projections speak through the darkened skies for the disenfranchised, the invisible citizen, for the men and women of long, sleepless nights. Night is the perfect time for Wodiczko. It is when assumptions are thrown into question scattered with flashes of visionary clarity and anxious doubt. Like the semiconscious apparitions of a fitful sleep, there is always a powerful ambiguity. Was it simply a dream? Did something actually occur? Will things be different in the light of day?
>
> (Phillips, 1992: 49)

CONCLUSION: THE WORDS ON THE STREET

In this short essay I have attempted an interpretation of various forms of Night Discourse or Culture Jamming in relation to the space of the street. The subversive scrawls that appear at night enter into a public discussion with other, more sanctioned, forms of discourse. The work of public artists from anonymous graffitists to an international figure such as Wodiczko reinscribes the urban topography producing new meanings and messages – not by negating the dominant messages of monumentality and capitalist consumption ((post)modernism from above) but by entering into a dialectical conversation with them. Billboard bandits create new synthetic meanings, not by ripping down the messages of Coca-Cola or Ford, but by engaging with them, letting the multiplicity of original meanings remain but in a fractured and tension-ridden form. Similarly Wodiczko's projections live off the meanings inscribed in the surfaces that surround the urban public realm – the facades of the monumental. In both these cases the new synthetic meanings (born out of the tensions between the original and the refinement) are products of a denial of eternity and Truth – both are ephemeral messages that are inscribed under the cover of night while the city sleeps. They are, metaphorically, the establishment's nightmares and repressed memories coming back to haunt it.

NOTES

1   I use the word writing here to denote all kinds of inscription, some legible, some not, some forming words and some just images.

REFERENCES

Acconci, V. (1990) 'Public Space in Private Time', *Critical Enquiry,* 16 (summer).

Arendt, H. (1958) *The Human Condition,* Chicago: University of Chicago Press.

Bauman, Z. (1994) 'Desert Spectacular', in K. Tester (ed.), *The Flaneur,* London: Routledge, pp. 138–57.

Benjamin, W. (1968) 'The Work of Art in the Age of Mechanical Reproduction', in H. Arendt (ed.), *Iluminations,* New York: Schocken, pp. 219–34.

Berman, M. (1988) *All That is Solid Melts into Air: The Experience of Modernity* (2nd ed.), Harmondsworth: Penguin.

Bourdieu, P. (1984) *Distinction: A Social Critique of the Judgement of Taste,* trans. R. Nice, Cambridge, MA: Harvard University Press.

Cresswell, T. (1996) *In Place/Out of Place: Geography, Ideology and Transgression,* Minneapolis: University of Minnesota Press.

Davis, M. (1992) *City of Quartz: Excavating the Future in Los Angeles,* New York: Vintage.

de Certeau, M. (1984) *The Practice of Everyday Life,* trans. S. Rendall, Berkeley, CA.: University of California Press.

Dery, M. (1993) *Culture Jamming: Hacking, Slashing and Sniping in the Empire of Signs,* vol. 25, Westfield, NJ: Open Media.

Deutsche, R. (1988) 'Uneven Development: Public Art in New York City', *October* 47: 3–52.

Duncan, J. (1990) *The City as Text: The Politics of Landscape Interpretation in the Kandyan Kingdom,* Cambridge: Cambridge University Press.

Eco, U. (1967) 'Towards a Semiological Guerrilla Warfare', in U. Eco (ed.), *Travels in Hyperreality,* New York: Harcourt Brace Jovanovich, pp. 135–44.

Fyfe, N. and Bannister, J. (1996) 'City Watching: Closed Circuit Television Surveillance in Public Spaces', *Area* 28/1: 37–46.

Hebdige, D. (1992) 'The Machine is unheimlich: Wodiczko's Homeless Vehicle Project', in K. Wodiczko (ed.), *Public Address,* Minneapolis: Walker Art Center, pp. 55–75.

Hoffman, B. (1992) 'Law for Art's Sake in the Public Realm', in W. J. T. Mitchell (ed.), *Art and the Public Sphere,* Chicago: University of Chicago Press, pp. 113–46.

Jacobs, J. (1961) *The Death and Life of Great American Cities,* New York: Vintage.

Jacobs, K. (1992) 'Night Discourse', in S. Heller and K. Jacobs (eds), *Angry Graphics,* Layton, Utah: Gibbs Smith, pp: 8–15.

North, M. (1992) 'The Public as Sculpture: From Heavenly City to Mass Ornament', in W. J. T. Mitchell (ed.), *Art and the Public Sphere,* Chicago: University of Chicago Press, pp. 9–47.

Phillips, P. C. (1992) 'Images of Repossession', in K. Wodiczko (ed.), *Public Address,* Minneapolis: Walker Art Center, pp: 43–54.

Posener, J. (1986) *Louder than Words,* London: Pandora.

Raven, A. (1993) 'Introduction', in A. Raven (ed.), *Art in the Public Interest,* New York: Da Capo Press, pp. 1–28.

Schivelbusch, W. (1988) *Disenchanted Night: The Industrialisation of Light in the Nineteenth Century*, trans. A.Davies, New York: Berg Publishers.

Sennett, R. (1977) *The Fall of Public Man*, Cambridge: Cambridge University Press.

Smith, N. (1993) 'Homeless/Global: Scaling Places', in J. Bird, B. Curtis, T. Putnam, G. Robertson and L. Tickner (eds) *Mapping the Futures: Local Cultures, Global Change*, London: Routledge, pp. 87–119.

Valentine, G. (1989) 'The Geography of Women's Fear', *Area* 21(4): 385–90.

Wodiczko, K. (1992) *Public Address*, Minneapolis: Walker Art Center.

Wodiczko, K. (with Crimp, D., Deutsche, R., Lajer-Burcharth, E.) (1986) 'Conversations with Krzysztof Wodiczko', *October,* 38 (Winter 1986): 22–51.

INDEX

•

Note: page numbers in *italic* type refer to figures. Page numbers followed by 'n' represent footnotes.